AF361652

Marine Nitrogen Fixation and Phytoplankton Ecology

Marine Nitrogen Fixation and Phytoplankton Ecology

Editors

Sang Heon Lee
P.S. Bhavya
Bo Kyung Kim

MDPI • Basel • Beijing • Wuhan • Barcelona • Belgrade • Manchester • Tokyo • Cluj • Tianjin

Editors
Sang Heon Lee
Pusan National University
Korea

P.S. Bhavya
Technical University of Munich
Germany

Bo Kyung Kim
Korea Polar Research Institute
Korea

Editorial Office
MDPI
St. Alban-Anlage 66
4052 Basel, Switzerland

This is a reprint of articles from the Special Issue published online in the open access journal *Water* (ISSN 2073-4441) (available at: https://www.mdpi.com/journal/water/special_issues/marine_n2).

For citation purposes, cite each article independently as indicated on the article page online and as indicated below:

LastName, A.A.; LastName, B.B.; LastName, C.C. Article Title. *Journal Name* **Year**, *Volume Number*, Page Range.

ISBN 978-3-0365-4667-4 (Hbk)
ISBN 978-3-0365-4668-1 (PDF)

Contents

About the Editors

Sang Heon Lee

Professor Sang Heon Lee received his doctoral degree in oceanography from the University of Alaska, Fairbanks, US. He worked as a senior research scientist at the Korea Polar Research Institute from 2007 to 2010. Currently, he has been working as a full professor at Pusan National University since September 2010. His major research interest is on marine ecosystems, mainly phytoplankton in various oceans from the Antarctic Ocean to the Arctic Ocean.

P.S. Bhavya

Dr. Bhavya received her doctoral degree in marine biogeochemistry from the Physical Research Laboratory, Ahmedabad, India. She was a recipient of the Technical University of Munich Foundation Fellowship 2020–2021. Currently, she has been working as a research scientist at Leibniz Institute for Baltic Sea Research since March 2021. Her major research interest is on primary production and nitrogen fixation rate estimation using 13C and 15N tracer techniques, respectively. She also worked on the impacts of anthropogenic interferences and climatological variations on the marine carbon and nitrogen cyclings.

Bo Kyung Kim

Dr. Bo Kyung Kim has been working as a Post-Doc of the Division of Polar Ocean Sciences, Korea Polar Research Institute, South Korea, since 2017. She received her doctoral degree in Oceanography from Pusan National University, South Korea. Her area of expertise is in the field of primary production by phytoplankton using stable isotopes and biomolecules (amino acids, protein, carbohydrate, and lipid) under various environmental conditions in Polar Oceans.

Editorial

Marine Nitrogen Fixation and Phytoplankton Ecology

Sang Heon Lee [1],*, Panthalil S. Bhavya [2] and Bo Kyung Kim [3]

[1] Department of Oceanography, Pusan National University, Busan 46241, Korea
[2] Aquatic System Biology, TUM School of Life Sciences Weihenstephan, Technical University of Munich, 80333 Munich, Germany; bhavya.panthalil@tum.de
[3] Division of Polar Ocean Sciences, Korea Polar Research Institute, Incheon 21990, Korea; bkkim@kopri.re.kr
* Correspondence: sanglee@pusan.ac.kr; Tel.: +82-051-510-3931

Citation: Lee, S.H.; Bhavya, P.S.; Kim, B.K. Marine Nitrogen Fixation and Phytoplankton Ecology. *Water* **2022**, *14*, 1638. https://doi.org/10.3390/w14101638

Received: 16 May 2022
Accepted: 18 May 2022
Published: 20 May 2022

Publisher's Note: MDPI stays neutral with regard to jurisdictional claims in published maps and institutional affiliations.

Marine Nitrogen Fixation and Phytoplankton Ecology

Many oceans are currently undergoing rapid changes in environmental conditions such as warming temperature, acidic water condition, coastal hypoxia, etc. Obvious warming and acidification in various oceans, from polar oceans to tropical oceans, was well reported in the fifth Assessment Report (AR5) by the Intergovernmental Panel on Climate Change [1]. These climate-driven environmental changes could lead to dramatic alterations in the biology and ecology of phytoplankton as major primary producers and biogeochemical drivers and subsequently impact the growth and survival of other marine organisms [2–5]. Consequently, the entire marine ecosystem and global biogeochemical cycles would be very different from what we have now.

Marine phytoplankton are an important indicator of marine ecosystem changes in response to climate-induced environmental change [2,4–6], since they are major primary producers that consolidate solar energy into organic matter and transfer it to marine ecosystems throughout the food web. Recently, increasing numbers of roles of small phytoplankton as a major contributor to primary production have been reported in various oceans, and it has been found that small phytoplankton could become more prominent under an ocean warming scenario [2,4,6–9]. The ecological and biogeochemical traits of small phytoplankton are very different from those of large phytoplankton [3,4,9]. Therefore, it is urgent to verify the different biological and chemical properties of small phytoplankton and understand their ecological roles under ongoing environmental changes.

It is widely known that nitrogenous nutrients are key components of primary production in the ocean, and the only biological source of such nutrients is diazotrophic N_2 fixation. Similar to primary producers, N_2 fixers (diazotrophs) are also vulnerable to changing environmental conditions. It was found that the polar regions can be introduced to diazotrophic activity under warming conditions, and the increased N availability can lead to elevated primary productivity [10–12]. However, if ocean acidification continues in the future, the diazotrophic activity is likely to decrease [13]. The documentation and processing of information on N_2 fixation is highly important as its role in the N cycle of the oceans is critical for preparing future projections on the effects of global environmental changes on the biogeochemical balance of the ecosystems. The employment of enriched isotopic tracers of dinitrogen ($^{15}N_2$) [14], natural abundance studies of the N_2 isotopes in particulate and dissolved forms of N, and the introduction of a simple enzyme-based assay, the acetylene reduction method, have opened possibilities expand our knowledge of biological N_2 fixation in the global oceans. The measured N_2 fixation rates in a *Trichodesmium* bloom in the Arabian Sea showed the highest depth-integrated values, ranging from ∼0.1 to 34 mmol N m^{-2} d^{-1} [15]. The highest depth-integrated N_2 fixation rates in non-bloom conditions are obtained in the western tropical South Pacific (638 ± 1689 µmol N m^{-2} d^{-1}), which are higher than those of the subtropical North Atlantic (182 ± 479 µmol N m^{-2} d^{-1}) and North Pacific (118 ± 101 µmol N m^{-2} d^{-1}). N_2

fixation rates in the eastern South Pacific are measured to be 86 ± 99 μmol N m^{-2} d^{-1}, whereas the southern Indian Ocean (<20 μmol N m^{-2} d^{-1}, [12]) rates are low.

Recent genetic studies on microbial communities report niches of diazotrophic activities which were previously unknown. Most measurements on N_2 fixation rates are from bulk samples, which might have possibly comprised cyanobacteria and other diazotrophs. The infusion of the cell-specific N_2 fixation method using a nanoscale mass spectrometer gave a high-definition perspective of cellular-level N_2 fixation; more importantly, it provides knowledge about diazotrophy on an individual-species basis [16]. This novel method helps to identify the species that are capable of N_2 fixation, and to determine their role in transferring fixed N_2 to the autotrophs in association with them. However, information on global N_2 fixation rates and cell-specific N_2 fixation so far is scant, and inconsistent in different spatio-temporal scales due to a lack of sufficient measurements. To tackle the perplexing response of diazotrophs, a detailed assessment of the diazotrophic community response toward the changing environmental conditions needs to be recorded thoroughly. Considering the fundamental roles of phytoplankton in marine ecosystems and global biogeochemical cycles, it is important to understand phytoplankton ecology and N_2 fixation as a potential N source in various oceans.

This Special Issue covers a wide range of geographic study regions from pole to pole and from coastal systems to open oceans, including Terra Nova Bay, Ross Sea in the Antarctic Ocean, Northern Bering Sea, Chukchi Sea, Canada Basin, and Kongsfjorden, Svalbard in the Arctic Ocean, Western South China Sea, Northern East China Sea, Northwestern Pacific Ocean, and Jaran Bay in South Korea. In this Special Issue, we present a total of 11 articles offering ecological and biogeochemical baselines as indicators for the changes in marine environments and ecosystems driven by global climate changes. In particular, articles on the compositions of intracellular biochemical components such as proteins, lipids, and carbohydrates of phytoplankton could provide important information for their physiological conditions and the nutritional value of organic matter available to grazers [17,18]. Recently, phytoplankton-derived transparent exopolymer particles (TEPs) are known for making a considerable contribution to the organic matter pools and thus marine biogeochemical cycles in aquatic environments [19]. Ref. [20] investigated monthly TEPs concentration and particulate organic carbon (POC) concentration in Jaran Bay, a large shellfish aquaculture site in a southern coastal region of Korea. They found that the contribution of TEPs ranged from 2.4% to as high as 78.0% of the POC concentration, which indicated that TEPs-C could be a significant contributor to the POC pool in a coastal bay. Since little information on the monthly variation in TEPs is available, their investigation on the TEPs could be a very important baseline in a coastal bay system. Moreover, in Jaran Bay, ref. [21] also observed the seasonal and spatial variations in the biochemical compositions of phytoplankton. They found that the dominant biochemical component was carbohydrates ($51.8 \pm 8.7\%$), followed by lipids ($27.3 \pm 3.8\%$) and proteins ($20.9 \pm 7.4\%$). Large phytoplankton and the $P \times (PO_4{}^{3-}1/16 \times NO_3{}^-)$ and $NH_4{}^+$ concentrations were identified as major controlling factors for food material (FM) in Jaran Bay. Over a year at Jang Bogo Station (JBS) in Antarctica, ref. [18] measured bi-weekly biochemical compositions of particulate organic matter (POM) and concentrations of TEPs. The high composition of lipids and proteins indicated a good food source in summer, whereas stably low concentrations of carbohydrates and lipids were utilized for long-term energy storage in the survival of phytoplankton in winter. They found that TEPs have a longer residence time than POC, and the contribution of TEPs-C to the POC pool could be important in the Ross Sea. The biochemical composition of POM deriving mainly from phytoplankton in the Chukchi Sea, Arctic Ocean were presented by [22]. They investigated the biochemical components of phytoplankton and their spatial pattern. Carbohydrates were the predominant macromolecules, accounting for 42.6% in the Chukchi Shelf and 60.5% in the Canada Basin, followed by lipids and proteins. Based on their study, the biochemical compositions of phytoplankton could be considerably different in the regions of the Arctic Ocean. In a similar region, ref. [23] estimated the bioavailable fraction of POM through enzymatic hydrolysis that can be utilized by higher trophic levels. Based on their results, nutrient, temperature, meltwater and different size classes of phytoplankton (micro and picophytoplankton) were the main factors of the compositional variations and the spatial distributions. More studies on the changes in the biochemical compositions of phytoplankton should be conducted under future environmental changes.

In terms of elemental composition and primary productivity driven by phytoplankton, ref. [24] determined the combined physiological–elemental ratio changes in two phytoplankton species, *Scrippsiella trochoidea* (Dinophyceae) and *Heterosigma akashiwo* (Raphidophyceae). They found higher average ratios of particulate organic nitrogen (PON) to chlorophyll-a (Chl-*a*) and POC to Chl-*a* in *S. trochoidea* than those of *H. akashiwo*. However, the authors observed similar ratios of POC/PON of the two microalgae. These results can be used to develop physiological models for phytoplankton, with implications for the marine biogeochemical cycle. In Kongsfjorden's high-latitude open fjord systems, ref. [25] found that the turbidity associated with glacier meltwater impacted the penetration depth of light and that nutrients could cause the lower productivity rates of phytoplankton. They found that picophytoplankton was largely based on regenerated nutrients, even more productive than that suggested by their biomass contribution and their nitrogen uptake. For a better understanding of the biochemical traits of small phytoplankton, ref. [26] conducted field measurements in the biologically productive northern Bering and Chukchi seas. The contributions of small phytoplankton to the total primary production were 38.0% (SD = ±19.9%) and 25.0% (SD = ±12.8%) in 2016 and 2017, respectively. They found that small phytoplankton synthesize different biochemical compositions with nitrogen-rich POC from large phytoplankton.

The three articles below on phytoplankton communities and cyanobacterial contributions provide significant ecological predictions under expected warming ocean conditions. In the North Pacific Ocean, ref. [27] determined the picocyanobacterial contribution and the total primary production. The average picocyanobacterial contributions to the carbon uptake rates were 45.2% in the tropical Pacific region and 70.2% in the subtropical and temperate Pacific region, respectively. In addition, their contributions to the nitrogen uptake rates were significantly higher than those of carbon uptake rates. Based on high-performance liquid chromatography (HPLC) pigment analysis, ref. [28] investigated spatiotemporal variations in phytoplankton community compositions in the northern East China Sea (ECS), the largest marginal sea in the north-western Pacific Ocean. Overall, the two major phytoplankton groups were diatoms (32.0%) and cyanobacteria (20.6%) in the northern ECS, and the two groups were negatively correlated. In the western South China Sea, ref. [29] investigated the distinct seasonal variation in phytoplankton community structure related to different oceanographic conditions and observed a major shift from a diatom-dominated regime in winter to a cyanobacteria-dominated system in summer. The authors found that the increased overall abundance of phytoplankton and cyanobacteria during the summer was caused by upwelling and enriched eddy activity, whereas the abundant symbiotic cyanobacteria–diatom association during the winter was mainly due to the influence of the cool temperature. Long-term monitoring of the phytoplankton communities and the picocyanobacterial contributions should be conducted for a better understanding of the ecological impacts of the global warming scenario, with a focus on the ecological roles of picocyanobacteria.

This Special Issue covers various articles on N_2 fixation and aspects of marine phytoplankton ecology, such as biodiversity, distribution, biomass, photosynthetic traits, biochemical compositions, productivity, etc., in various oceans, including polar oceans. Finally, we hope that this Special Issue provides ecological and biogeochemical baselines that broaden our existing knowledge on the current and ongoing changes in marine ecosystems in response to global climate change.

Author Contributions: Conceptualization, S.H.L., P.S.B. and B.K.K.; writing—review and editing, S.H.L., P.S.B. and B.K.K.; supervision, S.H.L. All authors have read and agreed to the published version of the manuscript.

Funding: Sang Heon Lee was financially supported by the project titled "Improvements of ocean prediction accuracy using numerical modeling and artificial intelligence technology" funded by the Ministry of Oceans and Fisheries, Korea (PM62850). P.S. Bhavya was supported by Leibniz Institute for Baltic Sea Research and N-Amazon project (DFG 08/2020-07/2022) by German Research Foundation (Deutsche Forschungsgemeinschaft). Support for Bo Kyung Kim was provided by the Korean government (MOF; 1525011760).

Acknowledgments: We thank all the authors who submitted their papers to this Special Issue. We are especially indebted to all reviewers for providing their valuable review comments for individual articles of this volume.

Conflicts of Interest: The authors declare no conflict of interest.

References

1. IPCC. Climate change 2014: Synthesis report. In *Contribution of Working Groups I, II and III to the Fifth Assessment Report of the Intergovernmental Panel on Climate Change*; IPCC: Geneva, Switzerland, 2014; ISBN 978-92-9169-142-5.
2. Li, W.K.W.; McLaughlin, F.A.; Lovejoy, C.; Carmack, E.C. Smallest Algae Thrive as the Arctic Ocean Freshens. *Science* **2009**, *326*, 539. [CrossRef] [PubMed]
3. Lee, S.H.; Yun, M.S.; Kim, B.K.; Saitoh, S.I.; Kang, C.K.; Kang, S.H.; Whitledge, T. Latitudinal Carbon Productivity in the Bering and Chukchi Seas during the Summer in 2007. *Cont. Shelf Res.* **2013**, *59*, 28–36. [CrossRef]
4. Kang, J.J.; Jang, H.K.; Lim, J.H.; Lee, D.; Lee, J.H.; Bae, H.; Lee, C.H.; Kang, C.K.; Lee, S.H. Characteristics of Different Size Phytoplankton for Primary Production and Biochemical Compositions in the Western East/Japan Sea. *Front. Microbiol.* **2020**, *11*, 560102. [CrossRef] [PubMed]
5. Sun, Y.; Youn, S.H.; Kim, Y.; Kang, J.J.; Lee, D.; Kim, K.; Jang, H.K.; Jo, N.; Yun, M.S.; Song, S.K.; et al. Interannual Variation in Phytoplankton Community Driven by Environmental Factors in the Northern East China Sea. *Front. Mar. Sci.* **2022**, *9*, 1–19. [CrossRef]
6. Lee, S.H.; Ryu, J.; Lee, D.; Park, J.W.; Kwon, J.-I.; Zhao, J.; Son, S. Spatial Variations of Small Phytoplankton Contributions in the Northern Bering Sea and the Southern Chukchi Sea. *GISci. Remote Sens.* **2019**, *56*, 794–810. [CrossRef]
7. Morán, X.A.G.; López-Urrutia, Á.; Calvo-Díaz, A.; LI, W.K.W. Increasing Importance of Small Phytoplankton in a Warmer Ocean. *Glob. Chang. Biol.* **2010**, *16*, 1137–1144. [CrossRef]
8. Bhavya, P.S.; Lee, J.H.; Lee, H.W.; Kang, J.J.; Lee, J.H.; Lee, D.; An, S.H.; Stockwell, D.A.; Whitledge, T.E.; Lee, S.H. First in Situ Estimations of Small Phytoplankton Carbon and Nitrogen Uptake Rates in the Kara, Laptev, and East Siberian Seas. *Biogeosciences* **2018**, *15*, 5503–5517. [CrossRef]
9. Bhavya, P.S.; Kang, J.J.; Jang, H.K.; Joo, H.; Lee, J.H.; Lee, J.H.; Park, J.W.; Kim, K.; Kim, H.C.; Lee, S.H. The Contribution of Small Phytoplankton Communities to the Total Dissolved Inorganic Nitrogen Assimilation Rates in the East/Japan Sea: An Experimental Evaluation. *J. Mar. Sci. Eng.* **2020**, *8*, 854. [CrossRef]
10. Blais, M.; Tremblay, J.É.; Jungblut, A.D.; Gagnon, J.; Martin, J.; Thaler, M.; Lovejoy, C. Nitrogen Fixation and Identification of Potential Diazotrophs in the Canadian Arctic. *Glob. Biogeochem. Cycles* **2012**, *26*, GB3022. [CrossRef]
11. Sipler, R.E.; Gong, D.; Baer, S.E.; Sanderson, M.P.; Roberts, Q.N.; Mulholland, M.R.; Bronk, D.A. Preliminary Estimates of the Contribution of Arctic Nitrogen Fixation to the Global Nitrogen Budget. *Limnol. Oceanogr. Lett.* **2017**, *2*, 159–166. [CrossRef]
12. Shiozaki, T.; Kodama, T.; Furuya, K. Large-scale impact of the island mass effect through nitrogen fixation in the western South Pacific Ocean. *Geophys. Res. Lett.* **2014**, *41*, 6413–6419. [CrossRef]
13. Hong, H.; Shen, R.; Zhang, F.; Wen, Z.; Chang, S.; Lin, W.; Kranz, S.A.; Luo, Y.-W.; Kao, S.-J.; Morel, F.M.M.; et al. The complex effects of ocean acidification on the prominent N_2-fixing cyanobacterium Trichodesmium. *Science* **2017**, *357*, 527–531. [CrossRef] [PubMed]
14. Montoya, J.P.; Voss, M.; Kahler, P.; Capone, D.G. A simple, high-precision, high-sensitivity tracer assay for N_2 fixation. *Appl. Environ. Microbiol.* **1996**, *62*, 986–993. [CrossRef] [PubMed]
15. Gandhi, N.; Singh, A.; Prakash, S.; Ramesh, R.; Raman, M.; Sheshshayee, M.S.; Shetye, S. First direct measurements of N_2 fixation during a Trichodesmium bloom in the eastern Arabian Sea. *Glob. Biogeochem. Cycles* **2011**, *25*, 1–10. [CrossRef]
16. Musat, N.; Musat, F.; Weber, P.K.; Pett-Ridge, J. Tracking Microbial Interactions with NanoSIMS. *Curr. Opin. Biotechnol.* **2016**, *41*, 114–121. [CrossRef] [PubMed]
17. Lee, J.H.; Kang, J.J.; Jang, H.K.; Jo, N.; Lee, D.; Yun, M.S.; Lee, S.H. Major Controlling Factors for Spatio-Temporal Variations in the Macromolecular Composition and Primary Production by Phytoplankton in Garolim and Asan Bays in the Yellow Sea. *Reg. Stud. Mar. Sci.* **2020**, *36*, 101269. [CrossRef]
18. Park, S.; Park, J.; Yoo, K.C.; Yoo, J.; Kim, K.; Jo, N.; Jang, H.K.; Kim, J.; Kim, J.; Kim, J.; et al. Seasonal Variations in the Biochemical Compositions of Phytoplankton and Transparent Exopolymer Particles (Teps) at Jang Bogo Station (Terra Nova Bay, Ross Sea), 2017–2018. *Water* **2021**, *13*, 2173. [CrossRef]
19. Passow, U. Transparent Exopolymer Particles (TEP) in Aquatic Environments. *Prog. Oceanogr.* **2002**, *55*, 287–333. [CrossRef]
20. Lee, J.H.; Lee, W.C.; Kim, H.C.; Jo, N.; Jang, H.K.; Kang, J.J.; Lee, D.; Kim, K.; Lee, S.H. Transparent Exopolymer Particle (TEPs) Dynamics and Contribution to Particulate Organic Carbon (POC) in Jaran Bay, Korea. *Water* **2020**, *12*, 1057. [CrossRef]

21. Lee, J.H.; Lee, W.C.; Kim, H.C.; Jo, N.; Kim, K.; Lee, D.; Kang, J.J.; Sim, B.R.; Kwon, J.-I.; Lee, S.H. Temporal and Spatial Variations of the Biochemical Composition of Phytoplankton and Potential Food Material (Fm) in Jaran Bay, South Korea. *Water* **2020**, *12*, 3093. [CrossRef]
22. Choe, K.; Yun, M.; Park, S.; Yang, E.; Jung, J.; Kang, J.; Jo, N.; Kim, J.; Kim, J.; Lee, S.H. Spatial Patterns of Macromolecular Composition of Phytoplankton in the Arctic Ocean. *Water* **2021**, *13*, 2495. [CrossRef]
23. Kim, B.K.; Joo, H.M.; Jung, J.; Lee, B.; Ha, S.Y. In Situ Rates of Carbon and Nitrogen Uptake by Phytoplankton and the Contribution of Picophytoplankton in Kongsfjorden, Svalbard. *Water* **2020**, *12*, 2903. [CrossRef]
24. Liu, X.; Liu, Y.; Noman, M.A.; Thangaraj, S.; Sun, J. Physiological Changes and Elemental Ratio of Scrippsiella Trochoidea and Heterosigma Akashiwo in Different Growth Phase. *Water* **2021**, *13*, 132. [CrossRef]
25. Kim, B.K.; Jung, J.; Lee, Y.; Cho, K.H.; Gal, J.K.; Kang, S.H.; Ha, S.Y. Characteristics of the Biochemical Composition and Bioavailability of Phytoplankton-Derived Particulate Organic Matter in the Chukchi Sea, Arctic. *Water* **2020**, *12*, 2355. [CrossRef]
26. Park, J.W.; Kim, Y.; Kim, K.W.; Fujiwara, A.; Waga, H.; Kang, J.J.; Lee, S.H.; Yang, E.J.; Hirawake, T. Contribution of Small Phytoplankton to Primary Production in the Northern Bering and Chukchi Seas. *Water* **2022**, *14*, 235. [CrossRef]
27. Lee, H.W.; Noh, J.H.; Choi, D.H.; Yun, M.; Bhavya, P.S.; Kang, J.J.; Lee, J.H.; Kim, K.W.; Jang, H.K.; Lee, S.H. Picocyanobacterial Contribution to the Total Primary Production in the Northwestern Pacific Ocean. *Water* **2021**, *13*, 1610. [CrossRef]
28. Kim, Y.; Youn, S.H.; Oh, H.J.; Kang, J.J.; Lee, J.H.; Lee, D.; Kim, K.; Jang, H.K.; Lee, J.; Lee, S.H. Spatiotemporal Variation in Phytoplankton Community Driven by Environmental Factors in the Northern East China Sea. *Water* **2020**, *12*, 2695. [CrossRef]
29. Ding, C.; Sun, J.; Narale, D.D.; Liu, H. Phytoplankton Community in the Western South China Sea in Winter and Summer. *Water* **2021**, *13*, 1209. [CrossRef]

Article

Seasonal Variations in the Biochemical Compositions of Phytoplankton and Transparent Exopolymer Particles (TEPs) at Jang Bogo Station (Terra Nova Bay, Ross Sea), 2017–2018

Sanghoon Park [1], Jisoo Park [2], Kyu-Cheul Yoo [2], Jaeill Yoo [2], Kwanwoo Kim [1], Naeun Jo [1], Hyo-Keun Jang [1], Jaehong Kim [1], Jaesoon Kim [1], Joonmin Kim [1] and Sang-Heon Lee [1,*]

[1] Department of Oceanography, Pusan National University, Busan 46241, Korea; mossinp@pusan.ac.kr (S.P.); goanwoo7@pusan.ac.kr (K.K.); nadan213@pusan.ac.kr (N.J.); jhk7947@naver.com (H.-K.J.); king9527@naver.com (J.K.); jaesoonkim1123@naver.com (J.K.); zoommini@pusan.ac.kr (J.K.)

[2] Korea Polar Research Institute, Incheon 21990, Korea; jspark@kopri.re.kr (J.P.); kcyoo@kopri.re.kr (K.-C.Y.); jiyoo@kopri.re.kr (J.Y.)

* Correspondence: sanglee@pusan.ac.kr

Citation: Park, S.; Park, J.; Yoo, K.-C.; Yoo, J.; Kim, K.; Jo, N.; Jang, H.-K.; Kim, J.; Kim, J.; Kim, J.; et al. Seasonal Variations in the Biochemical Compositions of Phytoplankton and Transparent Exopolymer Particles (TEPs) at Jang Bogo Station (Terra Nova Bay, Ross Sea), 2017–2018. *Water* **2021**, *13*, 2173. https://doi.org/10.3390/w13162173

Academic Editor: Oscar Romero

Received: 10 June 2021
Accepted: 5 August 2021
Published: 8 August 2021

Publisher's Note: MDPI stays neutral with regard to jurisdictional claims in published maps and institutional affiliations.

Abstract: The biochemical composition of particulate organic matter (POM) mainly originates from phytoplankton. Transparent exopolymer particles (TEPs) depend on environmental conditions and play a role in the food web and biogeochemical cycle in marine ecosystems. However, little information on their characteristics in the Southern Ocean is available, particularly in winter. To investigate the seasonal characteristics of POM and TEPSs, seawater samples were collected once every two weeks from November 2017 to October 2018 at Jang Bogo Station (JBS) located on the coast of Terra Nova Bay in the Ross Sea. The total chlorophyll-*a* (Chl-*a*) concentrations increased from spring (0.08 ± 0.06 µg L^{-1}) to summer (0.97 ± 0.95 µg L^{-1}) with a highest Chl-*a* value of 2.15 µg L^{-1}. After sea ice formation, Chl-*a* rapidly decreased in autumn (0.12 ± 0.10 µg L^{-1}) and winter (0.01 ± 0.01 µg L^{-1}). The low phytoplankton Chl-*a* measured in this study was related to a short ice-free period in summer. Strong seasonal variations were detected in the concentrations of proteins and lipids (one-way ANOVA test, $p < 0.05$), whereas no significant difference in carbohydrate concentrations was observed among different seasons (one-way ANOVA test, $p > 0.05$). The phytoplankton community was mostly composed of diatoms ($88.8\% \pm 11.6\%$) with a large accumulation of lipids. During the summer, the POM primarily consisted of proteins. The composition being high in lipids and proteins and the high caloric content in summer indicated that the phytoplankton would make a good food source. In winter, the concentrations of proteins decreased sharply. In contrast, relatively stable concentrations of carbohydrates and lipids have been utilized for respiration and long-term energy storage in the survival of phytoplankton. The TEPS values were significantly correlated with variations in the biomass and species of the phytoplankton. Our study site was characterized by dominant diatoms and low Chl-*a* concentrations, which could have resulted in relatively low TEP concentrations compared to other areas. The average contributions of TEP-C to the total POC were relatively high in autumn ($26.9\% \pm 6.1\%$), followed by those in summer ($21.9\% \pm 7.1\%$), winter ($13.0\% \pm 4.2\%$), and spring ($9.8\% \pm 3.1\%$).

Keywords: phytoplankton; macromolecular composition; transparent exopolymer particles; Ross Sea; polar night

1. Introduction

The Ross Sea is one of the most productive regions in the Southern Ocean, with strong seasonal variations in the biomass and production of phytoplankton [1]. The high phytoplankton biomass commonly observed in coastal areas during spring is linked to the stabilization of the water column and the release of micronutrients during ice melting [2]. However, phytoplankton growth is limited by micronutrient availability in ice-free waters under conditions of high irradiance in summer [3,4]. Furthermore, the changes in complex hydrographic features such as sea ice formation during the transition between late autumn and winter are likely to substantially influence the distribution of phytoplankton [5] and consequently the biochemical processes in the water columns [6–8]. Recently, sea ice concentration has increased significantly in the Ross Sea [9] and the annual ice-free period has shortened [10], which could potentially reduce water stratification and irradiance throughout the surface water. These changes can influence the composition and biomass of phytoplankton [11,12] and consequently the biogeochemical cycles and the food web structure in the Southern Ocean [13].

Through photosynthesis, phytoplankton synthesize inorganic nutrients into various organic compounds, which mainly consist of carbohydrates, proteins, and lipids [14]. The macromolecular composition of phytoplankton can change depending upon various environmental factors, such as nutrient concentrations, light intensity, and the species composition of the phytoplankton [14–16]. In the Southern Ocean, phytoplankton growth is generally limited by irradiance in spring [17] and by iron in summer [3,4,18]. Previous studies on the macromolecular composition of phytoplankton in the Ross Sea have been conducted primarily during spring and summer seasons [19–22]. Only a few studies [23] on the biochemical characteristics of phytoplankton have been previously conducted throughout a single year in the Southern Ocean. The Jang Bogo Station (JBS), located on the coast of Terra Nova Bay (TNB) in the Ross Sea, is one of the overwintering science stations in Antarctica. This study was conducted for approximately one year from November 2017 to October 2018 to monitor seasonal variations in the biochemical compositions of particulate organic matter (POM) and transparent exopolymer particles (TEPs) strongly related to phytoplankton.

TEPs are carbon-rich gel organic particles stainable with Alcian Blue, a specific dye for acidic polysaccharides, and phytoplankton are considered to be the main source of TEPs and their precursors [24,25]. TEPs are formed naturally by self-assembly of dissolved precursors, which appreciably contribute to the dissolved organic matter pool in the water column [26]. The formation of TEPs is critical as a significant pathway in which dissolved organic matter is transformed into POM [26]. Furthermore, TEPs are considered to play an important role in carbon cycling [27]. Because of their great stickiness, TEPs increase the aggregation rate of particles, forming marine snow, and enhance the carbon sinkage into deep waters in the marine system [28,29]. The distribution of TEPs is affected by environmental factors such as nutrient conditions [30,31] and the species and growth conditions of phytoplankton [32–34]. However, little information on TEPs' abundances and regulating ecological factors is currently available for the Southern Ocean [25,35,36]. Most previous studies on TEPs have described their formation and dynamics either under experimental conditions [25,37] or during phytoplankton bloom seasons [36,38]. Our study is the first to show seasonal variation in TEPs at the JBS in the Ross Sea. The present study aimed to (1) identify the seasonal characterization of phytoplankton, (2) investigate the seasonal variations in the macromolecular composition of phytoplankton and TEPs concentrations at the JBS throughout the year, and (3) understand the roles of macromolecules and TEPs' contributions to the organic carbon pool in Antarctic coastal environments.

2. Materials and Methods

2.1. Study Area and Water Sampling

The JBS is located on the coast of TNB in the Ross Sea, Antarctica (74°37′39.59″ S, 164°14′25.75″ E) (Figure 1). Surface seawater samples for biochemical parameters from a continuously flowing water tank at the station were collected biweekly from 3 November 2017 to 26 October 2018. In situ seawater temperature and salinity were measured by YSI professional plus. To verify seasonal differences, our entire study period was divided into four seasons: spring (November to December), summer (January to February), autumn (March to April), and winter (May to October).

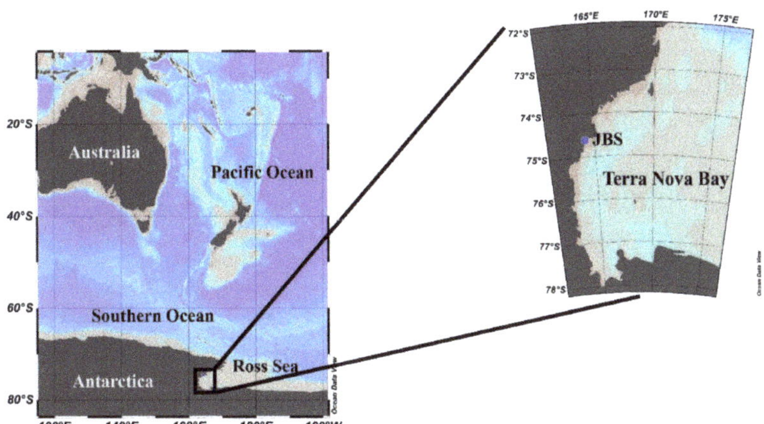

Figure 1. Location of the Jang Bogo Station (JBS) in Terra Nova Bay, Ross Sea.

2.2. Nutrient Analysis

Water samples for major inorganic nutrients (nitrite + nitrate, ammonium, silicate, and phosphate) were filtered through a 47 mm GF/F (0.7 μm pore size), and the filtered water samples (45 mL) were stored frozen at −80 °C in a 50 mL conical tube until analysis. The concentrations of inorganic nutrients were analyzed using an automated nutrient analyzer (Quaatro, Seal Analytical, Norderstedt, Germany) based on the manufacturer's instructions at the National Institute of Fisheries Science (NIFS).

2.3. Chlorophyll-a Analysis

Seawater samples (0.5 L) for the total chlorophyll-*a* (Chl-*a*) concentration were filtered through 0.7 μm pore size glass fiber filters (GF/F; Whatman, Maidstone, UK, 25 mm). Samples (1 L) for size-fractionated Chl-*a* concentration were passed sequentially through 20 and 2 μm Nuclepore membrane filters (20 μm; GVS, Sanford, Sioux Falls, SD, USA, 47 mm; 2 μm; Whatman, Maidstone, UK, 47 mm) and 0.7 μm pore size Whatman GF/F filters (47 mm). Chl-*a* was extracted with 90% acetone for 24 h in the dark at 4 °C following [39], and the Chl-*a* concentrations were measured using a Trilogy fluorometer (Turner Designs, San Jose, CA, USA) after calibration.

2.4. Particulate Organic Carbon and Nitrogen Analysis

For the particulate organic carbon (POC), nitrogen (PON), and δ^{13}C analyses of POM, 0.5 L of each seawater sample was filtered through Whatman GF/F filters (25 mm). The GF/F filters were kept frozen at −80 °C until mass spectrometric analysis in the stable isotope laboratory of the University of Alaska Fairbanks after HCl fuming overnight to remove carbonate.

2.5. Phytoplankton Community

Water samples (2–3 L) for photosynthetic pigment analysis were filtered through Whatman GF/F filters (47 mm). The GF/F filters were wrapped in aluminum foil to prevent photooxidation and stored at −80 °C until the extraction and analysis of pigments at the home laboratory in Pusan National University, Korea. The phytoplankton pigments were extracted in 100% acetone (5 mL) with canthaxanthin (100 μL) as an internal standard for 24 h in the dark at 4 °C and analyzed using a high-performance liquid chromatography (HPLC- Agilent Infinite 1260, Santa Clara, CA, USA) system. The contribution of each phytoplankton class to the total Chl-*a* concentration was estimated from the ratio of the different pigments to Chl-*a* based on the CHEMTAX program [40–42]. The ratio of initial pigment to the total Chl-*a* concentrations for the determination of the phytoplankton classes at the JBS was based on diverse phytoplankton groups obtained around the Ross Sea [43].

2.6. Macromolecular Compositions

Seawater samples (0.5 L) for macromolecular components (carbohydrates, proteins, and lipids) of total phytoplankton were filtered through Whatman GF/F filters (47 mm). The filters were immediately frozen and stored at $-80\ °C$ until analysis at the home laboratory. The content of total carbohydrates was determined by the phenol–sulfuric acid method from [44] with a glucose standard (1 mg mL^{-1}, 108 Sigma, St. Louis, MO, USA), and the protein concentration of POM was analyzed according to [45] with a protein standard solution (2 mg mL^{-1}, Sigma). For lipid extraction and assay, the methods described by [46] and [47] were used with the chloroform–methanol mixture (1:2 v/v). Tripalmitin solution was used as a standard for the protein concentration. The detailed method for the determination of macromolecular compositions was described by [48].

The sum of carbohydrate, protein, and lipid concentrations of POM was referred to as the food material (FM) [49]. The calorific value (Kcal g^{-1}) and contents (Kcal m^{-3}) for each macromolecule component were calculated using the equation from [50] with conversion factors of 0.041, 0.055, and 0.095 for carbohydrates, protein, and lipid, respectively.

2.7. TEP Analysis

TEPs sampling began on 17 December, which was later than the other samplings because Alcian Blue solution, which is prohibited for air shipping, was needed for the TEP sampling but was delayed by a research vessel, *Araon*. The missing TEP data would not have affected our results in this study since the sampling time before 17 December was biologically stagnant with a low Chl-*a* biomass. Seawater samples (0.3 L) for TEPs were filtered at a low constant vacuum onto 0.4 μm pore sized polycarbonate membrane filters (ADVANTEC; 25 mm, Toyo Roshi Kaisha, Tokyo, Japan) and stained with Alcian Blue solution. The concentrations of TEPs were estimated according to the method of [35], and their units were used as gum xanthan equivalents (μg Xeq. L^{-1}). The filters for TEP concentration were extracted with 80% sulfuric acid for 3 h, and the absorbance of the TEPs was measured by a spectrophotometer at 787 nm using Xanthan gum as a standard. Estimations of TEP carbon (TEP-C) for converting from micrograms of gum xanthan equivalents to micrograms of carbon were obtained from the conversion factor of 0.51 μg Xeq. L^{-1} reported by [51].

2.8. Data Treatment and Statistical Analyses

All statistical analyses were performed using the statistical software SPSS 12.0 K for Windows. Pearson's correlation analysis at an alpha value <0.01 was used to estimate the correlation between Chl-*a*, TEP concentrations and nutrients measured in the water samples. One-way ANOVA at an alpha value <0.05 was applied to identify differences in macromolecular concentrations, phytoplankton compositions of different sizes, and TEP-C contributions for each season. Principal component analysis (PCA) was performed using the prcomp function in the R stats package (R studio, version 1.1.463, Inc., Boston, MA, USA) to evaluate the relationship among the environmental factors (e.g., salinity, temperature, nitrite + nitrate, ammonium, and species of phytoplankton) and biochemical variables (e.g., POC, Chl-*a*, POM originating from phytoplankton, and TEP) and to identify relatively significant factors during the study period. Eigenvalues (>1.0) were considered to determine the number of principal components and the rotated eigenvectors were the results using the Varimax method with Kaiser normalization.

3. Results

3.1. Physical and Chemical Conditions in the Study Area

The landfast sea ice in our study area completely melted on 13 February 2018 and formed again on 24 February 2018. During the overall study period, surface water temperature and salinity ranged from 0.74 to −1.12 °C (mean ± S.D. = −0.69 ± 0.51 °C) and 30.88 to 32.51 (31.91 ± 0.38), respectively. Temperature increased and salinity decreased toward summer, largely caused by melting sea ice. Little seasonal variation in temperature (−1.06 ± 0.06 °C) and salinity (32.06 ± 0.03) was detected from May to October (Figure 2).

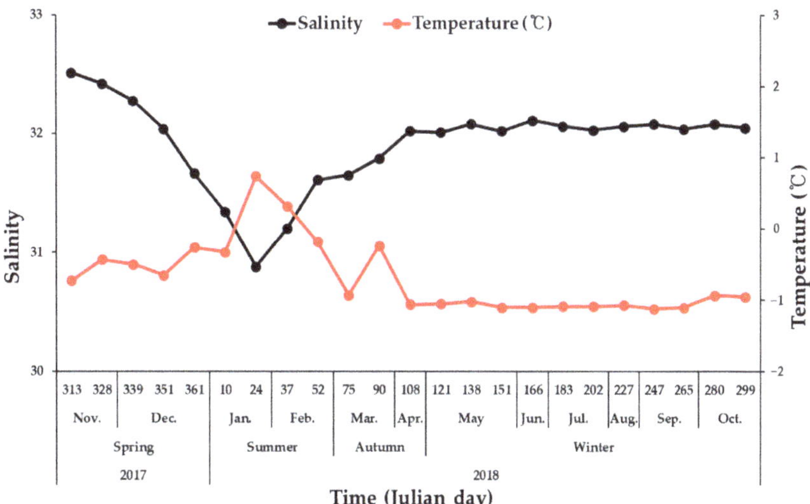

Figure 2. Salinity and temperature variations at the JBS during our sampling period.

Seasonal variations in major inorganic nutrient concentrations are shown in Figure 3. The concentrations of nitrate + nitrite, silicate, and phosphate were lowest in summer (11.0, 39.2, and 0.9 µM for nitrate + nitrite, silicate, and phosphate, respectively), whereas the ammonium concentration was lowest in winter 2018 (0.5 µM). In our study, all major inorganic nutrients had strong relationships with each other except for ammonium (r = 0.97, phosphate vs. nitrate + nitrite; r = 0.98, phosphate vs. silicate; r = 0.98, and nitrate + nitrite vs. silicate, $p < 0.01$, n = 24).

3.2. Chl-a Concentrations

The total Chl-*a* concentrations ranged from 0.01 to 2.15 µg L^{-1} (0.21 ± 0.50 µg L^{-1}) at the JBS during our study period from November 2017 to October 2018. The total Chl-*a* concentrations increased from early December 2017 and were highest (2.15 µg L^{-1}) in February 2018 (Figure 4). The concentration decreased rapidly as sea ice started to form between February and March, 2018. The average total Chl-*a* concentrations were 0.08 ± 0.06, 0.97 ± 0.95, 0.12 ± 0.10, and 0.01 ± 0.01 µg L^{-1} during spring, summer, autumn, and winter, respectively. The seasonal variations in the contributions of different size-fractionated Chl-*a* levels to the total Chl-*a* concentration are shown in Figure 5. The average contribution of micro-sized phytoplankton (>20 µm) to the total Chl-*a* in November 2017 to October 2018 was 39.7% ± 10.7%, and those of nano-sized (2–20 µm) and pico-sized (<2 µm) phytoplankton were 39.5% ± 10.5% and 20.8% ± 5.9%, respectively. Based on statistical analysis (one-way ANOVA test, $p > 0.05$), there were no significant differences in the average contributions of each size-fractionated Chl-*a* class over different seasons.

3.3. POC and PON

The POC and PON concentrations were 49.3–239.5 µg L^{-1} (92.5 ± 46.6 µg L^{-1}) and 7.0–42.7 µg L^{-1} (14.0 ± 10.4 µg L^{-1}), respectively. A strong linear relationship was found between POC and PON concentrations (PON = POC × 0.2192, r^2 = 0.98, $p < 0.01$). Unlike the Chl-*a* concentration, the POC concentration did not largely decrease after bloom (45.6% and 89.5% for POC and Chl-*a*, respectively) and was persistent throughout the winter period

(Figure 4). The average C/N ratios of POM were 7.1 ($\pm$1.0), 5.9 ($\pm$0.4), 8.3 ($\pm$0.7), and 10.6 ($\pm$0.9) during spring, summer, autumn, and winter, respectively.

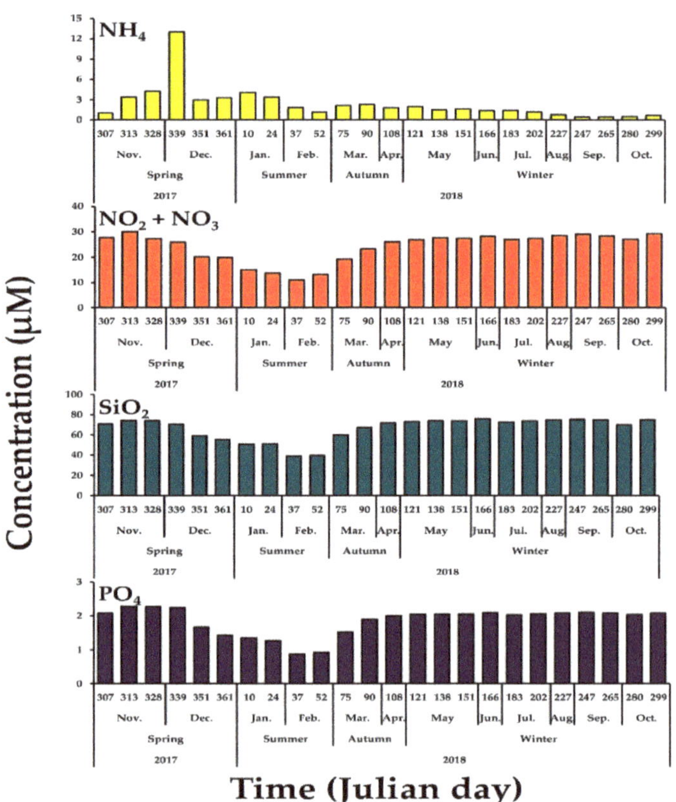

Figure 3. Variations in each major inorganic nutrient concentration at the JBS during our sampling period.

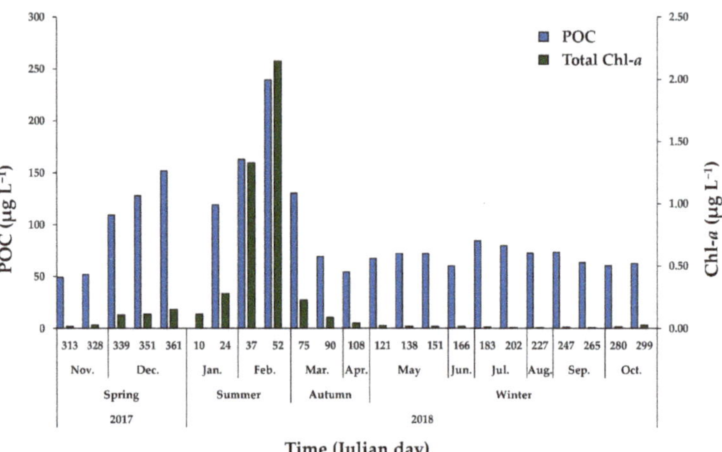

Figure 4. Variations in POC and total Chl-*a* concentrations at the JBS during our sampling period.

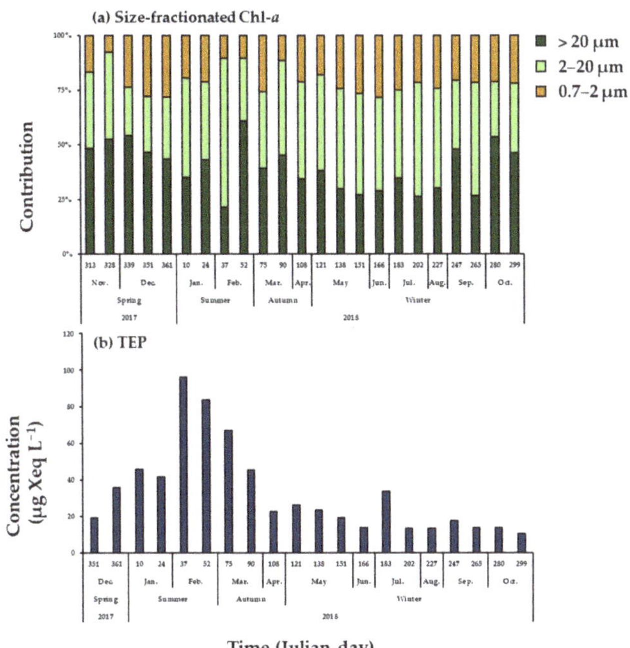

Figure 5. Contributions of different size-fractionated Chl-*a* concentrations to the total Chl-*a* concentration (**a**) and trans parent exopolymer particles (TEPs) concentrations (**b**) at the JBS during our sampling period.

3.4. Phytoplankton Community Composition

A seasonal variation in the phytoplankton community is shown in Figure 6. Seasonally, diatoms accounted for 95.2% ± 6.6% of the phytoplankton community during winter from early July to late October in 2018, whereas dinoflagellates were mostly the second dominant group in autumn (14.9% ± 5.9%). *Phaeocystis antarctica* was dominant during spring (10.9% ± 12.4%), and cryptophytes were observed to be relatively high in summer (5.2% ± 5.3%). Other phytoplankton communities were rarely observed in this study.

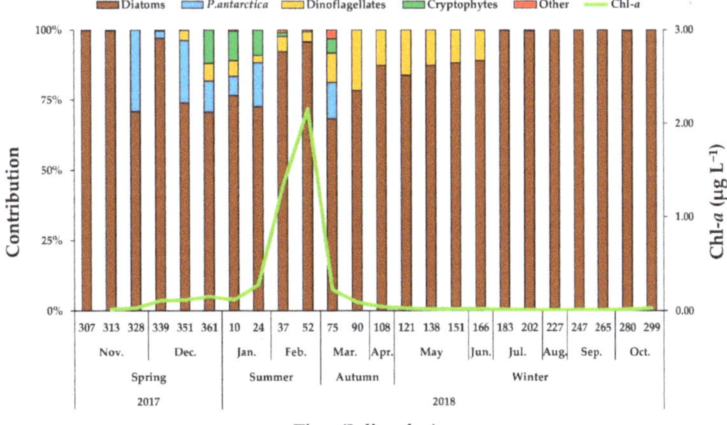

Figure 6. Contribution of phytoplankton communities at the JBS during our sampling period.

3.5. Macromolecular Composition of Phytoplankton

Strong seasonal variations in the concentrations of proteins and lipids (one-way ANOVA test, $p < 0.05$) were found, whereas no significant difference in carbohydrate concentrations was observed during the season (one-way ANOVA test, $p > 0.05$) in this study (Figure 7a). During the observation period, the overall carbohydrate, protein, and lipid concentrations of POM were 50.3–167.5 µg L^{-1} (123.7 ± 31.7 µg L^{-1}), 0–226.1 µg L^{-1} (39.3 ± 68.9 µg L^{-1}), and 51.1–141.9 µg L^{-1} (73.8 ± 27.1 µg L^{-1}), respectively (Figure 7a). The average contributions of carbohydrates, proteins, and lipids to the total POM during spring were 53.0% ± 11.0%, 9.6% ± 11.5%, and 37.3% ± 8.4%, respectively (Figure 7b). The average contributions of each component in summer were 33.3% ± 5.8%, 39.1% ± 5.2%, and 27.6% ± 2.2%, respectively. In autumn and winter, the carbohydrate, protein, and lipid compositions were 58.7% ± 15.0%, 10.5% ± 12.2%, and 30.8% ± 5.1% and 67.4% ± 5.5%, 0.1% ± 0.3%, and 32.5% ± 5.4%, respectively. The contribution of proteins was observed to be high in summer but decreased sharply after autumn. Based on the sum of each macromolecular composition, the highest averaged FM concentration was observed in summer (437.8 ± 76.0 µg L^{-1}), followed by autumn (239.6 ± 45.4 µg L^{-1}), spring (203.3 ± 84.8 µg L^{-1}), and winter (181.2 ± 34.5 µg L^{-1}) (Figure 7a). In comparison, the averaged caloric contents of FM had the highest value in summer (2.7 ± 0.5 Kcal m^{-3}) followed by autumn (1.4 ± 0.4 Kcal m^{-3}), spring (1.3 ± 0.5 Kcal m^{-3}), and winter (1.1 ± 0.2 Kcal m^{-3}) (Figure 8). Proteins were rarely detected whereas lipids and carbohydrates remained relatively constant during winter.

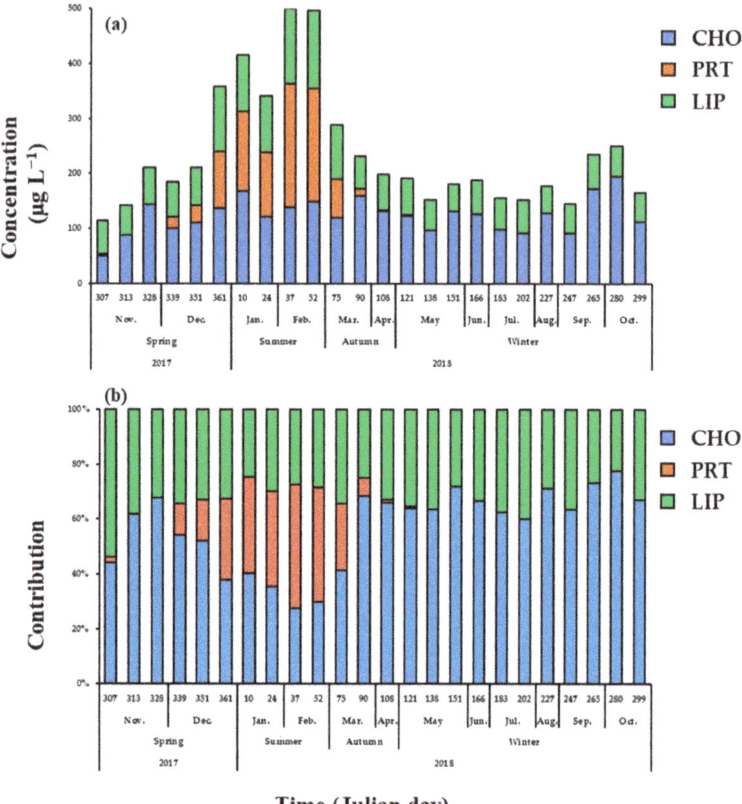

Time (Julian day)

Figure 7. Variations in concentrations (**a**) and relative contributions (**b**) of carbohydrate (CHO), protein (PRT), and lipid (LIP) concentrations (**a**) at the JBS during our sampling period.

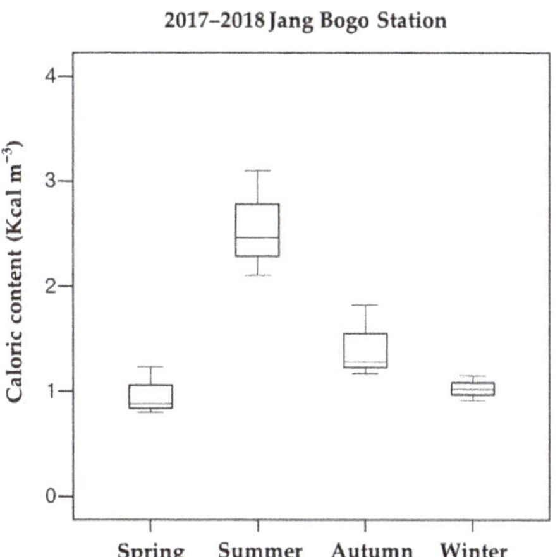

Figure 8. Seasonal variations in the caloric content of FM at the JBS during our sampling period.

3.6. TEP Concentration

The TEP concentration ranged from 10.2 to 96.1 μg Xeq. L^{-1} (32.7 ± 24.6 μg Xeq. L^{-1}) during our observation period from December 2017 to October 2018 (Figure 5b). The highest TEP concentration was observed during February, and the TEP concentrations were nearly stable during the winter period after April (18.3 ± 6.9 μg Xeq. L^{-1}). Overall, the TEP concentrations were significantly correlated with Chl-*a* and fucoxanthine concentrations in this study (Figure 9).

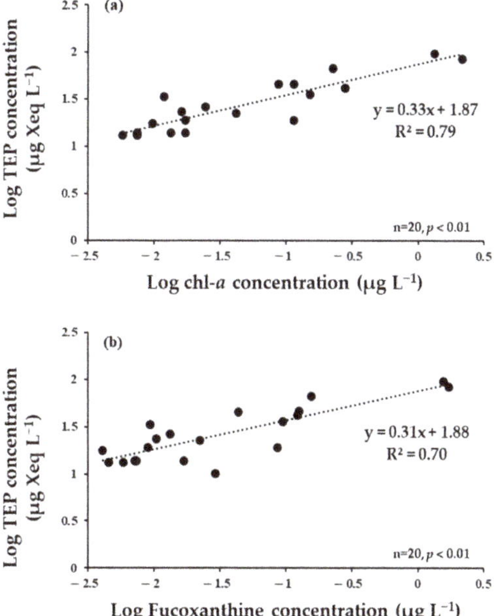

Figure 9. Correlations between log TEP concentrations and log Chl-*a* (**a**) and between log TEP concentrations and log fucoxanthine (**b**) at the JBS during our sampling period.

The concentrations of TEP-C ranging from 7.6 to 62.8 μg Xeq. L^{-1} (24.5 ± 18.4 μg Xeq. L^{-1}) were significantly different among each season (one-way ANOVA test, $p < 0.05$). The average contributions of TEP-C to the total POC were relatively higher in autumn (26.9% ± 6.1%), followed by summer (21.9% ± 7.1%), winter (13.0% ± 4.2%), and spring (9.8% ± 3.1%) (Figure 10).

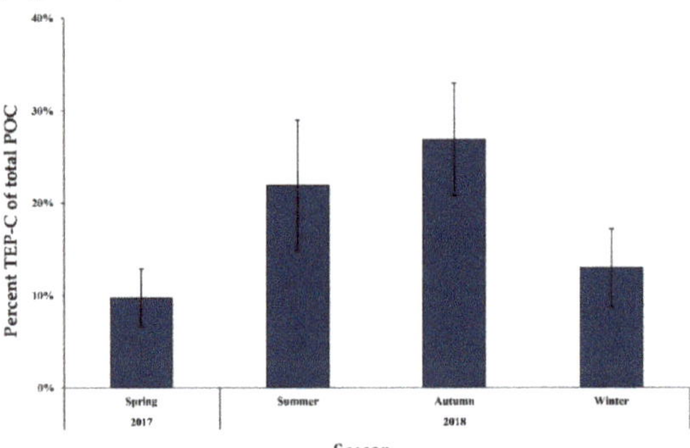

Figure 10. Seasonal variations in percent TEP-C of total POC at the JBS during our sampling period.

3.7. PCA

PCA was conducted to determine major environmental factors controlling macro-molecular concentrations, and TEPs primarily originated from phytoplankton during our sampling period (Figure 11). As a result of the PCA, the first and second principal components explained 64.8% and 16.8%, respectively, of the biochemical variables relative to the environmental factors during our research period. The first principal component was positively correlated with total Chl-*a* concentrations, POC, TEPs, lipids, proteins, diatoms, and dinoflagellates (eigenvectors of 0.957, 0.869, 0.911, 0.854, 0.894, 0.975, and 0.881, respectively) but negatively correlated with $NO_3 + NO_2$ (eigenvectors of -0.773). The second principal component was positively correlated with NH_4, water temperature, cryptophytes, and *P. antarctica* (eigenvectors of 0.909, 0.704, 0.851, and 0.828, respectively) but had negative loadings for salinity (eigenvectors of -0.719). However, we did not find a correlation between carbohydrates with either the first or second principal components.

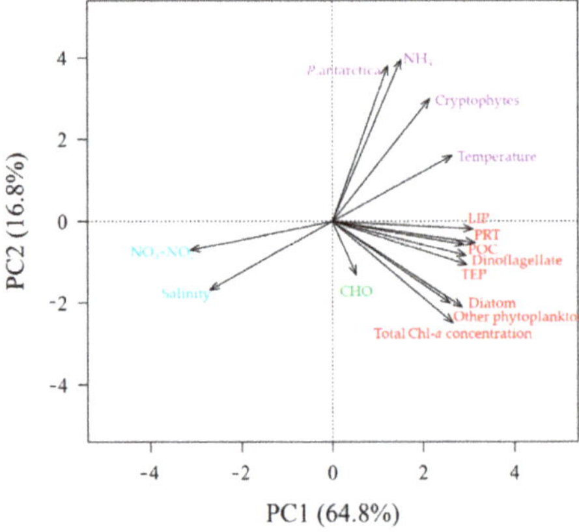

Figure 11. Principal component analysis based on environmental factors and biochemical variables.

4. Discussion

4.1. Environmental Conditions

The seasonal variations in temperature and salinity at the JBS were lower than those previously reported from other studies in the Antarctic coastal environment [52,53]. In spring and summer, relatively large variations in temperature and salinity were observed with the melting of sea ice, but they were nearly stable during the winter period, as expected. The ice-free period in our study was only 12 days, which is a relatively shorter period compared with a previous study in which the ice-free period lasted more than a month [23].

The concentrations of major inorganic nutrients were generally abundant throughout the study period, although relatively lower concentrations were found during the phytoplankton growth season in summer. Because of the consistently high concentrations, the growth of phytoplankton in Antarctic coastal waters is not limited by major inorganic nutrients [54,55]. We estimated nutrient utilization following [55] by subtracting the average concentration in summer from the average concentration in winter, which was largely not affected by sea ice melting or biological uptake. Although nitrate and phosphate in our study (14.7 and 1.0 μM, respectively) were within the range of net nutrient utilization values (8.9–23.0 and 0.9–1.6 μM, respectively) measured in a previous study in TNB [55], the utilization of silicate was noticeably higher in our study (29.0 μM) than in [55] (12.8–23.0 μM). Silicate is necessary for the growth of diatoms [55], which is consistent with our pigment results for diatom-dominant phytoplankton communities observed at JBS in our study.

4.2. Seasonal Variations in Biomass and Major Species Compositions of Phytoplankton

Large variations in both Chl-*a* and POC concentrations were observed in spring and summer, but they were nearly stable in winter. Generally, Chl-*a* concentrations in the Ross Sea reach their maximum in spring and decline thereafter, and then the second peak can be observed again in summer [17,56]. In this study, one short peak of Chl-*a* concentration was distinct in summer. This difference could be due to different locations between this and other studies. This study was conducted at the JBS, which is in a coastal area. Consistently, the authors of [57] reported that a phytoplankton bloom appeared in coastal waters of the Ross Sea in late summer. In addition, the maximum biomass of phytoplankton was reported in February, although interannual variability in biomass was observed in the Ross Sea according to [58]. The maximum concentration of Chl-*a* at the JBS in 2018 was relatively low compared to those from previous studies in other Antarctic coastal regions in summer [52,53,59,60]. Consistently, the maximum concentration of POC in summer also showed a relatively lower value compared with those of previous studies conducted in coastal Antarctica [20,61,62]. Our maximum Chl-*a* concentration (2.15 μg L^{-1}) was also relatively lower than that of [23], which was previously conducted in 2015 (4.29 μg L^{-1}). In addition, the average total Chl-*a* concentrations were lower for all seasons during this study compared with those in 2015 [23]. This could have been due to different growing season periods between the two studies at the JBS. This study had a relatively shorter ice-free growing season (approximately 12 days) than that in 2015 (>1 month). The considerable contribution of nano-sized phytoplankton (>40%) in spring and summer persisted through winter in this study whereas micro-sized phytoplankton (>40%) continued to dominate in the autumn and winter in 2015 [23]. These results could support the results in a previous study reporting that the phytoplankton community structure in winter could be determined in a previous season [5].

Seasonal variations in Chl-*a* concentrations were also reflected in the POC and PON concentrations in this study, suggesting that the majority of POM in Antarctic coastal waters originated directly from phytoplankton [20]. Similarly, Fabiano et al. [61] found that the sea ice melting process in TNB was a key factor influencing the distribution and composition of POM originating from phytoplankton. The C/N ratio in summer in this study (6.02 ± 0.48) was very similar to that in previous studies (6.2–6.5) conducted in TNB during summer [20,61]. This indicates a typical POM dominated by fresh algal material [63]. Furthermore, Fabiano and Pusceddu [20] found that the C/N ratio continued to increase until the end of their study period in middle February. We also found an increasing C/N ratio from autumn to winter, which suggests that N-based materials decreased more than C-based materials in POM. This is consistent with the macromolecular compositions of POM with dominant carbohydrates and lipids in the winter season, as we discuss later.

According to Smith and Asper [64], major phytoplankton communities include diatoms and *P. antarctica* in the Ross Sea. In particular, both groups occur in the spring, whereas diatoms dominate summer phytoplankton assemblages [65]. During the entire study period in 2018, the phytoplankton community was mostly composed of diatoms (88.8% ± 11.6%), followed by dinoflagellates (5.2% ± 6.3%), *P. antarctica* (4.2% ± 8.1%), and cryptophytes (1.6% ± 3.6%). Although the composition of *P. antarctica* was highest in spring in this study, it accounted for only 10.9% of the phytoplankton community. The proportion of dinoflagellates increased in autumn and winter during this study. This was similar to previous results, which accounted for approximately 10%–20% of the composition of dinoflagellates during summer in TNB [66]. Arrigo et al. [67] found different N/P ratios depending on the major phytoplankton community in the Ross Sea. The N/P ratios were 9.7 and 19.2 for diatom- and *P. antarctica*–dominated communities, respectively [67]. The average N/P ratios were 12.7 (±0.9) and 12.2 (±1.6) in spring and summer at the JBS, respectively, which were similar to the value (11.0 ± 0.9) from a previous study with diatom blooms during late summer in the Ross Sea [68].

4.3. Seasonal Variations of Macromolecular Composition

The macromolecular compositions of POM mainly originating from phytoplankton are affected by various factors, such as nutrients [15,69,70], light conditions [16,71], species composition [14,72,73], and different growth phases of phytoplankton [74,75]. Considering the incomplete removal of macronutrients (e.g., nitrate) at the end of the growth phase, iron and irradiance could be assumed to exert major controls on the productivity of phytoplankton in the Ross Sea [4,76–79]. The light intensity is sufficiently strong in spring [80], but POM cannot be synthesized by phytoplankton in the water column due to the blocking of light by sea ice. This could cause the low composition of proteins and high composition of lipids during spring in our study. The highest contribution of proteins observed in summer represented high growth rates and physiologically healthy conditions of phytoplankton [16]. The assemblages and species of phytoplankton could also have significant impacts on seasonal variations in macromolecular composition [81,82]. Previous studies have claimed that the dominance of *P. antarctica* biomass could be a key reason for the high carbohydrate concentrations [83,84]. In contrast, polar diatoms have been observed to have a large accumulation of lipids, partially related to temperature effects by altering the allocation of carbon into storage products, such as cytoplasmic lipid droplets [85,86]. This may explain the relatively higher contents of lipids in this study with diatom-predominant communities compared with those in other studies [84,87] during the spring and summer seasons. Indeed, the PCA results showed that lipids had the highest correlations with diatoms (r = 0.87, *p* < 0.01) (Table 1).

In addition, proteins were positively correlated with diatoms in our study (r = 0.94, *p* < 0.01), which is consistent with previous results from the Ross Sea (r = 0.59, *p* < 0.01) [88]. These results suggest that polar diatoms in the Southern Ocean have physiological mechanisms to increase protein concentrations to compensate for slow enzyme rates at cold temperatures [89].

In addition, carbohydrates linked to *P. antarctica* rapidly decrease during the degradation process, while carbohydrates linked to diatoms are retained more within their cellular matrix [90]. This indicates that carbohydrates remain persistent throughout the winter season at our study site. Ahn et al. [75] linked the growth phases of phytoplankton with seasonal variations in macromolecular composition. During the exponential phase corresponding to the summer in our study, the POM was mainly composed of proteins (39.1% ± 5.2%) as a functional reservoir of cell growth and division [91]. In contrast, cell storage compounds such as carbohydrates and lipids tended to increase in the stationary periods [92], corresponding to the autumn and winter in our study, with high carbohydrate (58.7% ± 15.0% and 67.4% ± 5.5%, respectively) and lipid contents (30.8% ± 5.1% and 32.5% ± 5.4%, respectively).

Table 1. Pearson's correlation analysis for the environmental factors and biochemical variables.

	1	2	3	4	5	6	7	8	9	10	11	12	13	14	15
1. Salinity	1	−0.93 **	0.88 **	−0.58 **	−0.61 **	−0.52 *	−0.72 **	−0.60 **	−0.68 **	-	−0.82 **	−0.50 *	-	−0.81 **	−0.77 **
2. Temperature		1	−0.88 **	0.61 **	0.63 **	0.55 *	0.66 **	0.62 **	0.64 **	-	0.71 **	-	-	0.80 **	0.73 **
3. $NO_3^- + NO_2^-$			1	−0.59 **	−0.86 **	−0.76 **	−0.87 **	−0.82 **	−0.86 **	−0.50 *	−0.68 **	−0.71 **	-	−0.94 **	−0.92 **
4. NH_4^+				1	-	-	-	-	-	0.76 **	0.68 **	-	-	-	-
5. POC					1	0.87 **	0.80 **	0.87 **	0.80 **	-	0.47 *	0.76 **	-	0.90 **	0.92 **
6. Chl-*a*						1	0.83 **	0.97 **	0.77 **	-	-	0.80 **	-	0.88 **	0.83 **
7. TEPs							1	0.89 **	0.95 **	-	0.49 *	0.90 **	-	0.89 **	0.87 **
8. Diatoms								1	0.87 **	-	-	0.85 **	-	0.94 **	0.87 **
9. Dinoflagellates									1	-	0.54 *	0.90 **	-	0.88 **	0.86 **
10. *P. antarctica*										1	0.62 **	-	-	-	-
11. Cryptophytes											1	-	-	0.61 **	0.67 **
12. Other phytoplankton												1	-	0.78 **	0.78 **
13. CHO													1	-	-
14. PRT														1	0.97 **
15. LIP															1

"-" indicates not significant, * $p < 0.05$, ** $p < 0.01$.

The relative contributions of different macromolecule compositions in phytoplankton can be an indicator of the quality of phytoplankton as primary food resources [75]. Proteins have a relatively higher efficiency in carbon transfer to herbivores than other macromolecular components [16,93]. Carbohydrates and lipids are necessary components for all cell membranes as well as major energy reservoirs in phytoplankton bodies [94,95]. In particular, glucan in carbohydrates is mobilized during respiration to maintain growth in the dark [94]. In terms of quality aspects, the calorie content of POM represents the condition of energy sources. Although the overall Chl-*a* concentrations were lower than those in other studies, the overall calorie content (1.42 ± 0.68 Kcal m^{-3}) in this study was similar to that from previous studies measured primarily in spring or summer [21,63,87]. The calorie content was high even during the polar night period, when photosynthesis was assumed to rarely occur. Although phytoplankton had a very low productivity in polar dark winters, they could maintain high caloric contents by having higher carbohydrate and lipid contents (Figure 8). Carbohydrates and lipids are maintained by respiration and long-term energy storage [96] for the survival of phytoplankton in winter.

4.4. Seasonal Variations of TEPs

Several previous studies on the distribution of TEPs in the Southern Ocean have previously been conducted in the Antarctic Peninsula [29,35,97], Drake Passage [97], and Ross Sea [36]. However, these previous studies were conducted during spring or summer. This is the first report of the annual variability in TEP concentrations in the Southern Ocean. The TEP concentrations obtained at the JBS during this study ranged from 10.2 to 96.1 μg Xeq. L^{-1}, which generally falls within the range observed in other studies from various oceans around the world [98–101]. However, the overall TEP concentrations obtained in this study were relatively lower than those previously reported for the Southern Ocean (Table 2).

Table 2. Comparison of the TEP concentrations between this and previous studies from the Southern Ocean.

Region	Season	Depth (m)	TEP Range (Mean)	TEPs/Chl-*a*	TEPs–Chl-*a* Relationship	References
			(μg Xeq. L^{-1})	(Mean)	(Log Converted)	
Anvers Island	summer	Surface	10–407 (207)	12–708 (123)	Not related	Passow et al. (1995) [35]
Ross Sea	summer	0–150	0–2800 (308)	89.1	y = 3.63x + 1.01	Hong et al. (1997) [36]
Bransfield Strait	summer	0–100	0–346 (57)	51	y = 0.32x + 1.63	Corzo et al. (2005) [97]
Gerlache Strait	summer	0–100	0–283	32.7	y = 0.67x + 1.52	Corzo et al. (2005) [97]
Northern Weddell Sea	summer	Surface	39.2–177.6 (102.3)	79.3	y = 0.35x + 1.90	Zamanillo et al. (2019) [34]
Antarctic Peninsula	summer	0–200	0–48.9 (15.4)	0–1492 (40.9)	y = 0.38x + 1.08	Ortega-Retuerta et al. (2009) [29]
Jang Bogo Station	summer	Surface	45.15–96.1 (73.0)	38.9–404.3 (165.9)	y = 0.33x + 1.87	This Study
Jang Bogo Station	winter	Surface	10–33.5 (17.9)	371.1–2807.5 (1477.9)	Not related	This Study

The species composition, biomass, and growth conditions of phytoplankton and environmental factors could have a great influence on the concentrations of TEPs [25,32,34]. Based on the PCA results (Figure 11), the seasonal variation in TEPs was primarily affected by diatoms and dinoflagellates in our study period. The TEP concentrations were positively correlated with the abundance of diatoms (r = 0.89, $p < 0.01$) and dinoflagellates (r = 0.95, $p < 0.01$) (Table 1). Similarly, previous studies reported that TEP concentrations in the Southern Ocean were significantly related to the abundance of diatoms (r = 0.67, $p < 0.01$) and dinoflagellates (r = 0.67, $p < 0.01$) [34]. In contrast, we found that TEP concentrations had no significant relationship with *P. antarctica* in our study, which conflicts with a previous finding that *P. antarctica* is a major producer of TEPs in the Southern Ocean [36], probably because the phytoplankton community was not dominated by *P. antarctica* in this study. In fact, TEP production by phytoplankton depends on the species composition of phytoplankton [25,32]. Hong et al. [36] observed that the concentrations of TEPs measured during a *P. antarctica*–dominant bloom were higher than those produced by coastal diatom blooms in the Ross Sea. The relatively lower TEP concentrations in this study could have resulted from the diatom-dominant phytoplankton community at the JBS. Additionally, the lower Chl-*a* concentrations observed in this study could be another reason for the lower TEP concentration since a significant relationship was found between the TEPs and Chl-*a* concentrations during this study period (Figure 10a). The relationship between TEPs and Chl-*a* can be expressed as TEPs = a(Chl-*a*)^b [97]. The slope, *a*, of the log-converted TEP and Chl-*a* relationship in our study (0.33) is comparable to the previously reported values of 0.32, 0.38, and 0.35 for the Antarctic Peninsula, Bransfield Strait, and the Northern Weddell Sea in the Southern Ocean, respectively [29,34,97]. In comparison, Hong et al. [36] reported an extremely high value (3.63) when *P. antarctica* was dominant in the Ross Sea.

TEPs frequently make up a substantial fraction of the POC in the seawater column [100]. To estimate TEP-C, we used the lowest conversion factor (0.51 µg Xeq. L^{-1}) commonly used for the diatom-dominated phytoplankton community [51,102]. Considerably large variations in the contribution of TEP-C to the POC stocks were observed seasonally during our observation period (Figure 9), which was different from the seasonal pattern in the TEP concentrations. TEP concentrations were highest in summer, but the contribution of TEP-C to the POC was highest in autumn (26.9% ± 6.1%). This suggests that TEPs could have a longer residence time than POC in the water column [29]. Previous studies have shown relatively faster degradation of protein compared to other components in the Antarctic Ocean [23,84]. Indeed, a decrease in the protein components of POC from the spring growth season to the summer and winter seasons was observed in this study (Figure 7). Therefore, TEP-C could stay longer and accumulate in comparison to phytoplankton or bacterial cells in the water column [29]. Although there have been few studies on TEP-C in the Southern Ocean, our average contribution of TEP-C to the total POC concentration in summer (21.9% ± 7.1%) was lower than that in the northern Weddell Sea (38.8% ± 12.3%) [34] but comparable to that on the Antarctic Peninsula (18%) [29]. The lower TEP-C contributions to the total POC suggest that relatively more particulate-based organic carbon materials were available in the water column during our observation period.

Author Contributions: Conceptualization, S.P., J.P., K.-C.Y., and S.-H.L.; methodology, S.P., K.K., N.J., and H.-K.J.; software, S.P.; validation, S.P. and S.-H.L.; formal analysis, S.P and S.-H.L.; investigation, S.P., K.-C.Y., and J.Y.; writing—original draft preparation, S.P.; writing—review and editing, S.-H.L.; visualization, S.P., K.K., N.J., H.-K.J., J.K. (Jaehong Kim), J.K. (Jaesoon Kim), and J.K. (Joonmin Kim); supervision, S.-H.L. All authors have read and agreed to the published version of the manuscript.

Funding: This research was funded by the National Research Foundation of Korea (NRF) grant funded by the Korea government (MSIT; NRF-2019R1A2C1003515) and the Korea Polar Research Institute (KOPRI; PE21110).

Institutional Review Board Statement: Not applicable.

Informed Consent Statement: Not applicable.

Data Availability Statement: Not applicable.

Acknowledgments: We sincerely thank members of the fifth overwintering team at the JBS for collecting seawater samples safely for 1 year. We would like to thank the anonymous reviewers and the handling editors for their constructive and valuable recommendations.

Conflicts of Interest: The authors declare no conflict of interest.

References

1. Nelson, D.M.; DeMaster, D.J.; Dunbar, R.B.; Smith, W.O. Cycling of organic carbon and biogenic silica in the Southern Ocean: Estimates of water-column and sedimentary fluxes on the Ross Sea continental shelf. *J. Geophys. Res. C Ocean.* **1996**, *101*, 18519–18532. [CrossRef]
2. Smith, W.O.; Nelson, D.M. Phytoplankton bloom produced by a receding ice edge in the ross sea: Spatial coherence with the density field. *Science* **1985**, *227*, 163–166. [CrossRef]
3. Sedwick, P.N.; Ditullio, G.R. Regulation of algal blooms in Antarctic shelf waters by the release of iron from melting sea ice. *Geophys. Res. Lett.* **1997**, *24*, 2515–2518. [CrossRef]
4. Sedwick, P.N.; Di Tullio, G.R.; Mackey, D.J. Iron and manganese in the Ross Sea, Seasonal iron limitation in Antarctic: Seasonal iron limitation in Antarctic shelf waters. *J. Geophys. Res. Ocean.* **2000**, *105*, 11321–11336. [CrossRef]
5. Krell, A.; Schnack-Schiel, S.B.; Thomas, D.N.; Kattner, G.; Zipan, W.; Dieckmann, G.S. Phytoplankton dynamics in relation to hydrography, nutrients and zooplankton at the onset of sea ice formation in the eastern Weddell Sea (Antarctica). *Polar Biol.* **2005**, *28*, 700–713. [CrossRef]
6. Garrison, D.L.; Buck, K.R. Surface-layer sea ice assemblages in Antarctic pack ice during the austral spring: Environmental conditions, primary production and community structure. *Mar. Ecol. Prog. Ser.* **1991**, *75*, 161–172. [CrossRef]
7. Smetacek, V.; Scharek, R.; Gordon, L.I.; Eicken, H.; Fahrbach, E.; Rohardt, G.; Moore, S. Early spring phytoplankton blooms in ice platelet layers of the southern Weddell Sea, Antarctica. *Deep Sea Res. Part A Oceanogr. Res. Pap.* **1992**, *39*, 153–168. [CrossRef]
8. Brierley, A.S.; Thomas, D.N. Ecology of Southern Ocean pack ice. *Adv. Mar. Biol.* **2002**, *43*, 171–276.
9. Parkinson, C.L.; Cavalieri, D.J. Antarctic sea ice variability and trends, 1979–2010. *Cryosphere* **2012**, *6*, 871–880. [CrossRef]
10. Eayrs, C.; Holland, D.; Francis, D.; Wagner, T.; Kumar, R.; Li, X. Understanding the Seasonal Cycle of Antarctic Sea Ice Extent in the Context of Longer-Term Variability. *Rev. Geophys.* **2019**, *57*, 1037–1064. [CrossRef]
11. Goffart, A.; Catalano, G.; Hecq, J.H. Factors controlling the distribution of diatoms and Phaeocystis in the Ross Sea. *J. Mar. Syst.* **2000**, *27*, 161–175. [CrossRef]
12. Grotti, M.; Soggia, F.; Ianni, C.; Frache, R. Trace metals distributions in coastal sea ice of Terra Nova Bay, Ross Sea, Antarctica. *Antarct. Sci.* **2005**, *17*, 289–300. [CrossRef]
13. Smith, W.O.; Shields, A.R.; Peloquin, J.A.; Catalano, G.; Tozzi, S.; Dinniman, M.S.; Asper, V.A. Interannual variations in nutrients, net community production, and biogeochemical cycles in the Ross Sea. *Deep Res. Part II Top. Stud. Oceanogr.* **2006**, *53*, 815–833. [CrossRef]
14. Finkel, Z.V.; Follows, M.J.; Liefer, J.D.; Brown, C.M.; Benner, I.; Irwin, A.J. Phylogenetic diversity in the macromolecular composition of microalgae. *PLoS ONE* **2016**, *11*, e0155977. [CrossRef]
15. Ben-Amotz, A.; Tornabene, T.G.; Thomas, W.H. Chemical Profile of Selected Species of Microalgae With Emphasis on Lipids. *J. Phycol.* **1985**, *21*, 72–81. [CrossRef]
16. Lee, S.H.; Kim, H.J.; Whitledge, T.E. High incorporation of carbon into proteins by the phytoplankton of the Bering Strait and Chukchi Sea. *Cont. Shelf Res.* **2009**, *29*, 1689–1696. [CrossRef]
17. Smith, W.O.; Marra, J.; Hiscock, M.R.; Barber, R.T. The seasonal cycle of phytoplankton biomass and primary productivity in the Ross Sea, Antarctica. *Deep Res. Part II Top. Stud. Oceanogr.* **2000**, *47*, 3119–3140. [CrossRef]
18. Olson, R.J.; Sosik, H.M.; Chekalyuk, A.M.; Shalapyonok, A. Effects of iron enrichment on phytoplankton in the Southern Ocean during late summer: Active fluorescence and flow cystometric analyses. *Deep Res. Part II Top. Stud. Oceanogr.* **2000**, *47*, 3181–3200. [CrossRef]
19. Fabiano, M.; Povero, P.; Danovaro, R. Participate organic matter composition in Terra Nova Bay (Ross Sea, Antarctica) during summer 1990. *Antarct. Sci.* **1996**, *8*, 7–13. [CrossRef]
20. Fabiano, M.; Pusceddu, A. Total and hydrolizable particulate organic matter (carbohydrates, proteins and lipids) at a coastal station in Terra Nova Bay (Ross Sea, Antarctica). *Polar Biol.* **1998**, *19*, 125–132. [CrossRef]
21. Pusceddu, A.; Cattaneo-Vietti, R.; Albertelli, G.; Fabiano, M. Origin, biochemical composition and vertical flux of particulate organic matter under the pack ice in Terra Nova Bay (Ross Sea, Antarctica) during late summer 1995. *Polar Biol.* **1999**, *22*, 124–132. [CrossRef]
22. Baldi, F.; Marchetto, D.; Pini, F.; Fani, R.; Michaud, L.; Lo Giudice, A.; Berto, D.; Giani, M. Biochemical and microbial features of shallow marine sediments along the Terra Nova Bay (Ross Sea, Antarctica). *Cont. Shelf Res.* **2010**, *30*, 1614–1625. [CrossRef]
23. Kim, K.; Park, J.; Jo, N.; Park, S.; Yoo, H.; Kim, J.; Lee, S.H. Monthly Variation in the Macromolecular Composition of Phytoplankton Communities at Jang Bogo Station, Terra Nova Bay, Ross Sea. *Front. Microbiol.* **2021**, *12*, 618999. [CrossRef]
24. Alldredge, A.L.; Passow, U.; Logan, B.E. The abundance and significance of a class of large, transparent organic particles in the ocean. *Deep Sea Res. Part I Ocean. Res. Pap.* **1993**, *40*, 1131–1140. [CrossRef]
25. Passow, U. Production of transparent exopolymer particles (TEPS) by phyto- and bacterioplankton. *Mar. Ecol. Prog. Ser.* **2002**, *236*, 1–12. [CrossRef]
26. Passow, U. Formation of transparent exopolymer particles, TEPS, from dissolved precursor material. *Mar. Ecol. Prog. Ser.* **2000**, *192*, 1–11. [CrossRef]
27. Mari, X.; Passow, U.; Migon, C.; Burd, A.B.; Legendre, L. Transparent exopolymer particles: Effects on carbon cycling in the ocean. *Prog. Oceanogr.* **2017**, *151*, 13–37. [CrossRef]

28. Passow, U.; Shipe, R.F.; Murray, A.; Pak, D.K.; Brzezinski, M.A.; Alldredge, A.L. The origin of transparent exopolymer particles (TEPS) and their role in the sedimentation of particulate matter. *Cont. Shelf Res.* **2001**, *21*, 327–346. [CrossRef]
29. Ortega-Retuerta, E.; Reche, I.; Pulido-Villena, E.; Agustí, S.; Duarte, C.M. Uncoupled distributions of transparent exopolymer particles (TEPS) and dissolved carbohydrates in the Southern Ocean. *Mar. Chem.* **2009**, *115*, 59–65. [CrossRef]
30. Mari, X.; Rassoulzadegan, F.; Brussaard, C.P.D.; Wassmann, P. Dynamics of transparent exopolymeric particles (TEPS) production by Phaeocystis globosa under N- or P-limitation: A controlling factor of the retention/export balance. *Harmful Algae* **2005**, *4*, 895–914. [CrossRef]
31. Beauvais, S.; Pedrotti, M.L.; Egge, J.; Iversen, K.; Marrasé, C. Effects of turbulence on TEPS dynamics under contrasting nutrient conditions: Implications for aggregation and sedimentation processes. *Mar. Ecol. Prog. Ser.* **2006**, *323*, 47–57. [CrossRef]
32. Hung, C.C.; Guo, L.; Santschi, P.H.; Alvarado-Quiroz, N.; Haye, J.M. Distributions of carbohydrate species in the Gulf of Mexico. *Mar. Chem.* **2003**, *81*, 119–135. [CrossRef]
33. Passow, U. Transparent exopolymer particles (TEPS) in aquatic environments. *Prog. Oceanogr.* **2002**, *55*, 287–333. [CrossRef]
34. Zamanillo, M.; Ortega-Retuerta, E.; Nunes, S.; Estrada, M.; Sala, M.M.; Royer, S.J.; López-Sandoval, D.C.; Emelianov, M.; Vaqué, D.; Marrasé, C.; et al. Distribution of transparent exopolymer particles (TEPS) in distinct regions of the Southern Ocean. *Sci. Total Environ.* **2019**, *691*, 736–748. [CrossRef]
35. Passow, U.; Alldredge, A.L. A dye-binding assay for the spectrophotometric measurement of transparent exopolymer particles (TEPS). *Limnol. Oceanogr.* **1995**, *40*, 1326–1335. [CrossRef]
36. Hong, Y.; Smith, W.O.; White, A.M. Studies on transparent exopolymer particles (TEPS) produced in the ross sea (Antarctica) and by *Phaeocystis antarctica* (Prymnesiophyceae). *J. Phycol.* **1997**, *33*, 368–376. [CrossRef]
37. Sugimoto, K.; Fukuda, H.; Baki, M.A.; Koike, I. Bacterial contributions to formation of transparent exopolymer particles (TEPS) and seasonal trends in coastal waters of Sagami Bay, Japan. *Aquat. Microb. Ecol.* **2007**, *46*, 31–41. [CrossRef]
38. Huertas, E.; Navarro, G.; Rodríguez-Gálvez, S.; Prieto, L. The influence of phytoplankton biomass on the spatial distribution of carbon dioxide in surface sea water of a coastal area of the Gulf of Cádiz (southwestern Spain). *Can. J. Bot.* **2005**, *83*, 929–940. [CrossRef]
39. Parsons, T.R.; Maita, Y.; Lalli, C.M. *A Manual of Chemical and Biological Methods for Seawater Analysis*; Pergamon Press: New York, NY, USA, 1984.
40. Mackey, M.D.; Mackey, D.J.; Higgins, H.W.; Wright, S.W. CHEMTAX—A program for estimating class abundances from chemical markers: Application to HPLC measurements of phytoplankton. *Mar. Ecol. Prog. Ser.* **1996**, *144*, 265–283. [CrossRef]
41. Wright, S.W.; Thomas, D.P.; Marchant, H.J.; Higgins, H.W.; Mackey, M.D.; Mackey, D.J. Analysis of phytoplankton of the Australian sector of the Southern Ocean: Comparisons of microscopy and size frequency data with interpretations of pigment HPLC data using the "CHEMTAX" matrix factorisation program. *Mar. Ecol. Prog. Ser.* **1996**, *144*, 285–298. [CrossRef]
42. Wright, S.W.; Van den Enden, R.L. Phytoplankton community structure and stocks in the East Antarctic marginal ice zone (BROKE survey, January-March 1996) determined by CHEMTAX analysis of HPLC pigment signatures. *Deep Res. Part II Top. Stud. Oceanogr.* **2000**, *47*, 2363–2400. [CrossRef]
43. DiTullio, G.R.; Geesey, M.E.; Jones, D.R.; Daly, K.L.; Campbell, L.; Smith, W.O. Phytoplankton assemblage structure and primary productivity along 170° W in the South Pacific Ocean. *Mar. Ecol. Prog. Ser.* **2003**, *255*, 55–80. [CrossRef]
44. Dubois, M.; Gilles, K.A.; Hamilton, J.K.; Rebers, P.A.; Smith, F. Colorimetric Method for Determination of Sugars and Related Substances. *Anal. Chem.* **1956**, *28*, 350–356. [CrossRef]
45. Lowry, O.H.; Rosebrough, N.J.; Farr, A.L.; Randall, R.J. Protein measurement with the Folin phenol reagent. *J. Biol. Chem.* **1951**, *193*, 265–275. [CrossRef]
46. Bligh, E.G.; Dyer, W.J. A rapid method of total lipid extraction and purification. *Can. J. Biochem. Physiol.* **1959**, *37*, 911–917. [CrossRef]
47. Marsh, J.B.; Weinstein, D.B. Simple charring method for determination of lipids. *J. Lipid Res.* **1966**, *7*, 574–576. [CrossRef]
48. Bhavya, P.S.; Kim, B.K.; Jo, N.; Kim, K.; Kang, J.J.; Lee, J.H.; Lee, D.; Lee, J.H.; Joo, H.T.; Ahn, S.H.; et al. A Review on the Macromolecular Compositions of Phytoplankton and the Implications for Aquatic Biogeochemistry. *Ocean Sci. J.* **2019**, *54*, 1–14. [CrossRef]
49. Danovaro, R.; Dell'Anno, A.; Pusceddu, A.; Marrale, D.; Della Croce, N.; Fabiano, M.; Tselepides, A. Biochemical composition of pico-, nano- and micro-particulate organic matter and bacterioplankton biomass in the oligotrophic Cretan Sea (NE Mediterranean). *Prog. Oceanogr.* **2000**, *46*, 279–310. [CrossRef]
50. Winberg, G.G. *Symbols, Units and Conversion Factors in Study of Fresh Waters Productivity*; International Biological Programme: London, UK, 1971.
51. Engel, A.; Passow, U. Carbon and nitrogen content of transparent exopolymer particles (TEPS) in relation to their Alcian Blue adsorption. *Mar. Ecol. Prog. Ser.* **2001**, *219*, 1–10. [CrossRef]
52. Schloss, I.R.; Abele, D.; Ferreyra, G.A.; González, O.; Moreau, S.; Demers, S. Response of Potter Cove phytoplankton dynamics to long term climate trends. In Proceedings of the Open Science Conference XXXI SCAR, Buenos Aires, Argentina, 3–6 August 2010.
53. Lee, S.H.; Joo, H.M.; Joo, H.T.; Kim, B.K.; Song, H.J.; Jeon, M.; Kang, S.H. Large contribution of small phytoplankton at Marian Cove, King George Island, Antarctica, based on long-term monitoring from 1996 to 2008. *Polar Biol.* **2015**, *38*, 207–220. [CrossRef]
54. Schloss, I.R.; Ferreyra, G.A.; Ruiz-Pino, D. Phytoplankton biomass in Antarctic shelf zones: A conceptual model based on Potter Cove, King George Island. *J. Mar. Syst.* **2002**, *36*, 129–143. [CrossRef]

55. Rivaro, P.; Luisa Abelmoschi, M.; Grotti, M.; Ianni, C.; Magi, E.; Margiotta, F.; Massolo, S.; Saggiomo, V. Combined effects of hydrographic structure and iron and copper availability on the phytoplankton growth in Terra Nova Bay Polynya (Ross Sea, Antarctica). *Deep Res. Part I Oceanogr. Res. Pap.* **2012**, *62*, 97–110. [CrossRef]

56. Tremblay, J.E.; Smith, W.O. Chapter 8 Primary Production and Nutrient Dynamics in Polynyas. *Elsevier Oceanogr. Ser.* **2007**, *74*, 239–269. [CrossRef]

57. Arrigo, K.R.; Weiss, A.M.; Smith, W.O. Physical forcing of phytoplankton dynamics in the southwestern Ross Sea. *J. Geophys. Res. Ocean.* **1998**, *103*, 1007–1021. [CrossRef]

58. Peloquin, J.A.; Smith, W.O. Phytoplankton blooms in the Ross Sea, Antarctica: Interannual variability in magnitude, temporal patterns, and composition. *J. Geophys. Res. Ocean.* **2007**, *112*, 1–12. [CrossRef]

59. Trimborn, S.; Hoppe, C.J.M.; Taylor, B.B.; Bracher, A.; Hassler, C. Physiological characteristics of open ocean and coastal phytoplankton communities of Western Antarctic Peninsula and Drake Passage waters. *Deep Res. Part I Oceanogr. Res. Pap.* **2015**, *98*, 115–124. [CrossRef]

60. Kim, H.; Ducklow, H.W.; Abele, D.; Barlett, E.M.R.; Buma, A.G.J.; Meredith, M.P.; Rozema, P.D.; Schofield, O.M.; Venables, H.J.; Schloss, I.R. Inter-decadal variability of phytoplankton biomass along the coastalWest Antarctic Peninsula. *Philos. Trans. R. Soc. A Math. Phys. Eng. Sci.* **2018**, *376*, 20170174. [CrossRef]

61. Fabiano, M.; Povero, P.; Danovaro, R. Distribution and composition of particulate organic matter in the Ross Sea (Antarctica). *Polar Biol.* **1993**, *13*, 525–533. [CrossRef]

62. Mangoni, O.; Saggiomo, V.; Bolinesi, F.; Margiotta, F.; Budillon, G.; Cotroneo, Y.; Misic, C.; Rivaro, P.; Saggiomo, M. Phytoplankton blooms during austral summer in the Ross Sea, Antarctica: Driving factors and trophic implications. *PLoS ONE* **2017**, *12*, e0176033. [CrossRef]

63. Pusceddu, A.; Fabiano, M. Changes in the biochemical composition of Tetraselmis suecia and Isochrysis galbana during growth and decay. *Chem. Ecol.* **1996**, *12*, 199–212. [CrossRef]

64. Smith, W.O.; Asper, V.L. The influence of phytoplankton assemblage composition on biogeochemical characteristics and cycles in the southern Ross Sea, Antarctica. *Deep Res. Part I Oceanogr. Res. Pap.* **2001**, *48*, 137–161. [CrossRef]

65. Mangoni, O.; Modigh, M.; Conversano, F.; Carrada, G.C.; Saggiomo, V. Effects of summer ice coverage on phytoplankton assemblages in the Ross Sea, Antarctica. *Deep Res. Part I Oceanogr. Res. Pap.* **2004**, *51*, 1601–1617. [CrossRef]

66. Andreoli, C.; Tolomio, C.; Moro, I.; Radice, M.; Moschin, E.; Bellato, S. Diatoms and dinoflagellates in Terra Nova Bay (Ross Sea-Antarctica) during austral summer 1990. *Polar Biol.* **1995**, *15*, 465–475. [CrossRef]

67. Arrigo, K.R.; Robinson, D.H.; Worthen, D.L.; Dunbar, R.B.; DiTullio, G.R.; VanWoert, M.; Lizotte, M.P. Phytoplankton community structure and the drawdown of nutrients and CO2 in the Southern Ocean. *Science* **1999**, *283*, 365–367. [CrossRef]

68. Saggiomo, V.; Catalano, G.; Mangoni, O.; Budillon, G.; Carrada, G.C. Primary production processes in ice-free waters of the Ross Sea (Antarctica) during the austral summer 1996. *Deep Res. Part II Top. Stud. Oceanogr.* **2002**, *49*, 1787–1801. [CrossRef]

69. Kilham, S.S.; Kreeger, D.A.; Goulden, C.E.; Lynn, S.G. Effects of nutrient limitation on biochemical constituents of Ankistrodesmus falcatus. *Freshw. Biol.* **1997**, *38*, 591–596. [CrossRef]

70. Van Leeuwe, M.A.; Scharek, R.; De Baar, H.J.W.; De Jong, J.T.M.; Goeyens, L. Iron enrichment experiments in the Southern Ocean: Physiological responses of plankton communities. *Deep Res. Part II Top. Stud. Oceanogr.* **1997**, *44*, 189–207. [CrossRef]

71. Morris, I.; Glover, H.E.; Yentsch, C.S. Products of photosynthesis by marine phytoplankton: The effect of environmental factors on the relative rates of protein synthesis. *Mar. Biol.* **1974**, *27*, 1–9. [CrossRef]

72. Liebezeit, G. Particulate carbohydrates in relation to phytoplankton in the euphotic zone of the Bransfield Strait. *Polar Biol.* **1984**, *24*, 225–228. [CrossRef]

73. Moal, J.; Jezequel, V.M.; Harris, R.P.; Samain, J.F.; Poulet, S.A. Interspecific and intraspecific variability of the chemical composition of marine phytoplankton. *Oceanol. Acta* **1987**, *10*, 339–346.

74. Fabregas, J.; Herrero, C.; Cabezas, B.; Abalde, J. Mass culture and biochemical variability of the marine microalga *Tetraselmis suecica* Kylin (Butch) with high nutrient concentrations. *Aquaculture* **1985**, *49*, 231–244. [CrossRef]

75. Ahn, S.H.; Whitledge, T.E.; Stockwell, D.A.; Lee, J.H.; Lee, H.w.; Lee, S.H. The biochemical composition of phytoplankton in the Laptev and East Siberian seas during the summer of 2013. *Polar Biol.* **2019**, *42*, 133–148. [CrossRef]

76. Fitzwater, S.E.; Johnson, K.S.; Gordon, R.M.; Coale, K.H.; Smith, W.O. Trace metal concentrations in the Ross Sea and their relationship with nutrients and phytoplankton growth. *Deep Res. Part II Top. Stud. Oceanogr.* **2000**, *47*, 3159–3179. [CrossRef]

77. Smith, W.O.; Dinniman, M.S.; Klinck, J.M.; Hofmann, E. Biogeochemical climatologies in the Ross Sea, Antarctica: Seasonal patterns of nutrients and biomass. *Deep Res. Part II Top. Stud. Oceanogr.* **2003**, *50*, 3083–3101. [CrossRef]

78. Arrigo, K.R.; van Dijken, G.L. Phytoplankton dynamics within 37 Antarctic coastal polynya systems. *J. Geophys. Res. Ocean.* **2003**, *108*. [CrossRef]

79. Ryan-Keogh, T.J.; DeLizo, L.M.; Smith, W.O.; Sedwick, P.N.; McGillicuddy, D.J.; Moore, C.M.; Bibby, T.S. Temporal progression of photosynthetic-strategy in phytoplankton in the Ross Sea, Antarctica. *J. Mar. Syst.* **2017**, *166*, 87–96. [CrossRef]

80. McKnight, D.M.; Howes, B.L.; Taylor, C.D.; Goehringer, D.D. Phytoplankton dynamics in a stably stratified antarctic lake during winter darkness. *J. Phycol.* **2000**, *36*, 852–861. [CrossRef]

81. Brown, P.C.; Painting, S.J.; Cochrane, K.L. Estimates of phytoplankton and bacterial biomass and production in the northern and southern benguela ecosystems. *S. Afr. J. Mar. Sci.* **1991**, *11*, 537–564. [CrossRef]

82. Lee, J.H.; Lee, D.; Kang, J.J.; Joo, H.T.; Lee, J.H.; Lee, H.W.; Ahn, S.H.; Kang, C.K.; Lee, S.H. The effects of different environmental factors on the biochemical composition of particulate organic matter in Gwangyang Bay, South Korea. *Biogeosciences* **2017**, *14*, 1903–1917. [CrossRef]

83. Alderkamp, A.C.; Buma, A.G.J.; Van Rijssel, M. The carbohydrates of *Phaeocystis* and their degradation in the microbial food web. *Biogeochemistry* **2007**, *83*, 99–118. [CrossRef]

84. Kim, B.K.; Lee, S.H.; Ha, S.Y.; Jung, J.; Kim, T.W.; Yang, E.J.; Jo, N.; Lim, Y.J.; Park, J.; Lee, S.H. Vertical Distributions of Macromolecular Composition of Particulate Organic Matter in the Water Column of the Amundsen Sea Polynya During the Summer in 2014. *J. Geophys. Res. Ocean.* **2018**, *123*, 1393–1405. [CrossRef]

85. DiTullio, G.R.; Smith, W.O. Spatial patterns in phytoplankton biomass and pigment distributions in the Ross Sea. *J. Geophys. Res. C Ocean.* **1996**, *101*, 18467–18477. [CrossRef]

86. Fabiano, M.; Danovaro, R.; Chiantore, M.; Pusceddu, A. *Bacteria, Protozoa and Organic Matter Composition in the Sediments of Terra Nova Bay (Ross Sea)*; Ross Sea ecology; Faranda, F., Guglielmo, L., Ianora, A., Eds.; Springer: Berlin/Heidelberg, Germany; Springer: New York, NY, USA, 2000; pp. 159–169.

87. Kim, B.K.; Lee, J.H.; Yun, M.S.; Joo, H.T.; Song, H.J.; Yang, E.J.; Chung, K.H.; Kang, S.H.; Lee, S.H. High lipid composition of particulate organic matter in the northern Chukchi Sea, 2011. *Deep Res. Part II Top. Stud. Oceanogr.* **2015**, *120*, 72–81. [CrossRef]

88. Jo, N.; La, H.S.; Kim, J.H.; Kim, K.; Kim, B.K.; Kim, M.J.; Son, W.; Lee, S.H. Different Biochemical Compositions of Particulate Organic Matter Driven by Major Phytoplankton Communities in the Northwestern Ross Sea. *Front. Microbiol.* **2021**, *12*, 5. [CrossRef]

89. Young, J.N.; Goldman, J.A.L.; Kranz, S.A.; Tortell, P.D.; Morel, F.M.M. Slow carboxylation of Rubisco constrains the rate of carbon fixation during Antarctic phytoplankton blooms. *New Phytol.* **2015**, *205*, 172–181. [CrossRef]

90. Hitchcock, G.L. A comparative study of the size-dependent organic composition of marine diatoms and dinoflagellates. *J. Plankton Res.* **1982**, *4*, 363–377. [CrossRef]

91. Berdalet, E.; Latasa, M.; Estrada, M. Effects of nitrogen and phosphorus starvation on nucleic acid and protein content of *Heterocapsa* sp. *J. Plankton Res.* **1994**, *16*, 303–316. [CrossRef]

92. Myklestad, S.; Haug, A. Production of carbohydrates by the marine diatom *Chaetoceros affinis* var. *willei* (Gran) Hustedt. I. Effect of the concentration of nutrients in the culture medium. *J. Exp. Mar. Bio. Ecol.* **1972**, *9*, 125–136. [CrossRef]

93. Lindqvist, K.; Lignell, R. Intracellular partitioning of $^{14}CO_2$ in phytoplankton during a growth season in the northern Baltic. *Mar. Ecol. Prog. Ser.* **1997**, *152*, 41–50. [CrossRef]

94. Handa, N. Carbohydrate metabolism in the marine diatom *Skeletonema costatum*. *Mar. Biol.* **1969**, *4*, 208–214. [CrossRef]

95. Parrish, C.C. Time series of particulate and dissolved lipid classes during spring phytoplankton blooms in Bedford Basin, a marine inlet. *Mar. Ecol. Prog. Ser.* **1987**, *35*, 129–139. [CrossRef]

96. Taipale, S.J.; Galloway, A.W.E.; Aalto, S.L.; Kahilainen, K.K.; Strandberg, U.; Kankaala, P. Terrestrial carbohydrates support freshwater zooplankton during phytoplankton deficiency. *Sci. Rep.* **2016**, *6*, 30897. [CrossRef]

97. Corzo, A.; Rodríguez-Gálvez, S.; Lubian, L.; Sangrá, P.; Martínez, A.; Morillo, J.A. Spatial distribution of transparent exopolymer particles in the Bransfield Strait, Antarctica. *J. Plankton Res.* **2005**, *27*, 635–646. [CrossRef]

98. Ramaiah, N.; Yoshikawa, T.; Furuya, K. Temporal variations in transparent exopolymer particles (TEPS) associated with a diatom spring bloom in a subarctic ria in Japan. *Mar. Ecol. Prog. Ser.* **2001**, *212*, 79–88. [CrossRef]

99. Ortega-Retuerta, E.; Duarte, C.M.; Reche, I. Significance of bacterial activity for the distribution and dynamics of transparent exopolymer particles in the Mediterranean Sea. *Microb. Ecol.* **2010**, *59*, 808–818. [CrossRef]

100. Malpezzi, M.A.; Sanford, L.P.; Crump, B.C. Abundance and distribution of transparent exopolymer particles in the estuarine turbidity maximum of Chesapeake Bay. *Mar. Ecol. Prog. Ser.* **2013**, *486*, 23–35. [CrossRef]

101. Yamada, Y.; Yokokawa, T.; Uchimiya, M.; Nishino, S.; Fukuda, H.; Ogawa, H.; Nagata, T. Transparent exopolymer particles (TEPS) in the deep ocean: Full-depth distribution patterns and contribution to the organic carbon pool. *Mar. Ecol. Prog. Ser.* **2017**, *583*, 81–93. [CrossRef]

102. Lee, J.H.; Lee, W.C.; Kim, H.C.; Jo, N.; Jang, H.K.; Kang, J.J.; Lee, D.; Kim, K.; Lee, S.H. Transparent exopolymer particle (TEPs) dynamics and contribution to particulate organic carbon (POC) in Jaran bay, Korea. *Water* **2020**, *12*, 1057. [CrossRef]

Article

Transparent Exopolymer Particle (TEPs) Dynamics and Contribution to Particulate Organic Carbon (POC) in Jaran Bay, Korea

Jae Hyung Lee [1], Won Chan Lee [2], Hyung Chul Kim [2], Naeun Jo [1], Hyo Keun Jang [1], Jae Joong Kang [1], Dabin Lee [1], Kwanwoo Kim [1] and Sang Heon Lee [1,*]

[1] Department of Oceanography, Pusan National University, Busan 46241, Korea; tlyljh78@pusan.ac.kr (J.H.L.); nadan213@pusan.ac.kr (N.J.); janghk@pusan.ac.kr (H.K.J.); jaejung@pusan.ac.kr (J.J.K.); ldb1370@pusan.ac.kr (D.L.); goanwoo7@pusan.ac.kr (K.K.)

[2] Marine Environmental Impact Assessment Center, National Institute of Fisheries Science, Busan 46083, Korea; phdleewc@korea.kr (W.C.L.); hckim072@korea.kr (H.C.K.)

* Correspondence: sanglee@pusan.ac.kr

Received: 11 March 2020; Accepted: 4 April 2020; Published: 8 April 2020

Abstract: Transparent exopolymer particles (TEPs) are defined as acidic polysaccharide particles and they are influenced by various biotic and abiotic processes that play significant roles in marine biogeochemical cycles. However, little information on their monthly variation, relationship and contribution to particulate organic carbon (POC) is currently available particularly in coastal regions. In this study, the water samples were collected monthly to determine TEP concentrations and POC concentrations in a southern coastal region of Korea, Jaran Bay from April 2016 to March 2017. The TEP concentrations varied from 26.5 to 1695.4 µg Xeq L^{-1} (mean ± standard deviation (S.D.) = 215.9 ± 172.2 µg Xeq L^{-1}) and POC concentrations ranged from 109.9 to 1201.9 µg L^{-1} (mean ± S.D. = 399.1 ± 186.5 µg L^{-1}) during our observation period. Based on the ^{13}C stable isotope tracer technique, monthly carbon uptake rates of phytoplankton ranged from 3.0 to 274.1 mg C m^{-2} h^{-1} (mean ± S.D. = 34.5 ± 45.2 mg C m^{-2} h^{-1}). The cross-correlation analysis showed a lag-time of 2 months between chlorophyll a and TEP concentrations ($r = 0.86$, $p < 0.01$; Pearson's correlation coefficient). In addition, we observed a 2 month lag-phased correlation between TEP concentrations and primary production ($r = 0.73$, $p < 0.05$; Pearson's correlation coefficient). In Jaran Bay, the TEP contribution was as high as 78.0% of the POC when the TEP-C content was high and declined to 2.4% of the POC when it was low. These results showed that TEP-C could be a significant contributor to the POC pool in Jaran Bay.

Keywords: TEP; TEP-C; phytoplankton; chlorophyll a; POC; primary production; Jaran Bay

1. Introduction

Transparent exopolymer particles (TEPs) are defined as carbon-rich gel particles, mainly consisting of acidic polysaccharides, ubiquitous in the marine environment [1–3]. TEPs are formed naturally from dissolved precursor substances, which contribute considerably to the colloidal dissolved organic matter (DOM) pool in aquatic systems [2]. TEP formation is very important as the major pathway through which DOM is converted to particulate organic matter (POM) [2]. These two different forms of organic matter have specific characteristics and specific roles in the chemistry and biology of the ocean. TEPs are a significant component of particulate organic carbon (POC) in the ocean, with an approximate size range of 0.4 to 100 µm [4–6]. The presence of TEPs is influenced by environmental factors such as temperature, salinity stratification, nutrient conditions, and biological processes that include phytoplankton and bacteria production [2,4,7–9]. Generally, phytoplankton and bacteria have been considered the major producers of TEPs and precursors in aquatic ecosystems [2,3,10]. TEPs and precursors that produced by these aquatic organisms cannot easily be predicted from their natural abundance because phytoplankton and bacteria do not produce TEPs or precursors equally [2]. For instance, the formation of TEPs by bacteria varies with species composition and growth conditions. Some phytoplankton species produce TEPs during growth, stationary phase and senescence [3]. According to Passow [2], TEP production by phytoplankton varies as a function of the conditions related to light, growth rates, and major or minor nutrient depletion. Therefore, the production of TEPs and precursors depends on the physiological status and growth conditions of the individual organisms.

In recent years, TEPs have received considerable attention in terms of their influence on carbon cycling [5,11–17]. Given their sticky character, TEPs increase particle aggregation forming marine snow and enhancing carbon export to deep waters in pelagic/oceanic ecosystems [11,18]. In coastal regions, coinciding with high biological productivity and phytoplankton biomass, TEP concentrations are also found to be high [10,12,19–22]. The half-lives of TEPs observed in a coastal environment are less than 1 day when large marine snow aggregates are formed, whereas they are more than 2 days during the period without large aggregations [23]. Although TEPs play an important role in carbon cycling and budget in coastal seas [24], little information has been reported on them in coastal seas to date.

Jaran Bay, located on the coast of the South Sea in Korea, is an area of productive bivalve farming in Korea [25,26]. There is no doubt that the roles of TEPs could be significant, but nothing has been studied in Jaran Bay. Thus, the present study aimed to (1) investigate the monthly variations in TEP concentrations for one year for the first time in Jaran Bay on the southern coast of Korea, (2) compare TEP with POC stocks to understand how much TEPs contribute to POC as a kind of carbon budget and (3) determine the relationship between the phytoplankton (biomass and productivity) and TEPs in coastal environmental conditions in Jaran Bay.

2. Materials and Methods

The study site was located in Jaran Bay on the southern coast of Korea, which borders the South Sea (Figure 1). Water samples were obtained at seven different stations in Jaran Bay monthly from April 2016 to March 2017 (Figure 1). The study area has a shallow area within 20 m depth. Using a 5 L Niskin water sampler (General Oceanics Inc., Miami, FL, USA), water samples were collected from two *in-situ* different light depths (100 and 1%) to include the depth range of the euphotic zone. The light depths were calculated by a Secchi disk using a vertical attenuation coefficient (K_d = 1.7/Secchi depth) from Poole and Atkins [27]. The water temperature and salinity were measured with a YSI-30 (YSI Incorporated, Yellow Spring, OH, USA). The large zooplankton were removed by 333 μm mesh when the water samples were collected.

The chlorophyll a concentration was determined by filtering 200 mL of seawater on 0.7 μm pore sized Whatman glass fiber filters (GF/F; 25 mm, Whatman, Maidstone, UK) and the filters were kept frozen until analysis. Chlorophyll a pigments were extracted from the filters with 10 mL acetone (90% v:v) in the dark at 4 °C for 24 h and fluorescence was measured with a Turner Designs model 10-AU fluorometer (Turner Designs, San Jose, CA, USA). The method for the chlorophyll a calculation followed that of Parson et al. [28].

TEP concentrations were estimated following the method of Passow and Alldredge [29]. Fifty milliliters of water was filtered onto 0.4 μm pore sized Nuclepore polycarbonate membrane filters (ADVANTEC; 25 mm, Toyo Roshi Kaisha, Tokyo, Japan) at a low and constant vacuum (<10 cm Hg), because the live cells could be damaged during strong vacuum filtration [29]. The filters were stained with 0.5 mL of precalibrated (with a xanthan gum solution) Alcian Blue (8GX, Sigma) for 5 s, rinsed with ultrapure water, frozen immediately and returned to the laboratory at Pusan National University in South Korea for further analysis. The filters for TEP concentration were soaked in 80% sulfuric acid for 3 h, and absorbance was read at 787 nm in a spectrophotometer (Hitachi-UH 5300, Hitachi, Tokyo, Japan). TEPs were quantified by a standard curve prepared with xanthan gum particles, as described in Passow and Alldredge [29]. The detection limit of the measurements was 0.025 absorbance units and the coefficient of variation of the replicates was 12.4%. Three blanks (empty filters stained with Alcian Blue) were also prepared with every batch of filtered samples every day. TEP concentration was expressed in terms of xanthan gum equivalents (μg Xeq L^{-1}). Based on the conversion factor ranging from 0.51 to 0.88 μg Xeq L^{-1} for a comparison of TEP with POC stocks [12,21], we used the lowest conversion factor to estimate TEP-C conservatively, since it corresponds to diatom-dominated ecosystems [12]. The conversion factor of 0.51 μg Xeq L^{-1} was used to convert from micrograms of xanthan gum to micrograms of carbon (TEP-C). The water samples (300 mL) were obtained from 100% and 1% light depths and filtered on pre-combusted (4 h, 450 °C) GF/F filters (25 mm) for POC concentrations. The filters were immediately frozen at −20 °C until further mass spectrometric analysis within 3 months.

The estimation of the carbon uptake rates was performed using ^{13}C stable isotope labeling experiments. Seawater samples at 6 different light depths (100, 50, 30, 12, 5, and 1%) determined by a Secchi disk were transferred from the Niskin bottles to 1 L polycarbonate incubation bottles (NALGENE, Rochester, NY, USA), wrapped with neutral density light filters (LEE filters, Andover, England), to match the desired light levels. Then, the water samples were inoculated with a labeled carbon (NaH^{13}CO$_3$) solution, and the bottles were incubated in large polycarbonate incubators under natural light conditions for at least 4 h, or up to 6 h when it was cloudy. To maintain in situ environments, ambient surface seawater was continuously run through the incubators during incubation. The 4–6 h incubations were terminated by filtration. The incubated waters were well mixed, and the measured volume (0.3 L) was filtered through precombusted (450 °C) GF/F filters (25 mm).

All filter samples for POC concentration and the abundance of ^{13}C/^{15}N were kept frozen (−20 °C) until the mass spectrometric analysis in a Finnigan Delta+XL mass spectrometer in the stable isotope laboratory (University of Alaska Fairbanks, Fairbanks, AL, USA) after overnight HCl fuming to remove carbonate. The measurement uncertainties for δ^{13}C and δ^{15}N were ± 0.06‰ and ± 0.02‰, respectively. Finally, the carbon uptake rates were calculated based on Hama et al. [30].

Statistical tests were performed using the statistical software IBM SPSS Statistics (*t*-test, one-way ANOVA and Pearson's correlation coefficient).

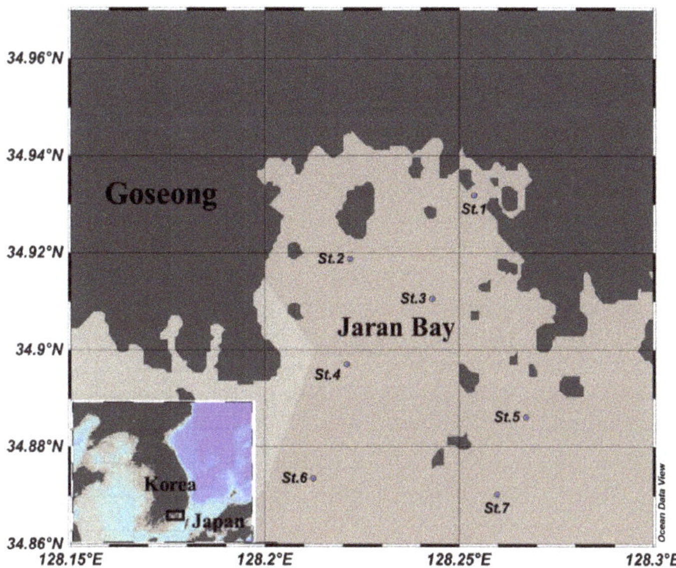

Figure 1. Study area map with the sampling stations in Jaran Bay, Korea.

3. Results

The monthly average water temperature and salinity were 6.4–28.7 °C and 28.7–34.0 during our sampling period, respectively (Table 1). The water temperature substantially varied, whereas salinity showed a relatively lower variation. The temporal variations in temperature and salinity showed patterns typical of those in mid-latitude regions. Water transparency was measured using Secchi disk depths, ranging between 2 and 10 m in Jaran Bay (Table 1). During the study period, the average euphotic depth was 8 ± 2 m, ranging from 5 to 18 m.

Table 1. Station-averaged environmental variables observed at the sampling stations in Jaran Bay. N.D. means no data.

Year	Month	Light Depth (%)	Temperature (°C)	Salinity	Secchi Depth (m)
2016	Apr.	100	13.0 ± 0.5	32.8 ± 0.6	6–10
		1	11.8 ± 0.5	33.1 ± 0.3	
	May	100	20.8 ± 0.5	31.7 ± 0.6	5–9
		1	18.0 ± 0.3	32.4 ± 0.6	
	Jun.	100	23.6 ± 0.9	29.3 ± 1.9	5–8
		1	20.7 ± 0.7	30.5 ± 2.9	
	Jul.	100	23.7 ± 1.1	29.6 ± 3.3	2–5
		1	23.4 ± 0.6	32.7 ± 0.7	
	Aug.	100	28.7 ± 0.8	29.1 ± 1.0	6–10
		1	23.4 ± 0.6	29.5 ± 3.2	
	Sep.	100	23.9 ± 0.4	N.D.	2–4
		1	23.8 ± 0.3	N.D.	
	Oct.	100	21.7 ± 0.2	N.D.	3–5
		1	21.8 ± 0.3	N.D.	
	Nov.	100	16.8 ± 0.8	29.8 ± 2.5	3–5
		1	16.9 ± 1.0	28.7 ± 3.2	
	Dec.	100	12.2 ± 1.1	32.2 ± 0.4	5–7
		1	12.1 ± 1.1	32.0 ± 0.3	
2017	Jan.	100	9.1 ± 0.6	32.0 ± 1.0	5–7
		1	9.2 ± 0.5	32.2 ± 0.5	
	Feb.	100	6.4 ± 0.7	32.9 ± 0.8	6–7
		1	6.4 ± 0.8	33.4 ± 0.3	
	Mar.	100	9.6 ± 0.4	33.8 ± 0.1	4–8
		1	9.6 ± 0.6	33.6 ± 0.3	

The chlorophyll a concentration at the euphotic depth ranged from 0.2 to 8.4 µg L^{-1} (mean ± S.D. = 1.8 ± 1.6 µg L^{-1}) in this study (Figure 2a). The chlorophyll a concentration was averaged from the two light depths at each station, because they were not significantly different (one-way ANOVA test, $p > 0.05$). The monthly chlorophyll a concentration averaged from the two light depths at the 7 stations ranged from 0.3 ± 0.1 µg L^{-1} in March 2017 to 4.9 ± 1.6 µg L^{-1} in October 2016 (Figure 2a). Distinct temporal variations in chlorophyll a concentrations at 100% and 1% light depths were observed in this study. Two bloom peaks were found during the study period. The first bloom was initiated at the 100% light depth in July 2016, followed by the bloom at the 1% light depth in August 2016. The second blooms were at both the 100% and 1% light depths in October 2016, which were relatively higher than the former blooms (Figure 2a).

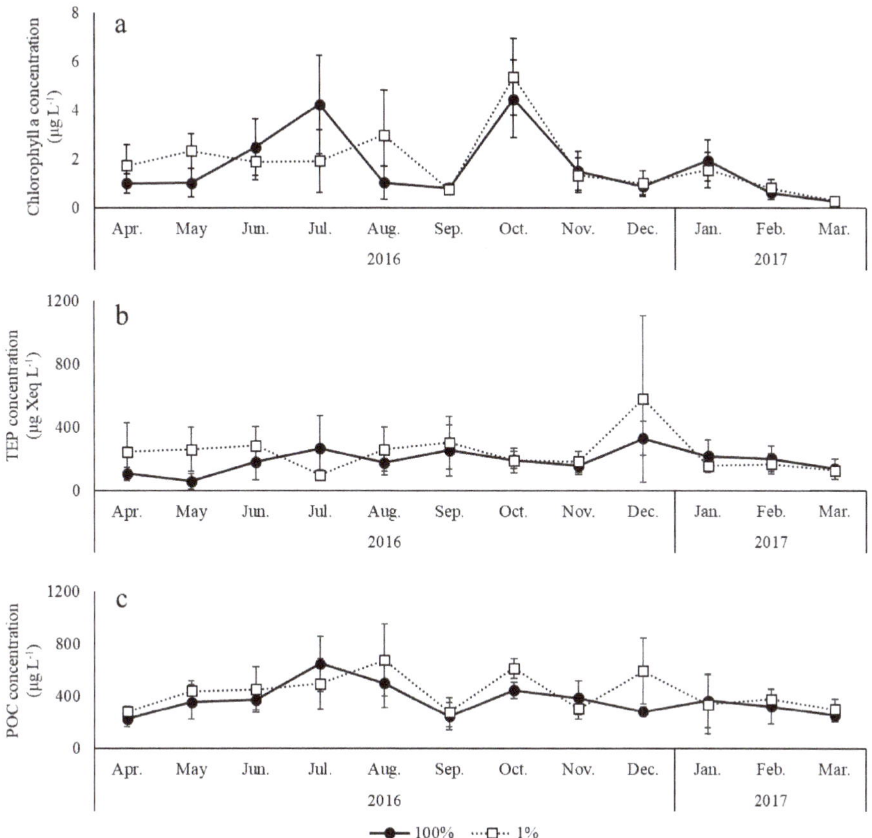

Figure 2. Monthly variations in the (**a**) transparent exopolymer particles (TEP) (μg Xeq L^{-1}), (**b**) chlorophyll a (μg L^{-1}) and (**c**) particulate organic carbon (POC) (μg L^{-1}) concentrations at the 100% and 1% light depths averaged from 7 different stations in Jaran Bay.

The overall TEP concentrations ranged from 26.5 to 1695.4 μg Xeq L^{-1} (mean ± S.D. = 215.9 ± 172.2 μg Xeq L^{-1}) in Jaran Bay from April 2016 to March 2017 in this study (Figure 2b). The TEP concentrations from the two light depths (100 and 1%) were different during the observation period (one-way ANOVA test, $p < 0.05$); the mean concentrations of TEP from the two light depths were 192.0 ± 119.3 μg Xeq L^{-1} and 239.9 ± 210.4 μg Xeq L^{-1}, respectively. The highest TEP concentrations monthly averaged from the 7 stations were observed in December 2016 (100% = 332.1 ± 105.9 μg Xeq L^{-1}; 1% = 581.9 ± 527.4 μg Xeq L^{-1}). After the winter peak in December, TEP concentrations decreased rapidly throughout the month (Figure 2b). TEP-C values were calculated by multiplying the TEP concentrations by the lowest conversion factor (0.51 μg Xeq L^{-1}), ranging from 13.5 to 864.7 μg C L^{-1} (mean ± S.D. = 110.1 ± 87.8 μg C L^{-1}) during the study period. The average values of TEP-C at the two light depths (100 and 1%) were 97.9 ± 60.8 μg C L^{-1} and 122.3 ± 107.3 μg C L^{-1}, respectively.

POC concentrations ranged from 109.9 to 1201.9 μg L^{-1}, with an average of 399.1 ± 186.5 μg L^{-1} in this study (Figure 2c). The mean concentration of POC was statistically higher at the 1% light depth than at the 100% light depth (one-way ANOVA test, $p < 0.05$). The average concentrations of POC were 368.0 ± 167.7 μg L^{-1} at the 100% light depth and 430.2 ± 199.8 μg L^{-1} at the 1% light depth. In this study, the highest POC concentrations at the euphotic depth were observed in the summer season (100% light depth: July 2016; 1% light depth: August 2016) in Jaran Bay (Figure 2c). Monthly TEP, chlorophyll a and POC concentrations were not significantly correlated (Pearson's correlation coefficient).

The average TEP-C contributions to POC were 21.4% ± 11.7% at the 100% light depth and 22.0% ± 11.2% at the 1% light depth (Figure 3). The TEP-C contributions to POC were not constant throughout the different months in this study (one-way ANOVA test, $p < 0.05$). The monthly average contribution of TEP-C to POC was relatively high in September (100% light depth: 35.3 ± 23.8%; 1% light depth: 36.2% ± 19.9%). In contrast, relatively lower TEP-C contributions to POC were found in May (12.8% ± 6.7%; 100% light depth) and July (10.1% ± 3.9%; 1% light depth) (Figure 3).

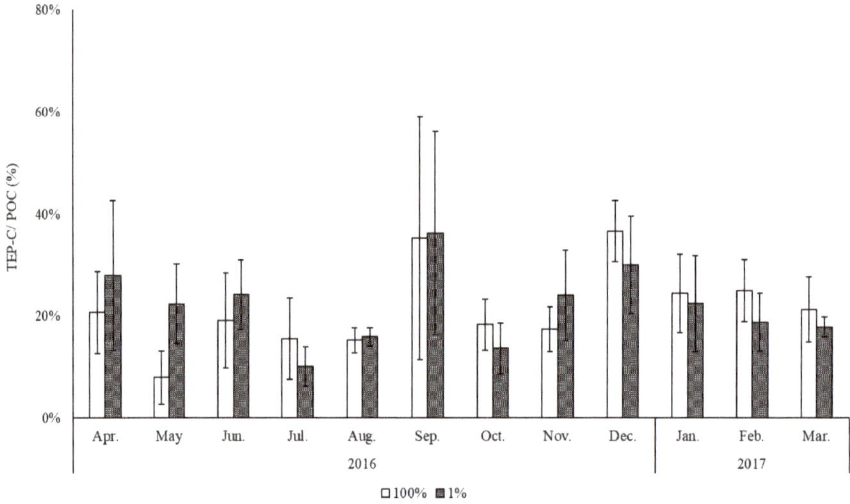

Figure 3. Monthly TEP-C contributions to POC at different light depths (100 and 1%) in Jaran Bay.

The monthly mean specific and absolute carbon uptake rates of the phytoplankton in the water column during this study are presented in Table 2. The specific carbon uptake rates were 0.001–0.112 h^{-1} at the 100% light depth and 0–0.004 h^{-1} at the 1% light depth. The absolute carbon uptake rates were 0.2–137.4 µg C L^{-1} h^{-1} and 0.0–1.3 µg C L^{-1} h^{-1} at the 100% light depth and 1% light depth, respectively (Table 2). The hourly specific and absolute carbon uptake rates were relatively higher in the surface waters than at other light depths. The monthly carbon uptake rates of the phytoplankton integrated from the surface to a 1% light depth (primary production of phytoplankton from the euphotic layer) ranged from 3.0 to 274.1 mg C m^{-2} h^{-1} (mean ± S.D. = 34.5 ± 45.2 mg C m^{-2} h^{-1}). The highest monthly primary production was recorded in the early summer (mean ± S.D. = 115.4 ± 97.3 mg C m^{-2} h^{-1}; July 2016) and the lowest value was measured in the summer (mean ± S.D. = 5.7 ± 2.3 mg C m^{-2} h^{-1}; August 2016) (Table 2).

Table 2. Station-averaged specific and absolute carbon uptake rates and primary production of phytoplankton at the 100% and 1% light depths in Jaran Bay.

Year	Month	Light Depth (%)	Specific Uptake Rate (h^{-1})	Absolute Uptake Rate ($\mu g\ C\ h^{-1}\ L^{-1}$)	Integral Primary Production ($mg\ C\ m^{-2}\ h^{-1}$)
2016	Apr.	100	0.0125 ± 0.0114	3.6 ± 3.9	18.3 ± 19.4
		1	0.0004 ± 0.0004	0.1 ± 0.1	
	May	100	0.0121 ±0.0112	3.9 ± 3.9	21.6 ± 5.2
		1	0.0010 ± 0.0014	0.4 ± 0.4	
	Jun.	100	0.0210 ± 0.0145	12.7 ± 10.4	43.2 ± 23.7
		1	0.0006 ± 0.0002	0.3 ± 0.2	
	Jul.	100	0.0532 ± 0.0407	52.8 ± 50.4	115.4 ± 97.3
		1	0.0010 ± 0.0004	0.5 ± 0.2	
	Aug.	100	0.0065 ± 0.0071	0.6 ± 0.3	5.7 ± 2.3
		1	0.0007 ± 0.0003	0.3 ± 0.2	
	Sep.	100	0.0080 ± 0.0046	4.6 ± 2.8	12.9 ± 7.2
		1	0.0003 ± 0.0002	0.2 ± 0.1	
	Oct.	100	0.0458 ± 0.0127	28.0 ± 12.4	93.3 ± 42.7
		1	0.0011 ± 0.0006	0.5 ± 0.2	
	Nov.	100	0.0242 ± 0.0089	8.5 ± 3.8	29.1 ± 19.7
		1	0.0008 ± 0.0005	0.2 ± 0.1	
	Dec.	100	0.0091 ± 0.0034	2.9 ± 1.3	16.9 ± 6.4
		1	0.0005 ± 0.0002	0.1 ± 0.1	
2017	Jan.	100	0.0198 ± 0.0046	6.3 ± 2.6	31.3 ± 13.5
		1	0.0009 ± 0.0005	0.2 ± 0.2	
	Feb.	100	0.0047 ±0.0015	2.0 ± 0.7	15.2 ± 7.4
		1	0.0003 ± 0.0002	0.1 ± 0.1	
	Mar.	100	0.0048 ± 0.0015	1.3 ± 0.3	10.5 ± 4.9
		1	0.0004 ± 0.0002	0.1 ± 0	

4. Discussion

The TEP concentrations measured in Jaran Bay during this study ranging from 26.5 to 1695.4 µg Xeq L^{-1} generally fall within the range observed in other studies from various coastal seas and bays to open oceans around the world (Table 3). However, our TEP concentrations were relatively higher than those observed in open ocean environments [6,31–33]. In general, TEP concentrations are higher in coastal waters than in adjacent open ocean waters [2,34]. Since TEPs are mainly formed by phytoplankton, the general pattern of higher TEP concentrations in coastal oceans are closely related to the phytoplankton biomass [18]. Similarly, the relatively higher chlorophyll a concentration in this study (0.3 to 4.9 µg L^{-1}) could be explained by the higher TEP concentrations compared to those of the open oceans (Figure 2a) [6,9]. Previous studies have shown that TEP production rates are largely involved in the physiological status of phytoplankton [2,10,23]. To investigate this relationship in this study, the specific and absolute carbon uptake rates of phytoplankton were used for their physiological conditions during our research period. Generally, the monthly mean specific and absolute carbon uptake rates were statistically higher at the 100% light depth than at the 1% light depth (one-way ANOVA test, $p < 0.05$). However, neither specific nor absolute carbon uptake rates were significantly correlated with the TEP concentrations in our study. One of the reasons for the higher TEP concentrations observed in our study compared with those in previous studies could be related to the phytoplankton community. Previous studies have reported that extracellular organic matter created from phytoplankton serves as a precursor for TEPs [18,35,36]. In particular, TEPs are considerably correlated with the abundance of diatoms, which are considered major producers of TEPs [2,23]. In this study, we found relatively high chlorophyll a concentrations and carbon uptake rates of phytoplankton. In addition, the dominant species were diatoms during this study based on the relative contributions of the major phytoplankton classes in a parallel study (unpublished data). The monthly contributions of diatoms to the total phytoplankton classes ranged from 19.8% to 96.7%, with an average of 66.4% ± 24.3% in Jaran Bay (unpublished data) [26,37]. It was also reported that the predominant species are diatoms in this area. Specific to these diatom-dominated ecosystems, the overall dominant macromolecular compositions of POC are carbohydrates (51.8 ± 8.7%) in Jaran Bay [38]. The carbohydrate-dominant phytoplankton could be a potential reason for the higher TEP concentrations observed in this study, since TEPs can be produced from dissolved carbohydrate polymers exuded by phytoplankton [2]. Another potential reason for the higher TEP concentration in this study is that large shellfish aquaculture farms were intensely implemented in our study areas. In coastal waters, mucus nets of filter feeders and anthropogenic sources, such as river inputs (urban-industrial release) and aquaculture farms (macroalgae and shellfish) could be major sources of TEPs [39,40]. There are no major river inputs into Jaran Bay, so no significant influence of urban-industrial release would be expected. Based on previous studies, cation availability might be one such important controlling factor on the TEP formation system [34]. The formation and sustenance of TEP are dependent on cation availability, as cations (particularly Ca^{2+} and Mg^{2+}) stabilize the structural integrity of TEP through cation bonding [34,41]. In Jaran Bay, cation availability (Ca^{2+}) might be high because of intensive shellfish (oysters and scallops) aquaculture conditions. Jaran Bay is a relatively shallow coastal bay with an average water depth of 10 m, which might sustain a high concentration of cations from the relatively easy resuspension of biodeposition on the top of the sediments underneath aquaculture farms. In fact, oyster soft tissues are generally reported for high mineral elements such as Ca^{2+} and Mg^{2+} [42]. However, this should be verified in further studies.

During the study period, the TEP/chlorophyll a ratio value ranged from 8 to 1233 µg Xeq µg Chl a^{-1}, averaging 221 ± 217 µg Xeq µg Chl a^{-1}. High levels of variability in TEP/chlorophyll a ratio values were observed in this study. These results were similar to those in former studies. The mean TEP/chlorophyll a ratio values were 206.8 and 281.5 in Otsuchi Bay and the Gulf of Cadiz, respectively [43–45].

TEPs often make up a large fraction of the POC in the water column [4]. Although TEPs play a key role in the sequestration of excess carbon to deeper waters in open oceans, TEP-C is also an important part in understanding the carbon budget as an additional carbon source in coastal marine food webs [2]. Previous studies have reported significant relationships between TEP and POC concentrations [9,17]. However, no significant correlation was found between TEP and POC concentrations in this study, which is similar to the result of Ortega-Retuerta et al. [22] in the coastal NW Mediterranean Sea. Considering the TEP-C and POC values reported herein, we estimated the TEP-C contribution to the POC pool in Jaran Bay during our study period. In the present study, we used the lowest conversion factor (0.51 µg Xeq L^{-1}), which corresponds to diatom-dominated phytoplankton ecosystems, to estimate TEP-C. The same conversion factor has been used to investigate the TEP-C contribution to the total organic carbon pool in the upper surface water column in open deep oceans [6,21]. In the present study, the estimated TEP-C contribution to the total POC ranged from 2.4 to 67.3%, with an average of 21.7% ± 11.4% (Table 4). In comparison, the TEP-C contribution to the total POC based on the highest conversion factor (0.88 µg Xeq L^{-1}) ranged from 4.1% to 78.0% (31.4% ± 13.8%). The contribution of TEP-C to the total POC in Jaran Bay during our research period was within the range observed in previous studies in different bays, ranging from 7% to 32%. Furthermore, Bhaskar and Bhosle [46] reported that TEP-C could constitute 7% of the total POC in Dona Paula Bay, and Malpezzi et al. [4] found that TEP-C averaged 32% of the total POC in the Chesapeake Bay. In contrast to the values in bays, the TEP-C contributions can be higher in relatively deeper environmental conditions [9,17,23]. Moreover, Parinos et al. [17] observed that TEP-C represented 70% of the POC, on average, in the NE Aegean Sea. Notably, Zamanillo et al. [9] reported that the monthly mean TEP contribution to POC was 73% of the POC pool in the southwestern Atlantic Shelf, and Ortega-Retuerta et al. [22] found that TEP-C averaged 77% of the POC in coastal NW Mediterranean Sea in early summer. Higher TEP concentrations, as discussed previously, but lower TEP-C contributions to the total POC pools, indicate relatively more POC-based carbon available under coastal or bay environmental conditions, compared to that under open ocean environmental conditions. TEP production and formation rates are largely dependent on physiological conditions, especially nutrient stress, of phytoplankton and bacteria [23].

In general, relatively better phytoplankton nutrient conditions in bay systems than in deeper ocean could be a major reason for the lower contribution of TEP-C to the total POC. Another potential explanation might be different degrees of overestimation from different pore sized filters for TEP and POC measurements. Typically, 0.4 µm pore-sized filters are used for TEP concentration, while 0.7 µm pore-sized filters are used for POC. Therefore, the TEP-C contribution to the total POC could potentially be overestimated by methodological sampling approaches. However, TEPs smaller than 0.7 µm might be more abundant in open ocean conditions than in coastal or bay environments with active TEP formation conditions, as discussed previously, which could result in increased overestimations of TEP-C in open oceans. Verification for this concept could be conducted in future studies.

Table 3. Comparison of the TEP concentrations between this and previous studies from various regions.

Region	Season	Depth (m)	TEP Range (Mean) (μg Xeq L^{-1})	Reference
Western Arctic Ocean	Summer	0–200	37–130 (120)	[6]
Eastern tropical and eastern subarctic, North Pacific Ocean	Summer	Above mixed layer depth	78–970	[33]
Western and central Pacific Ocean (subtropical and equatorial regions)	Summer and winter	0–200	5–40 (25)	[6]
Northeast coast of Japan	Spring and winter	0–15	136–2321	[43]
Tokyo bay	All year	0–20	14–1774	[47]
Changjiang estuary	Spring, summer and autumn	0–80	37–1227	[48]
Mediterranean Sea	Spring	Upper mixed layer	19–53 (29)	[32]
Coastal NW (rocky shore) Mediterranean Sea	All year	Surface	5–91	[49]
Baltic Sea	Summer	40	145–322	[50]
Bransfield Strait	Summer	0–100	0–346	[45]
Southern Ibrian coasts	Spring	Mixed layer	507–560	[51]
Neuse River estuary	All year	Surface	805–1801	[34]
Chesapeake bay	All year	0–24	37–2820	[4]
Northeast Atlantic Ocean	Spring	0–10	20–420	[31]
Western North Atlantic Ocean and Sargasso Sea	Spring	2–5	100–200	[52]
Jaran bay (Southern coast of Korea)	All year	Euphotic depth	27–1695 (222)	This study

In this study, monthly TEP and chlorophyll a concentrations were not significantly coupled (Figure 2a,b). Previous studies reported strong interrelationships between TEP and chlorophyll a concentrations for some time series [17,24,53,54]. In contrast, no strong relationships were found [46,55], or some correlations were observed for certain periods of the year [22,56]. Ortega-Retuerta et al. [22] observed a 4-month lag phase between chlorophyll a and TEP concentration peaks in the coastal NW Mediterranean Sea ($r = 0.48$, $p < 0.04$). In the present study, the cross-correlation analysis indicated a lag-time of two months between chlorophyll a and TEP concentration peaks to a significant degree ($r = 0.86$, $p < 0.01$; Pearson's correlation coefficient) (Figure 4a). Moreover, we observed a two-month lag-phased correlation between TEP concentrations and primary production ($r = 0.73$, $p < 0.05$; Pearson's correlation coefficient) (Figure 4b). Previous studies reported that the lag-phase between chlorophyll a and TEP concentration peaks could be caused by an enhancement in the TEP production rate under nutrient-limiting conditions of phytoplankton after phytoplankton blooms [23]. Based on the concentrations and molar ratios of nutrients [57–59]; no major nutrient limiting conditions were found in this study area except in May 2016 [26]. In coastal areas, TEP production and removal/consumption processes are influenced by various biotic factors (e.g., phytoplankton biomass and community structure, growth, grazing, and heterotrophy) and abiotic (e.g., stratification, current, tidal, land run-off, and sorption to sediments) processes [43,46]. In addition, nutrient limitations to phytoplankton growth enhance TEP production rates [23,60]. Therefore, the lag-phase patterns between chlorophyll a and TEP concentration and primary production and TEP concentration could not be controlled by any single factor.

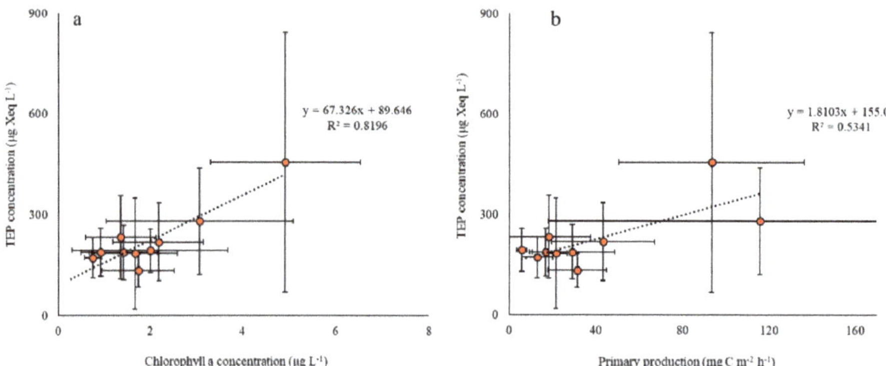

Figure 4. Lag-phased correlations between (**a**) chlorophyll a and TEP concentration and (**b**) primary production and TEP concentration.

Table 4. Comparison of the TEP-C contribution to the POC (%) between this and previous studies.

Region	Season	TEP-C/POC %	Reference
NE Aegean Sea	Early spring, summer and autumn	70	[17]
Coastal NW Mediterranean Sea	Early summer	77	[22]
Epipelagic Mediterranean Sea		75	
Deep Mediterranean Sea	Spring	50	[61]
North East Atlantic Ocean		85	
Open Atlantic Ocean	Spring and fall	66	[9]
Southwestern Atlantic Shelf	Spring	73	
Dona Paula Bay	All year	7	[46]
Chesapeake Bay	Early and late spring, fall and winter	32	[4]
Jaran bay (Southern coast of Korea)	All year	22	This study

5. Conclusions

This study reported the spatiotemporal dynamics and relative POC contributions of TEP in Jaran Bay, South Korea, based on monthly field measurement data. This study clearly showed that the large monthly variation in TEP is mainly driven by phytoplankton biomass, such as chlorophyll a concentration and their photosynthetic productivity, with 2-month lag phases in Jaran Bay. Although the TEP carbon contribution to the POC pool can be as high as up to approximately 70% at some sampling sites, the overall TEP contribution was 21.7% (± 11.4%) in Jaran Bay, which is consistent with the results of previous studies. Generally, high TEP concentrations but relatively lower TEP carbon contributions in coastal or bay environments suggest that POC-based carbon could be more available in shallow waters than in deep open oceans. However, some of the different sampling methods that we discussed should not be ignored because of the potential possibilities for the discrepancy between TEP concentrations and their carbon contributions. In this study, relatively higher TEP concentrations were observed in Jaran Bay than in other bays. This could be due to the diatom-dominant coastal environments in Jaran Bay, with overall monthly contributions of diatoms > 60% based on a parallel study. The relatively higher TEP concentration in Jaran Bay could be due to the many shellfish aquaculture farms with potentially high cation availability (Ca^{2+}), which should be verified in further studies. This result contributes to a comprehensive understanding of the seasonal dynamics of TEPs and their potential roles in the organic carbon pool in coastal or bay environments. In particular, this study contributes to providing the background for using TEPs to evaluate phytoplankton responses to ongoing changes in coastal ecosystems associated with global climate change.

Author Contributions: Conceptualization, J.H.L., W.C.L., H.C.K. and S.H.L.; Data curation, J.H.L.; Formal analysis, H.K.J.; Investigation, J.H.L., N.J., J.J.K., D.L. and K.K.; Methodology, H.K.J.; Validation, J.H.L., W.C.L. and H.C.K.; Visualization, J.H.L.; Writing—original draft, J.H.L.; Writing—review and editing, J.H.L., N.J. and S.H.L. All authors have read and agreed to the published version of the manuscript.

Funding: This research was supported by the grant (R2020050) from the National Institute of Fisheries Science (NIFS) and "Improvements of ocean prediction accuracy using numerical modeling and artificial intelligence technology", funded by the Ministry of Oceans and Fisheries, Republic of Korea.

Acknowledgments: We thank the anonymous reviewers who greatly improved an earlier version of manuscript.

Conflicts of Interest: The authors declare no conflict of interest.

References

1. Alldredge, A.L.; Passow, U.; Logan, B.E. The abundance and significance of a class of large, transparent organic particles in the ocean. *Deep Sea Res. Part I* **1993**, *40*, 1131–1140. [CrossRef]
2. Passow, U. Transparent exopolymer particles (TEP) in aquatic environments. *Prog. Oceanogr.* **2002**, *55*, 287–333. [CrossRef]
3. Ortega-Retuerta, E.; Reche, I.; Pulido-Villena, E.; Agustí, S.; Duarte, C.M. Uncoupled distributions of transparent exopolymer particles (TEP) and dissolved carbodydrates in the Southern Ocean. *Mar. Chem.* **2009**, *115*, 59–65. [CrossRef]
4. Malpezzi, M.A.; Sanford, L.P.; Crump, B.C. Abundance and distribution of transparent exopolymer particles in the estuarine turbidity maximum of Chesapeake Bay. *Mar. Ecol. Prog. Ser.* **2013**, *486*, 23–35. [CrossRef]
5. Mari, X.; Passow, U.; Migon, C.; Burd, A.B.; Legendre, L. Transparent exopolymer particles: Effects on carbon cycling in the ocean. *Prog. Oceanogr.* **2017**, *151*, 13–37. [CrossRef]
6. Yamada, Y.; Yokokawa, T.; Uchimiya, M.; Nishino, S.; Fukuda, H.; Ogawa, H.; Nagata, T. Transparent exopolymer particles (TEP) in the deep oceaen: Full-depth distribution patterns and contribution to the organic carbon pool. *Mar. Ecol. Prog. Ser.* **2017**, *583*, 81–93. [CrossRef]
7. Precali, R.; Giani, M.; Marini, M.; Grilli, F.; Ferrari, C.R.; Pečar, O.; Paschinic, E. Mucilaginous aggregates in the northern Adriatic in the period 1999–2002: Typology and distribution. *Sci. Total Environ.* **2005**, *353*, 10–23. [CrossRef]
8. Beauvais, S.; Pedrotti, M.L.; Egge, J.; Iversen, K.; Marrasé, C. Effects of turbulence on TEP dynamics under contrasting nutrient conditions: Implications for aggregation and sedimentation processes. *Mar. Ecol. Prog. Ser.* **2006**, *323*, 47–57. [CrossRef]
9. Zamanillo, M.; Ortega-Retuerta, E.; Nunes, S.; Rodríguez-Ros, P.; Dall'Osto, M.; Estrada, M.; Sala, M.M.; Simó, R. Main drivers of transparent exopolymer particle distribution across the surface Atlantic Ocean. *Biogeosciences* **2019**, *16*, 733–749. [CrossRef]
10. Passow, U.; Alldredge, A.L. Distribution, size and bacterial colonization of transparent exopolymer particles (TEP) in the ocean. *Mar. Ecol. Prog. Ser.* **1994**, *113*, 185–198. [CrossRef]
11. Passow, U.; Shipe, R.F.; Murray, A.; Pak, D.K.; Brzezinski, M.A.; Alldredge, A.L. The origin of transparent exopolymer particles (TEP) and their role in the sedimentation of particulate matter. *Cont. Shelf Res.* **2001**, *21*, 327–346. [CrossRef]
12. Engel, A.; Passow, U. Carbon and nitrogen content of transparent exopolymer particles (TEP) in relation to their Alcian Blue adsorption. *Mar. Ecol. Prog. Ser.* **2001**, *219*, 1–10. [CrossRef]
13. Mari, X.; Beauvais, S.; Lemée, R.; Pedotti, M.L. Non-Redfield C:N ratio of transparent exopolymetric particles in the northwestern Mediterranean Sea. *Limnol. Oceanogr.* **2001**, *46*, 1831–1836. [CrossRef]
14. Engel, A.; Thoms, S.; Riebesell, U.; Rochelle-Newall, E.; Zondervan, I. Polysaccharide aggregation as a potential sink of marine dissolved organic carbon. *Nature* **2004**, *428*, 929–932. [CrossRef]
15. De La Rocha, C.L.; Nowald, N.; Passow, U. Interactions between diatom aggregates, minerals, particulate organic carbon, and dissolved organic matter: Further implications for the ballast hypothesis. *Glob. Biogeochem. Cycles* **2008**, *22*. [CrossRef]
16. Gogou, A.; Repeta, D.J. Particulate-dissolved transformations as a sink for semi-labile dissolved organic matter: Chemical characterization of high molecular weight dissolved and surface-active organic matter in seawater and in diatom cultures. *Mar. Chem.* **2010**, *121*, 215–223. [CrossRef]

17. Parinos, C.; Gogou, A.; Krasakopoulou, E.; Lagaria, A.; Giannakourou, A.; Karageogis, A.P.; Psarra, S. Transparent Exopolymer Particles (TEP) in the NE Aegean Sea frontal area: Seasonal dynamics under the influence of Black Sea water. *Cont. Shelf Res.* **2017**, *149*, 112–123. [CrossRef]

18. Passow, U.; Alldredge, A.L.; Logan, B.E. The role of particulate carbohydrate exudates in the flocculation of diatom blooms. *Deep Sea Res. Part I* **1994**, *41*, 335–357. [CrossRef]

19. Klein, C.; Claquin, P.; Pannard, A.; Napoléon, C.; Le Roy, B.; Véron, B. Dynamic of soluble extracellular polymeric substances and transparent exopolymer particle pools in coastal ecosystems. *Mar. Ecol. Prog. Ser.* **2011**, *427*, 13–27. [CrossRef]

20. Sun, C.C.; Wang, Y.S.; Li, Q.P.; Yue, W.Z.; Wang, Y.T.; Sun, F.L.; Peng, Y.L. Distribution characteristics of transparent exopolymer particles in the Pearl River estuary, China. *J. Geophys. Res.* **2012**, *117*. [CrossRef]

21. Jennings, M.K.; Passow, U.; Wozniak, A.S.; Hansell, D.A. Distribution of transparent exopolymer particles (TEP) asross an organic carbon gradient in the western North Atlantic Ocean. *Mar. Chem.* **2017**, *190*, 1–12. [CrossRef]

22. Ortega-Retuerta, E.; Marrasé, C.; Muñoz-Fernández, A.; Sala, M.M.; Simó, R.; Gasol, J.M. Seasonal dynamics of transparent exopolymer particles (TEP) and their drivers in the coastal NW Mediterranean Sea. *Sci. Total Environ.* **2018**, *631*, 180–189. [CrossRef] [PubMed]

23. Logan, B.E.; Passow, U.; Alldredge, A.L.; Grossart, H.P.; Simon, M. Rapid formation and sedimentation of large aggregates is predictable from coagulation rates (half-lives) of transparent exopolymer particles (TEP). *Deep-Sea Res. Part II* **1995**, *42*, 203–214. [CrossRef]

24. Beauvais, S.; Pedrotti, M.L.; Villa, E.; Lemée, R. Transparent exopolymer particle (TEP) dynamics in relation to trophic and hydrological conditions in the NW Mediterranean Sea. *Mar. Ecol. Prog. Ser.* **2003**, *262*, 91–109. [CrossRef]

25. Cho, C.H.; Park, K. Eutrophication of bottom mud in shellfish farms, the Goseong-Jaran Bay. *Korean J. Fish. Aquat. Sci.* **1983**, *16*, 260–264.

26. Lee, J.H.; Kim, H.C.; Lee, T.; Lee, W.C.; Kang, J.J.; Jo, N.; Lee, D.; Kim, K.; Min, J.; Kang, S.; et al. Monthly variations in the intracellular nutrient pools of phytoplankton in Jaran Bay, Korea. *J. Coast. Res.* **2018**, *85*, 331–335. [CrossRef]

27. Poole, H.H.; Atkins, R.G. Photo-electric Measurements of Submarine Illumination throughout the Year. *J. Mar. Biol. Assoc. U. K.* **1929**, *16*, 297–324. [CrossRef]

28. Parsons, T.R.; Maita, Y.; Lalli, C.M. *A Manual of Biological and Chemical Methods for Seawater Analysis*; Pergamon Press: Oxford, UK, 1984.

29. Passow, U.; Alldredge, A.L. A dye-binding assay for the spectrophotometric measurement of transparent exopolymer particles (TEP). *Limnol. Oceanogr.* **1995**, *40*, 1326–1335. [CrossRef]

30. Hama, T.; Miyazaki, T.; Ogawa, Y.; Iwakuma, T.; Takahashi, M.; Otsuki, A.; Ichimura, S. Measurement of photosynthetic production of a marine phytoplankton population using a stable ^{13}C isotope. *Mar. Biol.* **1983**, *73*, 31–36. [CrossRef]

31. Leblanc, K.; Hare, C.E.; Feng, Y.; Berg, G.M.; DiTullio, G.R.; Neeley, A.; Benner, I.; Sprengel, C.; Beck, A.; Sanudo-Wilhemy, S.A.; et al. Distribution of calcifying and silicifying phytoplankton in relation to environmental and biogeochemical parameters during the late stages of the 2005 North East Atlantic Spring Bloom. *Biogeosciences* **2009**, *6*, 1–25. [CrossRef]

32. Ortega-Retuerta, E.; Duarte, C.M.; Reche, I. Significance of Bacterial Activity for the Distribution and Dynamics of Transparent Exopolymer Paticles in the Mediterranean Sea. *Microb. Ecol.* **2010**, *59*, 808–818. [CrossRef] [PubMed]

33. Wurl, O.; Miller, L.; Vagle, S. Production and fate of transparent exopolymer particles in the ocean. *J. Geophys. Res.* **2011**, *116*. [CrossRef]

34. Wetz, M.S.; Robbins, M.C.; Paerl, H.W. Transparent Exopolymer Particles (TEP) in a River-Dominated Estuary: Spatial-Temporal Distributions and an Assessment of Controls upon TEP Formation. *Estuaries Coasts* **2009**, *32*, 447–455. [CrossRef]

35. Alldredge, A.L.; Passow, U.; Haddock, H.D. The characteristics and transparent exopolymer particle (TEP) content of marine snow formed from thecate dinoflagellates. *J. Plankton Res.* **1998**, *20*, 393–406. [CrossRef]

36. Stoderegger, K.E.; Herndl, G.J. Production of exopolymer particles by marine bacterioplankton under contrasting turbulence conditions. *Mar. Ecol. Prog. Ser.* **1999**, *189*, 9–16. [CrossRef]

37. Park, J.S.; Yoon, Y.H.; Oh, S.J. Variational characteristics of phytoplankton community in the mouth parts of Gamak Bay, Southern Korea. *Korean J. Environ. Biol.* **2009**, *27*, 205–215.

38. Kim, H.C.; Lee, J.H.; Lee, W.C.; Hong, S.; Kang, J.J.; Lee, D.; Jo, N.; Bhavya, P.S. Decoupling of Macromolecular compositions of particulate organic matters between the water columns and the sediment in Geoje-Hansan Bay, South Korea. *Ocean Sci. J.* **2018**, *53*, 735–743. [CrossRef]

39. Herndl, G.J. Microbial Dynamics in Marine Aggregates. In *Seasonal Dynamics of Plankton Ecosystems and Sedimentation in Coastal Waters*; Symposium Proceedings, NurmiPrint OY: Nurmijärvi, Finland, 1995; pp. 81–105.

40. Goto, N.; Kawamura, T.; Mitamura, O.; Terai, H. Importance of extracellular organic carbon production in the total primary production by tidal-flat diatoms in comparison to phytoplankton. *Mar. Ecol. Prog. Ser.* **1999**, *190*, 289–295. [CrossRef]

41. Chin, W.C.; Orellana, M.V.; Verdugo, P. Spontaneous assembly of marine dissolved organic matter in polymer gels. *Nature* **1998**, *391*, 568–572. [CrossRef]

42. Lu, G.Y.; Ke, C.H.; Zhu, A.; Wang, W.X. Oyster-based national mapping of trace metals pollution in the Chinese coastal waters. *Environ. Pollut.* **2017**, *224*, 658–669. [CrossRef]

43. Ramaiah, N.; Yoshikawa, T.; Furuya, K. Temporal variations in transparent exopolymer particles (TEP) associated with a diatom spring bloom in a subarctic ria in Japan. *Mar. Ecol. Prog. Ser.* **2001**, *212*, 79–88. [CrossRef]

44. García, C.M.; Prieto, L.; Vargas, M.; Echevarría, F.; García-Lafuente, J.; Ruiz, J.; Rubín, J.P. Hydrodynamics and the spatial distribution of plankton and TEP in the Gulf of Cádiz (SE Iberian Peninsula). *J. Plankton Res.* **2002**, *24*, 817–833. [CrossRef]

45. Corzo, A.; Rodríguez-Gálvez, S.; Lubian, L.; Sangrá, P.; Martínez, A.; Morillo, J.A. Spatial distribution of transparent exopolymer particles in the Branfield Strait, Antarctica. *J. Plankton Res.* **2005**, *27*, 635–646. [CrossRef]

46. Bhaskar, P.V.; Bhosle, N.B. Dynamics of transparent exopolymeric particles (TEP) and particle-associated carbohydrates in the Dona Paula bay, west coast of India. *J. Earth Syst. Sci.* **2006**, *115*, 403–413. [CrossRef]

47. Ramaiah, N.; Furuya, K. Seasonal variations in phytoplankton composition and transparent exopolymer particles in a eutrophicated coastal environment. *Aquat. Microb. Ecol.* **2002**, *30*, 69–82. [CrossRef]

48. Guo, S.; Sun, X. Concentrations and distribution of transparent exopolymer particles in a eutrophic coastal sea: a case study of the Changjiang (Yangtze River) estuary. *Mar. Freshwater Res.* **2019**, *70*, 1389–1401. [CrossRef]

49. Iuculano, F.; Duarte, C.M.; Marbá, N.; Agustí, S. Seagrass as major source of transparent exopolymer particles in the oligotrophic Mediterranean coast. *Biogeosciences* **2017**, *14*, 5069–5075. [CrossRef]

50. Engel, A.; Meyerhöfer, M.; von Bröckel, K. Chemical and biological composition of suspended particles and aggregates in the Baltic Sea in summer (1999). *Estuar. Coast. Shelf S.* **2002**, *55*, 729–741. [CrossRef]

51. Prieto, L.; Navarro, G.; Cózar, A.; Echevarría, F.; García, C.M. Distribution of TEP in the euphotic and upper mesopelagic zones of the southern Iberian coasts. *Deep Sea Res. Part II* **2006**, *53*, 1314–1328. [CrossRef]

52. Aller, J.Y.; Radway, J.C.; Kilthau, W.P.; Bothe, D.W.; Wilson, T.W.; Vaillancourt, R.D.; Quinn, P.K.; Coffman, D.J.; Murray, B.J.; Knopf, D.A. Size-resolved characterization of the polysaccharidic and proteinaceous components of sea spray aerosol. *Atmos. Environ.* **2017**, *154*, 331–347. [CrossRef]

53. Scoullos, M.; Plavšić, M.; Karavoltsons, S.; Sakellari, A. Partitioning and distribution of dissolved copper, cadmium and organic matter in Mediterranean marine coastal areas: The case of a mucilage event. *Estuar. Coast. Shelf Sci.* **2006**, *67*, 484–490. [CrossRef]

54. Engel, A.; Piontek, J.; Metfies, K.; Endres, S.; Sprong, P.; Peeken, I.; Gäbler-Schwarz, S.; Nöthig, E.M. Inter-annual variability of transparent exopolymer particles in the Arctic Ocean reveals high sensitivity to ecosystem changes. *Sci. Rep.* **2017**, *7*, 4129. [CrossRef] [PubMed]

55. Taylor, J.D.; Cottingham, S.D.; Billinge, J.; Cunliffe, M. Seasonal microbial community dynamics correlate with phytoplankton-derived polysaccharides in surface coastal waters. *ISME J.* **2014**, *8*, 245–248. [CrossRef] [PubMed]

56. Dreshchinskii, A.; Engel, A. Seasonal variations of the sea surface microlayer at the Boknis Eck Times Series Station (Baltic Sea). *J. Plankton Res.* **2017**, *39*, 943–961. [CrossRef]

57. Redfield, A.C.; Ketchum, B.H.; Richards, F.A. *The Influence of Organisms on the Composition of Seawater, Comparative and Descriptive Oceanography*; Harvard University Press: Cambridge, MA, USA, 1963; pp. 26–77.

58. Dortch, Q.; Whitledge, T.E. Does nitrogen or silicon limit phytoplankton production in the Mississippi River plume and nearby regions? *Cont. Shelf Res.* **1992**, *12*, 1293–1309. [CrossRef]
59. Justíc, D.; Rabalais, N.N.; Turner, R.E.; Dortch, Q. Changes in nutrient structure of river-dominated coastal waters: Stoichiometric nutrient balance and its consequences. *Estuar. Coast. Shelf Sci.* **1995**, *40*, 339–356. [CrossRef]
60. Corzo, A.; Morillo, J.A.; Rodríguez, S. Production of transparent exopolymer particles (TEP) in culture of Chaetoceros calcitrans under nitrogen limitation. *Aquat. Microb. Ecol.* **2000**, *23*, 63–72. [CrossRef]
61. Ortega-Retuerta, E.; Mazuecos, I.P.; Reche, I.; Gasol, J.M.; Álvarez-Salgado, X.A.; Álvarez, A.; Montero, M.F.; Arístegui, J. Transparent exopolymer particle (TEP) distribution and in situ prokaryotic generation across the deep Mediterranean Sea and nearby North East Atlantic Ocean. *Prog. Oceanogr.* **2019**, *173*, 180–191. [CrossRef]

Article

Temporal and Spatial Variations of the Biochemical Composition of Phytoplankton and Potential Food Material (FM) in Jaran Bay, South Korea

Jae Hyung Lee [1], Won-Chan Lee [2], Hyung Chul Kim [2], Naeun Jo [3], Kwanwoo Kim [3], Dabin Lee [3], Jae Joong Kang [3], Bo-Ram Sim [2], Jae-Il Kwon [4] and Sang Heon Lee [3,*]

1   South Sea Fisheries Research Institute, National Institute of Fisheries Science, Yeosu-si 59780, Korea; jhlee88@korea.kr
2   Marine Environment Research Division, National Institute of Fisheries Science, Busan 46083, Korea; phdleewc@korea.kr (W.-C.L.); hckim072@korea.kr (H.C.K.); qhfka3465@gmail.com (B.-R.S.)
3   Department of Oceanography, Pusan National University, Busan 46241, Korea; nadan213@pusan.ac.kr (N.J.); goanwoo7@pusan.ac.kr (K.K.); ldb1370@pusan.ac.kr (D.L.); jaejung@pusan.ac.kr (J.J.K.)
4   Marine Disaster Research Center, KIOST, Busan 49111, Korea; jikwon@kiost.ac.kr
*   Correspondence: sanglee@pusan.ac.kr

Received: 22 September 2020; Accepted: 30 October 2020; Published: 4 November 2020

Abstract: Food material (FM) derived from biochemical components (e.g., proteins, lipids, and carbohydrates) of phytoplankton can provide important quantitative and qualitative information of the food available to filter-feeding animals. The main objective of this study was to observe the seasonal and spatial variations of the biochemical compositions of phytoplankton and to identify the major controlling factors of FM as a primary food source in Jaran Bay, a large shellfish aquaculture site in South Korea. Base d on monthly sampling conducted during 2016, significant monthly variations in the depth-integrated concentrations of major inorganic nutrients and chlorophyll *a* within the euphotic water column and a predominance (49.9 ± 18.7%) of micro-sized phytoplankton (>20 μm) were observed in Jaran Bay. Carb ohydrates were the dominant biochemical component (51.8 ± 8.7%), followed by lipids (27.3 ± 3.8%) and proteins (20.9 ± 7.4%), during the study period. The biochemical compositions and average monthly FM levels (411.7 ± 93.0 mg m^{-3}) in Jaran Bay were not consistent among different bays in the southern coastal region of South Korea, possibly due to differences in controlling factors, such as environmental and biological factors. Acco rding to the results from multiple linear regression, the variations in FM could be explained by the relatively large phytoplankton and the P* ($PO_4^{3-} - 1/16 \times NO_3^-$) and NH_4^+ concentrations in Jaran Bay. The macromolecular compositions and FM, as alternatives food source materials, should be monitored in Jaran Bay due to recent changes in nutrient concentrations and phytoplankton communities.

Keywords: phytoplankton; biochemical compositions; carbohydrates; proteins; lipids; Jaran Bay

---

1. Introduction

Bays are important aquatic systems that provide food resources for fisheries and aquaculture since they provide habitats and prey for various marine organisms. Rece ntly, mollusk farming, including bivalves, has contributed greatly to global farming production [1]. The present study site, Jaran Bay, is one of the largest shellfish aquaculture regions for oysters and scallops in South Korea [2], and these filter-feeding oysters and scallops feed mainly on water-dwelling phytoplankton for their growth and reproduction [3,4].

The growth and physiological conditions of phytoplankton can vary depending on environmental conditions [5–7]. In particular, phytoplankton synthesize biochemical components through photosynthesis and are therefore highly dependent on light conditions and quality [8–10], temperature [11], species composition [12,13] and nutrient availability [5,8,14]. Rece ntly, Lee et al. [5] reported that dissolved inorganic nitrogen loading from river discharge is a major factor that controls the photosynthetic biochemical compositions (e.g., carbohydrates, proteins and lipids) of phytoplankton in Gwangyang Bay. More over, the community structure and, consequently, biochemical composition of phytoplankton can be altered by differences in nutrient inputs due to river discharge [7]. Diff erences in the biochemical compositions of phytoplankton can lead to differences in nutritional qualities for potential consumers [5,15–17]. Ther efore, the biochemical compositions of phytoplankton, as natural food resources, are very important for phytoplankton-grazing herbivores. In agreement with this finding, Yun et al. [16] reported a strong positive relationship between the lipid composition in phytoplankton and protein content in the mesozooplankton community in the northern Chukchi Sea, indicating that a high lipid content in phytoplankton can be important for protein synthesis for zooplankton growth.

Food material (FM) is represented as the sum of the concentrations of proteins, lipids and carbohydrates [18,19]. FM indicates the quantity of food that is available to potential consumers [19] and is also used as a food index of food quality [20]. Seas onal and spatial variations in the quantity and quality of the natural diet available to filter feeders could be important for their grazing characteristics [20]. Nava rro and Thompson [20] observed that the seasonal trends in FM dynamics are closely correlated with the trends of the chlorophyll *a* concentration in Logy Bay, southeast Newfoundland, Canada. Rece ntly, Kang et al. [21] found that small-sized cells of phytoplankton could assimilate higher amounts of FM per unit of chlorophyll *a* concentration compared to large-sized cells of phytoplankton in the East/Japan Sea based on size fractionation filtering methods. Simi lar results from Gwangyang Bay, Korea, were also in agreement with this consistent observation [7].

Previously, most biochemical composition studies have been conducted once a year or, at most, seasonally [5,7,21]. Cons idering the importance of phytoplankton as a primary food source for filter-feeding aquaculture animals, the present study aimed to observe monthly and spatial variations in biochemical compositions as a food quality indicator of phytoplankton and to determine the major environmental controlling factors of FM available to shellfish, such as oysters and scallops, growing in Jaran Bay as a large aquaculture site in South Korea.

2. Materials and Methods

2.1. Water Sampling and Analysis

Using a 5 L Niskin sampler (General Oceanics Inc., Miami, FL, USA), water samples for the determination of the nutrient and chlorophyll *a* concentrations were obtained from three different light depths (e.g., 100, 30 and 1% of photosynthetic active radiation (PAR), determined by using a Secchi disk) at seven different stations (Figure 1). The study area Jaran Bay is a relatively shallow coastal bay with an average water depth of 10 m [2]. Samp ling was conducted monthly from January to December 2016. The depth-averaged values were obtained from the three light depths (e.g., 100, 30 and 1% PAR), and monthly observed values were obtained from all depths and stations.

The water samples (0.2 L) used for determining the dissolved inorganic nutrient concentrations were filtered through a 47 mm GF/F filters (0.7-µm pore size, Whatman, Maidstone, UK), and the filtrates were stored at −20 °C for further analysis using an Auto Analyzer (Quaatro, Bran+Luebbe, Germany) at the National Institute of Fisheries Science (NIFS), Korea. For determining the total chlorophyll *a* concentration as a proxy for biomass, water samples (0.2 L) were filtered through 25-mm GF/F filters (0.7-µm pore size, Whatman, Maidstone, UK). The water samples (0.6 L) were filtered sequentially through 47-mm Nucleopore filters (20- and 2-µm) and 47-mm GF/F filters (0.7-µm pore size, Whatman, Maidstone, UK) to determine size-fractionated chlorophyll *a* concentrations of different cell-sized phytoplankton communities [22,23]. The filters retained chlorophyll *a* and were immediately frozen and preserved at −70 °C for chlorophyll *a* extraction at the home laboratory at Pusan National University, South Korea. The chlorophyll *a* concentrations were measured using a previously calibrated

10-AU fluorometer (Turner Designs, San Jose, CA, USA) after extraction (approximately 24 h, 4 °C) with 90% acetone and centrifugation at 4480 g for 20 min [24].

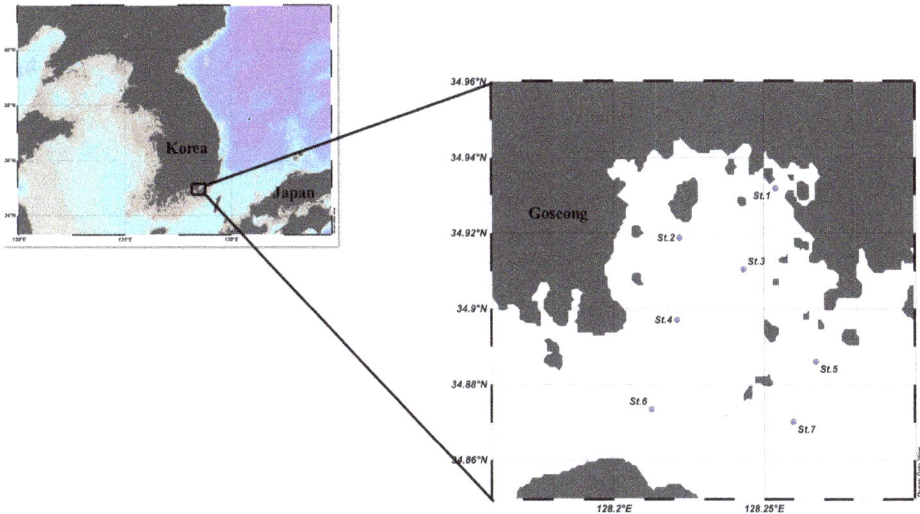

Figure 1. Sampling stations in Jaran Bay, South Sea of Korea.

The water samples that were used for determining the macromolecular compositions (e.g., carbohydrates, proteins and lipids) of particulate organic matter (POM) were filtered through 47 mm GF/F filters, and the filters were immediately preserved at −70 °C until further spectrophotometric analysis. The samples were filtered under a constant vacuum (<10 cm Hg) because live cells could be damaged during the strong vacuum filtration [23]. Carb ohydrate extraction was performed by following Dubois et al. [25]. The preground POM-retained filter paper was transferred to a polypropylene (PP) tube. Afte r the addition of 1 mL deionized water, 1 mL of a 5% phenol solution was added and allowed to rest for 40 min. Then , 5 mL of sulfuric acid (H_2SO_4) was added and allowed to stand for 10 min. Next , the solutions were centrifuged at 3430 g for 10 min. The absorbance of the supernatant was measured at 490 nm. A glucose solution (1 mg mL^{-1}, Sigma Aldrich) was used as the standard for determining the carbohydrate concentration.

For protein extraction, each preground sample filter was transferred to a 12-mL glass tube with 1 mL deionized water (DH_2O) and was added to 5 mL of an alkaline copper solution. Afte r the solution was well mixed, 0.5 mL of diluted Folin–Ciocalteu phenol reagent (1:1, *v/v*) were added and allowed to sit for 1 h 30 min at room temperature. Then , the solutions were centrifuged for 10 min at 2520 g. The absorbance of the supernatant was measured at 750 nm. Bovi ne serum albumin (2 mg mL^{-1}, Sigma Aldrich) was used as the standard for determining the protein concentration based on previous works in various oceans [5–7,16,17,21].

Last, the filters used for lipid extraction were transferred into a 16-mL glass tubes, ground with 3 mL of chloroform-methanol (1:2, *v/v*) and stored at 4 °C for 1 h. Afte r the solution was homogenized with 4 mL of DH_2O, the lower (chloroform) phase of the solution was dried at 40 °C for 48 h and then heated at 200 °C for 15 min with 2 mL of H_2SO_4. An additional 3 mL of DH_2O was added to the chloroform phase in the glass tubes and then they were allowed to rest for 10 min. The absorbance of the supernatant was measured at 375 nm, and a tripalmitin solution (Sigma Aldrich) was used as the standard for determining the lipid concentration. Afte r each extraction process, the concentration of each biochemical component was determined using a UV spectrophotometer (Hitachi-UH5300, Hitachi, Tokyo, Japan).

2.2. Statistical Analysis

Principal component analysis (PCA) was performed on our field-obtained data of the chemical and biological variables (i.e., nutrient concentrations and phytoplankton biomass) for their relative significance and interrelationship patterns among the various biochemical conditions measured during our sampling period. Bart lett's sphericity tests were used to determine the validity of the PCA ($p < 0.01$) [26,27]. Fact or analysis was conducted to obtain various factors selected by the principal component method with varimax rotation [28]. Due to the strong dependency between PO_4^{3-} and NO_3^- ($r = 0.56$, $p < 0.01$; Pearson's correlation coefficient), PO_4^{3-} was excluded but included P* ($PO_4^{3-} - 1/16 \times NO_3^-$) in the PCA. P * reflects the excess (or deficiency) of PO_4^{3-} versus NO_3^- [29,30].

To determine the major factors controlling the macromolecular composition and FM of POM, multiple linear regression analysis was conducted in this study based on the PCA results. The multiple linear regression equation of Pedhazur [31] is as follows:

$$Y = \alpha + b_1 X_1 + \cdots + b_k X_k + e \tag{1}$$

where Y denotes a dependent variable and the FM of POM is estimated from the independent variables (predictors), $X_1 \cdots X_k$. Para meter α is a constant, $b_1 \cdots b_k$ are the regression coefficients for the predictors (FM in this study), and e is an error term.

Insignificant variables for the controlling the FM variation were stepwise eliminated from the model by stepwise variable selection after multiple linear regression analysis. Stat istical analysis was performed with IBM SPSS software version 12.0 (SPSS Inc., Chicago, IL, USA). t statistics were conducted for testing the regression coefficients and values of the coefficient of determination (R^2) were obtained for measure of goodness of fit for the FM in this study.

3. Results

3.1. Monthly Concentrations of Nutrients and Chlorophyll a

The monthly depth-integrated nutrient concentrations within the euphotic water column from 100 to 1% light depths during the present study period are summarized in Table 1. The ranges of the NH_4^+, $NO_2^- + NO_3^-$, PO_4^{3-} and $Si(OH)_4^{2-}$ concentrations were 4.0–47.5, 10.9–80.0, 0.5–6.0 and 20.9–166.9 μm, respectively, in Jaran Bay from January to December 2016. The concentration ranges varied significantly during the observation period, and the highest concentrations were detected in September, except for the $Si(OH)_4^{2-}$ concentrations, which showed the largest peak in June and a secondary peak in September.

Table 1. Monthly variations in the water column-integrated major nutrient concentrations averaged from seven different stations in Jaran Bay.

	Integrated Nutrients			
	NH_4^+	$NO_2^- + NO_3^-$	DIP	SiO_2-Si
	mmol m^{-2}			
Jan.	8 ± 5	21 ± 19	3 ± 2	64 ± 46
Feb.	4 ± 3	8 ± 6	2 ± 1	21 ± 11
Mar.	4 ± 2	7 ± 6	1 ± 1	22 ± 7
Apr.	7 ± 2	12 ± 5	2 ± 1	47 ± 5
May	6 ± 1	7 ± 2	0.5 ± 0.2	74 ± 14
Jun.	8 ± 3	12 ± 8	1 ± 1	167 ± 48
Jul.	11 ± 4	15 ± 10	2 ± 1	161 ± 31
Aug.	9 ± 4	8 ± 4	2 ± 1	90 ± 46
Sep.	48 ± 19	33 ± 13	6 ± 2	146 ± 37
Oct.	6 ± 2	22 ± 23	1 ± 2	72 ± 70
Nov.	11 ± 4	44 ± 24	4 ± 2	102 ± 46
Dec.	9 ± 4	46 ± 31	4 ± 2	106 ± 66

The total monthly chlorophyll *a* concentration averaged from the three light depths at seven stations ranged from 0.77 µg L^{-1} in September to 4.89 µg L^{-1} in October, with an average of 2.13 µg L^{-1} (S.D. = ± 1.18 µg L^{-1}) (Figure 2). Base d on the different size-fractionated chlorophyll *a* concentrations (Figure 3), the compositions of the micro- (> 20 µm), nano- (2–20 µm) and pico-sized chlorophyll *a* concentrations (0.7–2 µm) varied significantly in Jaran Bay among the different months. The compositions of the micro-sized chlorophyll *a* concentrations ranged from the lowest value in April (23.8 ± 18.7%) to the highest value in January (77.8 ± 6.8%), whereas the nano-sized chlorophyll *a* compositions ranged from the lowest value in January (14.3 ± 7.0%) to the highest value in June (50.3 ± 21.3%). In comparison, the compositions of pico-sized chlorophyll *a* were lowest in January (7.9 ± 4.4%) and highest in April (46.0 ± 18.1%). Seas onally, the compositions of the micro-sized chlorophyll *a* concentrations steadily increased from spring (March–May) to winter (December–February), although significant monthly variations were present. In contrast, the compositions of the pico-sized chlorophyll *a* concentrations steadily decreased from spring to winter. The compositions of the nano-sized chlorophyll *a* concentrations were highest in summer (June–August) and lowest in winter. On average, micro-sized (>20 µm) cells contributed 49.9% (± 18.7%) of the total chlorophyll *a* concentration in Jaran Bay during our observation period. In comparison, the nano- and pico-sized chlorophyll *a* compositions contributed 28.5% (± 12.4%) and 21.6% (± 11.2%), respectively. A strong positive relationship was found between the micro-sized chlorophyll *a* concentrations and total chlorophyll *a* concentrations integrated from the euphotic water columns in this study (y = 1.31x + 5.74, r^2 = 0.82; Figure 4).

3.2. Spatial and Temporal Variations of the Macromolecular Compositions of POM

Figure 5 shows the average of three light depth values of each macromolecular composition of POM in Jaran Bay from January to December 2016. No distinctive spatial variations were detected in the macromolecular compositions among the different stations; however, they significantly varied among the different months. Carb ohydrates were the predominant biochemical component during our observation period from January to December, with monthly proportions of carbohydrates ranging from 40.9% to 66.4%. In comparison, the protein and lipid proportions were 11.1–31.0% and 22.5–35.1%, respectively. The lipid proportion appeared to decrease steadily from January to December. Seas onally, the carbohydrate proportion were relatively variable compared to the protein and lipid proportions. The carbohydrate proportion was lowest during summer (45.6 ± 1.4%) and highest during autumn (59.1 ± 10.9%). In comparison, the protein proportion was lowest during winter (17.5 ± 5.7%) and highest during summer (28.1 ± 2.7%), while the lipid proportion was lowest in autumn (23.1 ± 0.6%) and highest in winter (31.1 ± 3.7%).

The monthly FM concentrations ranged from 297 to 630 mg m^{-3}, with an average of 411.7 mg m^{-3} (S.D. = ± 93.0 mg m^{-3}), in this study (Table 2). No noticeable monthly variations were observed for the FM concentrations. Spat ial variations in the FM concentrations were not noticeable for the seven stations during the observation period except for March, April and June, which had considerably higher FM concentrations at several stations (Figure 6). The monthly calorific values and FM contents of FM averaged from the three light depths at the seven stations did not vary significantly and ranged from 5.5–6.3 Kcal g^{-1} and 1.7–3.7 Kcal m^{-3}, respectively, in Jaran Bay (Table 2).

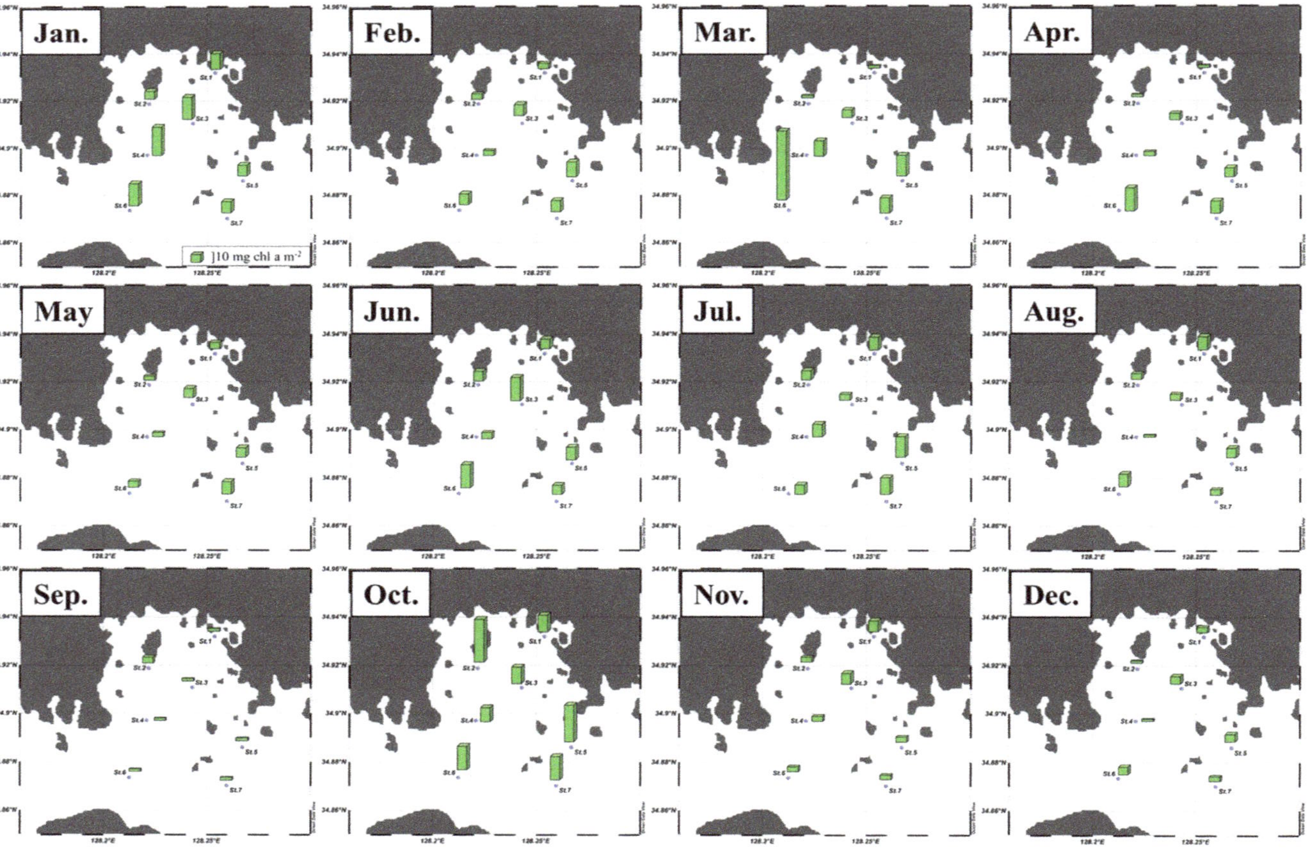

Figure 2. Water column-integrated chlorophyll *a* concentrations at the sampling stations in Jaran Bay.

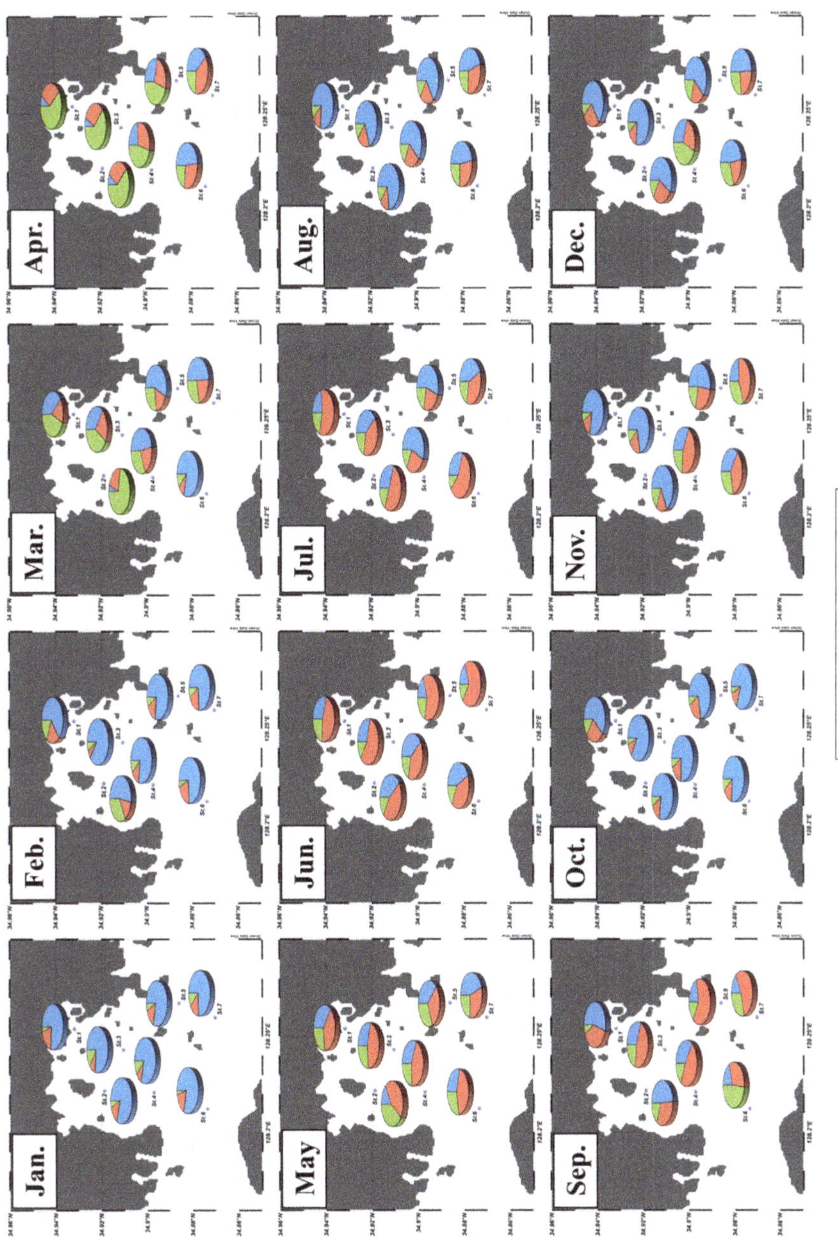

Figure 3. Different size compositions of chlorophyll *a* concentrations at the sampling stations in Jaran Bay.

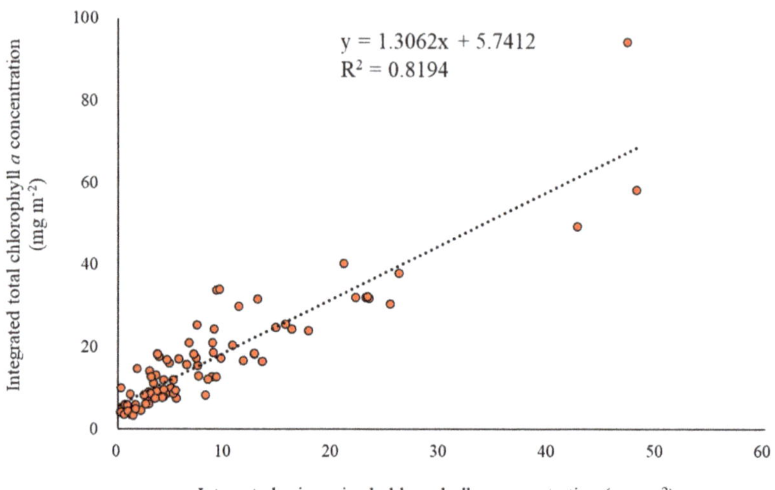

Figure 4. Relationship between the euphotic depth-integrated micro-sized chlorophyll *a* concentrations and the integrated total chlorophyll *a* concentrations in Jaran Bay.

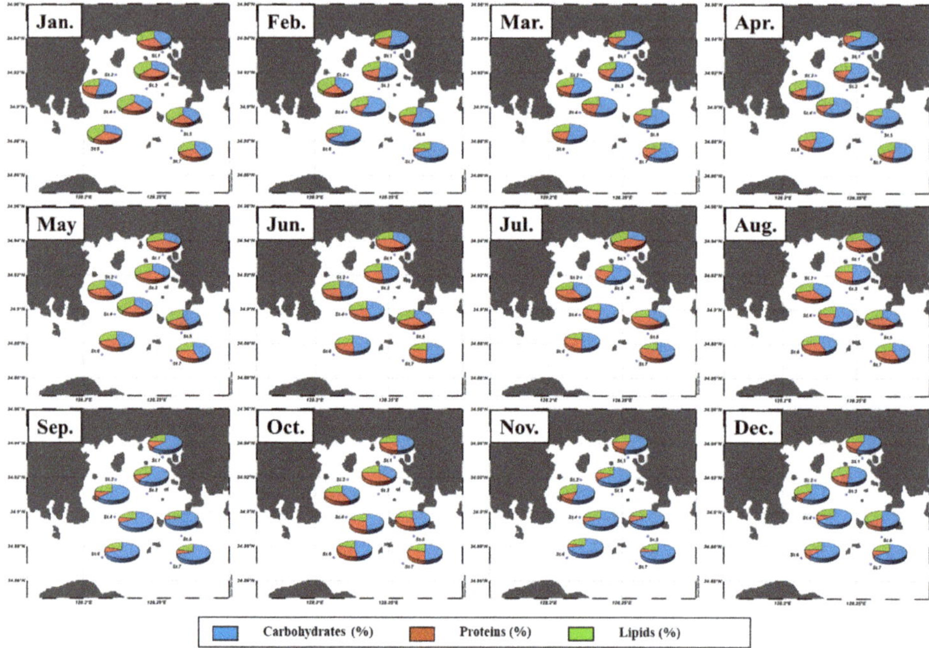

Figure 5. Biochemical compositions of POM relative to the total FM at the sampling stations in Jaran Bay.

Table 2. Monthly averaged compositions of different sized chlorophyll a concentrations and biochemical concentrations and compositions of POM averaged from seven different stations in Jaran Bay.

	Total chl a (μg L^{-1})	Micro (%)	Nano (%)	Pico (%)	CHO (μg L^{-1})	PRT (μg L^{-1})	LIP (μg L^{-1})	FM (μg L^{-1})	CHO (%)	PRT (%)	LIP (%)
Jan.	3.2 ± 1.0	78 ± 7	14 ± 7	8 ± 4	145 ± 64	81 ± 19	119 ± 34	345 ± 84	41 ± 9	24 ± 4	35 ± 8
Feb.	1.7 ± 0.6	72 ± 19	15 ± 8	13 ± 18	243 ± 56	65 ± 21	134 ± 44	442 ± 56	55 ± 10	15 ± 5	30 ± 8
Mar.	2.5 ± 1.6	45 ± 22	19 ± 5	37 ± 18	206 ± 48	68 ± 21	92 ± 26	368 ± 78	56 ± 6	18 ± 3	25 ± 5
Apr.	1.4 ± 0.7	24 ± 19	30 ± 6	46 ± 18	183 ± 37	53 ± 17	96 ± 31	332 ± 57	56 ± 8	16 ± 3	29 ± 7
May	1.4 ± 0.8	32 ± 16	38 ± 12	29 ± 12	163 ± 44	103 ± 37	129 ± 50	395 ± 101	42 ± 7	26 ± 7	32 ± 5
Jun.	2.5 ± 1.4	32 ± 19	50 ± 21	18 ± 6	239 ± 182	122 ± 64	131 ± 82	492 ± 317	47 ± 8	26 ± 5	27 ± 8
Jul.	3.2 ± 1.8	40 ± 15	41 ± 15	19 ± 6	269 ± 94	202 ± 108	158 ± 99	630 ± 250	45 ± 11	31 ± 8	24 ± 6
Aug.	1.6 ± 1.5	62 ± 16	21 ± 11	16 ± 8	168 ± 71	101 ± 34	101 ± 32	370 ± 125	45 ± 6	28 ± 5	28 ± 4
Sep.	0.8 ± 0.2	32 ± 15	45 ± 14	23 ± 18	255 ± 48	42 ± 11	85 ± 14	382 ± 53	66 ± 6	11 ± 3	23 ± 4
Oct.	4.9 ± 1.6	75 ± 11	17 ± 8	8 ± 3	239 ± 48	157 ± 43	117 ± 18	513 ± 83	47 ± 6	30 ± 6	23 ± 3
Nov.	1.4 ± 0.7	51 ± 20	28 ± 12	20 ± 9	240 ± 36	45 ± 19	89 ± 22	375 ± 45	64 ± 8	12 ± 5	24 ± 5
Dec.	0.9 ± 0.4	55 ± 14	23 ± 6	22 ± 13	172 ± 23	41 ± 14	85 ± 32	297 ± 49	58 ± 7	14 ± 4	28 ± 7

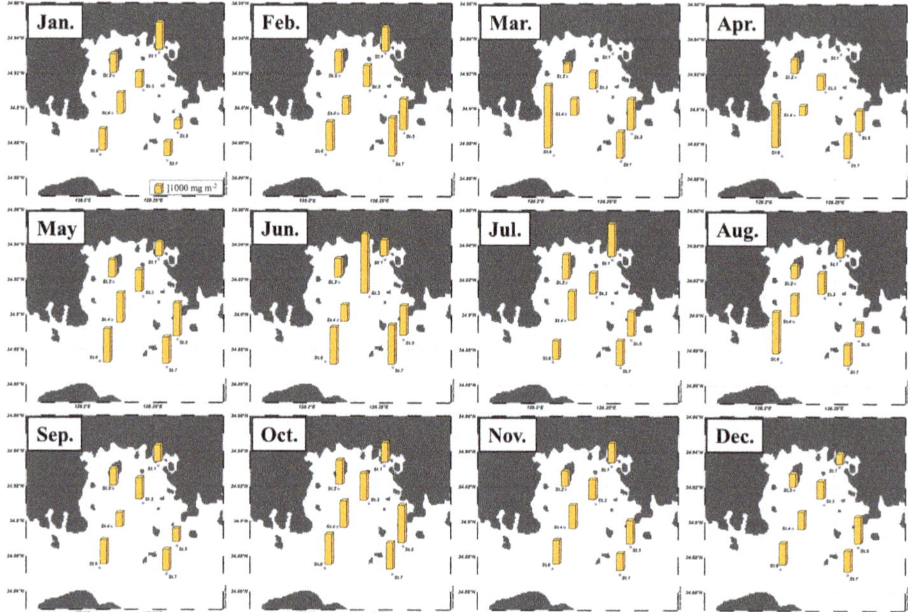

Figure 6. Water column-integrated the total FM concentrations at the sampling stations in Jaran Bay.

3.3. Principal Component Analysis (PCA)

The PCA results for our field-observed biochemical parameters are summarized in Table 3. Thre e PCs were selected for multiple linear regression analysis in this study. The variables shown in bold indicate the highest correlations among the 12 variables and the corresponding components. The nano-sized chlorophyll *a* concentrations, carbohydrates, proteins, lipids and FMs had the highest correlations with PC1, whereas the concentrations of NH_4^+, NO_3^-, P* and $Si(OH)_4^{2-}$ were highest correlated with PC2. For PC3, temperature and the micro- and pico-sized chlorophyll *a* concentrations showed the highest correlations. Base d on the PCA results in Table 3, multiple linear regression analysis was performed to obtain the major controlling factors for the variation in the FM in Jaran Bay (Table 4). The nano- and micro-sized chlorophyll *a* concentrations and P* and NH_4^+ concentrations were found to be the major factors for controlling the FM in Jaran Bay during our observation period (Table 4). The concentrations of nano- and micro-sized chlorophyll *a* and NH_4^+ had positive effects whereas the P* concentration had a negative impact on the FM in Jaran Bay during the study period. In other word, a total increase in the concentrations of nano- and micro-sized chlorophyll *a* and NH_4^+ could bring an increase in the FM. On the other hand, an increase in P* concentration could lead to a decrease in the FM.

Table 3. Principal component analysis (PCA) results in Jaran Bay during the observation period.

Variables in	Standardized Weight of Variables in Selected PC (t_{ik}; $I = 1, 2, \ldots, 12$ and $k = 1, 2$)			Loading of Variables (v_{ik})			Communalities
	PC 1	PC 2	PC 3	PC 1	PC 2	PC 3	
Temperature	−0.103	0.18	0.373	0.109	0.506	**0.587**	0.612
NH_4^+	−0.002	0.35	−0.06	−0.087	**0.874**	−0.065	0.775
NO_3^-	0.054	0.217	−0.028	0.154	**0.543**	0.047	0.321
P*	−0.052	0.289	−0.141	−0.371	**0.708**	−0.289	0.723
SiO_2-Si	−0.017	0.273	0.222	0.221	**0.719**	0.43	0.75
Micro	−0.102	−0.133	0.416	0.171	−0.276	**0.624**	0.495
Nano	0.204	0.046	0.033	**0.772**	0.118	0.33	0.719
Pico	−0.107	0.023	0.335	0.048	0.105	**0.489**	0.253
CHO	0.357	0.108	−0.357	**0.818**	0.219	−0.19	0.754
PRT	0.094	−0.068	0.277	**0.694**	−0.133	**0.627**	0.892
LIP	0.252	−0.053	−0.075	**0.805**	−0.147	0.177	0.702
FM	0.304	0.012	−0.099	**0.961**	0.013	0.209	0.967

Table 4. Regression analysis results for the FM in Jaran Bay (**: $p < 0.01$, n = 252).

Included Independent Variables	Regression Coefficient (b_k)	Standard Error of b_k	Standardized Regression Coefficient	t Statics	p Value	Adjusted R^2 (%)
Constant	337.872	12.08		27.969	0.000 **	
Nano-chlorophyll *a* concentration	112.476	8.16	0.617	13.784	0.000 **	0.544
Micro-chlorophyll *a* concentration	20.115	5.412	0.156	3.716	0.000 **	0.57
P*	−230.321	49.425	−0.305	−4.66	0.000 **	0.582
NH_4^+ concentration	19.321	5.362	0.225	3.603	0.000 **	0.602

4. Discussion

The monthly-averaged concentrations of the depth-integrated nutrient concentrations measured were within the ranges previously reported from regions near Jaran Bay [32–35]. The present study indicates that each nutrient concentration showed significant seasonal variations. For example, the DIN concentrations were relatively higher during the period from September to December, whereas the silicate concentrations were higher in June–September compared to other months (Table 1).

The monthly depth-integrated total chlorophyll *a* concentrations within the euphotic water column from 100% to 1% light depths ranged from 5.2 to 36.7 mg m^{-2} (mean ± S.D. = 17.0 ± 9.2 mg chl-*a* m^{-2}) during the study period from January to December 2016. The largest peak was observed in October immediately, followed by the nutrient peaks observed in September (Table 1). Howe ver, the seasonal chlorophyll *a* concentrations did not vary greatly and ranged from 15.8 to 17.8 mg m^{-2}. Gene rally, the spatial variation of the total chlorophyll *a* concentrations appeared to be low among the seven stations in Jaran Bay during the observation period except for March (Figure 2). Over all, the phytoplankton community was dominated by micro-sized phytoplankton based on the size-fractionated chlorophyll *a* concentration results during our observation period. Prev ious studies have reported that the predominant species in this area consisted of diatoms [23,36]. In general, the spatial and seasonal variations of the total chlorophyll *a* concentrations were strongly related to the micro-sized (> 20 µm) chlorophyll *a* concentrations (Figure 4). This finding suggests that micro-sized cells greatly contributed to the total chlorophyll *a* concentration in Jaran Bay. In other words, 49.9% (± 18.7%) of total chlorophyll *a* was from micro-sized cells (Figure 3) during our observation period.

The overall dominant macromolecular composition of POM was carbohydrates (51.8 ± 8.7%), followed by lipids (27.3 ± 3.8%) and proteins (20.9 ± 7.4%), during our observation period (Figure 5). The macromolecular compositions obtained from the present study fell in a similar range to those obtained from Geoje-Hansan Bay by Kim et al. [6], in which their study area was close to our research site. Howe ver, the compositions in Jaran and Geoje-Hansan bays were considerably different from those in Gwangyang Bay. The mean compositions in Gwangyang Bay were 26.4% (± 9.4%), 37.8% (± 16.1%), and 35.7% (± 13.9%) carbohydrates, proteins, and lipids, respectively [5]. Thes e differences may have been due to the influence of river-borne nutrients. The protein and lipid proportions are largely dependent on the input of dissolved inorganic nitrogen from the Seomjin River in Gwangyang Bay [5]. In comparison, there are no large river inputs in the Jaran and Geoje-Hansan bays. For coastal management plans, e.g., artificial dam construction, the potential influence of river inputs on the dominant cell size and photosynthetic end-products of phytoplankton should be considered [7].

Although the macromolecular compositions between Jaran and Geoje-Hansan Bays [6] in south Korea are similar, the monthly FM concentrations were relatively lower in Jaran Bay and ranged from 297 to 630 mg m^{-3} with an average of 411.7 mg m^{-3} (S.D. = ± 93.0 mg m^{-3}), than in Geoje-Hansan Bay, which had a range of 346–1280 mg m^{-3} (615.5 ± 291.7 mg m^{-3}; Table 5). Howe ver, the average monthly FM concentration (411.7 ± 93.0 mg m^{-3}) of POM in Jaran Bay during our observation period was similar to that in Gwangyang Bay (434.5 ± 175.5 mg m^{-3}) [5] despite the large difference in macromolecular compositions between the two bays. Base d on the fact that FM concentrations are derived from the total concentrations of carbohydrates, proteins and lipids [18,19] and that their relative compositions can be affected by various environmental and biological factors [5,8–14], different macromolecular compositions are unlikely to be strongly related to the FM concentrations of POM. Inst ead of the compositions of the chlorophyll *a* concentrations, which are often used to represent phytoplankton biomass, would be more appropriate for comparisons. Howe ver, no strong relationship between the FM concentrations and total chlorophyll *a* concentrations was found in the present study, although a strong correlation was found in Gwangyang Bay by [7]. Simi larly, no significant linear relationship was observed between the FM and total chlorophyll *a* concentrations among the different bays in South Korea (Table 5). The average chlorophyll *a* concentrations were 2.13 µg L^{-1} (S.D. = ± 1.18 µg L^{-1}, this study), 4.34 µg L^{-1} [6] and 3.45 µg L^{-1} [5] in the Jaran, Geoje-Hansan and Gwangyang Bays, (Table 5). Thes e bays are all in the South Sea of South Korea. In the Garolim-Asan Bay, Yellow Sea [37], the average chlorophyll *a* concentration (2.81 ± 2.12 µg L^{-1}) was within the low range (2.13–4.34 µg L^{-1}) among the three bays, but the average FM concentration (781.4 ± 228.2 mg m^{-3}) was highest among the bays in this study. The chlorophyll *a* concentration has been used as a proxy for biomass, but may not be completely representative of phytoplankton biomass since the chlorophyll *a* concentration is greatly influenced by light and nutrient conditions, physiological status and species composition of phytoplankton [38–41]. Inst ead of the chlorophyll *a* concentration, Lee et al. [5] and Kim et al. [7] suggested that the FM concentration of POM, mainly phytoplankton, could be an alternative proxy for food sources available to higher trophic levels in bay or coastal marine ecosystems. Ther efore, the FM concentration could have a quantitatively complementary value for the amount of various food material sources available to potential consumers in estuarine or bay ecosystems [7,21]. With respect to energy aspects, the calorific content, which depends on the different macromolecular compositions of the FM concentration, should be considered as representative of the physiological or ecological conditions of higher trophic levels of consumers [5,7,21].

Table 5. Comparison of the total chlorophyll *a* concentrations and FM concentrations of POM among different Korean bays.

Region	Period	Total Chlorophyll *a* Concentration (µg L^{-1})	FM Concentration (mg m^{-3})	Reference
Gwangyang Bay, Korea	Seasonally, 2012–2013	3.45 (±2.81)	434.5 (±175.5)	[5]
Geoje-Hansan Bay, Korea	Monthly, 2015	4.34 (±2.42)	615.5 (±291.7)	[6]
Garolim-Asan Bay, Korea	Seasonally, 2015–2016	2.81 (±2.12)	781.4 (±228.2)	[37]
Jaran Bay, Korea	Monthly, 2016	2.13 (±1.18)	411.7 (±93.0)	This study

According to the PCA results, spatiotemporal variations in FM are primarily governed by the nano-sized chlorophyll *a* concentrations, carbohydrates, proteins and lipids since FM is the sum of the concentrations of the three different macromolecules. Howe ver, the positive relationship between the nano-sized chlorophyll *a* concentration and FM would not be predictable. In Jaran Bay, the spatiotemporal change of the total chlorophyll *a* concentration was primarily controlled by the micro-sized chlorophyll *a* concentrations because of their high contribution to the total chlorophyll *a* concentration. In comparison, nano-sized chlorophyll *a* compositions contributed 28.5% (± 12.4%) of the total chlorophyll *a* concentration in this study, although their monthly contributions varied somewhat broadly and ranged from 14.3% (±7.0%) in January to 50.3% (±21.3%) in June (Figure 3). In PC2, the positive correlations among the major inorganic nutrient concentrations (e.g., NH$_4^+$, NO$_3^-$, P* and Si(OH)$_4^{2-}$) were reasonable. Temp erature and the micro- and pico-sized chlorophyll *a* concentrations in PC3 indicate positive correlations among the three variables in Jaran Bay (Table 3). PCA was used in this study for ranking their relative significance (Table 3) among our field-observed biochemical parameters for multiple linear regression analysis and deriving major controlling factors (Table 4) of the FM in our study site. In this approach, we could predict the FM in our study site based on the multiple linear regression analysis. Acco rding to the multiple linear regression, approximately 60% of the variation in FM could be explained by the nano- and micro-sized chlorophyll *a* concentrations and P* and NH$_4^+$ concentrations in Jaran Bay (Table 4). With this approach, the four major controlling factors were determined for the observed FM variations in Jaran Bay during our observation period from January to December 2016. Howe ver, the somewhat low prediction of up to 60% suggests that other potential factors in addition to our observed parameters should be investigated to improve the spatiotemporal variation in the FM in Jaran Bay. Sinc e this study was a pilot study, some of important parameters were not considered. For example, grazing effects from predators, such as aquaculture shellfish and zooplankton, could be highly correlated with FM, which is a main food source available to them.

5. Conclusions

A detailed spatiotemporal evaluation of the biochemical compositions and FM of POM of phytoplankton communities and a set of multiple linear regression analyses were conducted in Jaran Bay to understand their major controlling factors. Base d on this research, the variations in FM representing food source materials could be explained by large-cell-sized phytoplankton (>2 μm) and major inorganic nutrient concentrations. Kim et al. [42] observed progressive decreases in dissolved inorganic nutrients in the southern coastal region of South Korea in recent decades. A progressive decline of the chlorophyll *a* concentration has been consistently reported in several regions in the southern coastal region of South Korea [43]. At this point, we cannot assume that the changes of the species compositions or size compositions of phytoplankton are correlated with the decreases of the concentrations of nutrients and chlorophyll *a*. Howe ver, we may expect greater numbers of small-sized phytoplankton cells than of large cell-sized phytoplankton cells under these conditions. Thes e changes in nutrient concentrations and dominant phytoplankton communities could cause changes in FM and further alterations in potential consumers. Jara n Bay is one of the largest shellfish aquaculture sites in the South Sea of Korea. Furt her studies on the spatial and temporal variations in the macromolecular compositions and FM of POM in regard to various environmental conditions are needed to better understand the quality and quantity of the primary food source available to higher trophic animals.

Author Contributions: Conceptualization, J.H.L., W.-C.L., H.C.K. and S.H.L.; data curation: J.H.L., N.J. and K.K.; formal analysis: K.K. and B.-R.S.; investigation: J.H.L., N.J., K.K., D.L. and J.J.K.; methodology: N.J. and K.K.; validation: J.H.L., W.-C.L., H.C.K. and J.-I.K.; visualization: J.H.L.; writing—original draft: J.H.L. and S.H.L.; writing—review & editing: J.H.L., N.J. and S.H.L. All authors have read and agreed to the published version of the manuscript.

Funding: This research was supported by a grant (R2020048) from the National Institute of Fisheries Science (NIFS) and partly by a project titled 'Improvements of ocean prediction accuracy using numerical modeling and artificial intelligence technology' which are funded by the Ministry of Oceans and Fisheries, Republic of Korea.

Acknowledgments: We thank the anonymous reviewers who greatly improved an earlier version of manuscript.

Conflicts of Interest: The authors declare no conflict of interest.

References

1. Mathiesen, A.M. *The state of the World Fisheries and Aquaculture 2012*; Food and Agriculture Organization: Rome, Italy, 2012; p. 290.
2. Cho, C.H.; Park, K. Eutr ophication of bottom mud in shellfish farms, the Goseong-Jaran Bay. *Korean J. Fish . Aqua t. Sci .* **1983**, *16*, 260–264.
3. Xu, Q.; Yang, H. Food sources of three bivalves living in two habitats of Jiaozhou Bay (Qingdao, China): Indicated by lipid biomarkers and stable isotope analysis. *J. Shel lfish Res.* **2007**, *26*, 561–567. [CrossRef]
4. Umehara, A.; Asaoka, S.; Fujii, N.; Otani, S.; Yamamoto, H.; Nakai, S.; Okuda, T.; Nishijima, W. Biol ogical productivity evaluation at lower trophic levels with intensive Pacific oyster farming of *Grassostrea gigas* in Hiroshima Bay, Japan. *Aquaculture* **2018**, *495*, 311–319. [CrossRef]
5. Lee, J.H.; Lee, D.; Kang, J.J.; Joo, H.; Lee, J.H.; Lee, H.W.; Ahn, S.H.; Kang, C.K.; Lee, S.H. The effects of different environmental factors on the biochemical composition of particulate organic matter in Gwanyang Bay, South Korea. *Biogeosciences* **2017**, *14*, 1903–1917. [CrossRef]
6. Kim, H.C.; Lee, J.H.; Lee, W.C.; Hong, S.; Kang, J.J.; Lee, D.; Jo, N.; Bhavya, P.S. Deco upling of macromolecular compositions of particulate organic matters between the water columns and the sediment in Geoje-Hansan Bay, South Korea. *Ocean Sci. J .* **2018**, *53*, 735–743. [CrossRef]
7. Kim, Y.; Lee, J.H.; Kang, J.J.; Lee, J.H.; Lee, H.W.; Kang, C.K.; Lee, S.H. Rive r discharge effects on the contribution of small-sized phytoplankton to the total biochemical composition of POM in Gwangyang Bay, Korea. *Estuar. Coas t. Shel f Sci.* **2019**, *226*, 106293. [CrossRef]
8. Morris, I.; Glover, H.E.; Yentsch, C.S. Prod ucts of photosynthesis by marine phytoplankton: The effect of environmental factors on the relative rates of protein synthesis. *Mar. Biol .* **1974**, *21*, 1–9. [CrossRef]

9. Kowallik, W. Blue light effects on carbohydrate and protein metabolism. In *Blue Light Responses: Phenomena and Occurrence in Plants*; Senger, H., Ed.; CRC Press: Boca Raton, FL, USA, 1978; Volume 1, pp. 8–13.

10. Suárez, I.; Maranón, E. Phot osynthate allocation in a temperate sea over an annual cycle: The relationship between protein synthesis and phytoplankton physiological state. *J. Sea Res.* **2003**, *50*, 285–299. [CrossRef]

11. Mortensen, S.H.; Børsheim, K.Y.; Rainuzzo, J.; Knutsen, G. Fatt y acid and elemental composition of the marine diatom Chaetoceros gracilis Schütt. Effe cts of silicate deprivation, temperature and light intensity. *J. Exp . Mar . Biol . Ecol .* **1988**, *122*, 173–185. [CrossRef]

12. Liebezeit, G. Part iculate carbohydrate in relation to phytoplankton in the euphotic zone of the Bransfield Strait. *Polar Biol.* **1984**, *2*, 225–228. [CrossRef]

13. Moal, J.; Jezequel, V.M.; Harris, R.P.; Samain, J.F.; Poulet, S.A. Inte rspecific and intraspecific variability of the chemical-composition of marine-phytoplankton. *Oceanol. Acta* **1987**, *10*, 339–346.

14. Kilham, S.S.; Kreeger, D.A.; Goulden, C.E.; Lynn, S.G. Effe cts of nutrient limitation on biochemical constituents of *Ankistrodesmus falcatus. Freshw. Biol .* **1997**, *38*, 591–596. [CrossRef]

15. Lee, S.H.; Kim, H.J.; Whitledge, T.E. High incorporation of carbon into proteins by the phytoplankton of the Bering Strait and Chukchi Sea. *Cont. Shel f Res.* **2009**, *29*, 1689–1696. [CrossRef]

16. Yun, M.S.; Lee, D.B.; Kim, B.K.; Kang, J.J.; Lee, J.H.; Yang, E.J.; Park, W.G.; Chung, K.H.; Lee, S.H. Comp arison of phytoplankton macromolecular compositions and zooplankton proximate compositions in the northern Chukchi Sea. *Deep Sea Res. Part II* **2015**, *120*, 82–90. [CrossRef]

17. Jo, N.; Kang, J.J.; Park, W.G.; Lee, B.R.; Yun, M.S.; Lee, J.H.; Kim, S.M.; Lee, D.; Joo, H.; Lee, J.H.; et al. Seas onal variation in the biochemical compositions of phytoplankton and zooplankton communities in the southwestern East/Japan Sea. *Deep Sea Res. Part II* **2017**, *143*, 82–90. [CrossRef]

18. Navarro, J.M.; Clasing, E.; Urrutia, G.; Asencio, G.; Stead, R.; Herrera, C. Bioc hemical composition and nutritive value of suspended particulate matter over a tidal flat of Southern Chile. *Estuar. Coas t. Shel f Sci.* **1993**, *37*, 59–73. [CrossRef]

19. Danovaro, R.; Dell'Anno, A.; Pusceddu, A.; Marrale, D.; Croce, N.D.; Fabiano, M.; Tselepides, A. Bioc hemical composition of pico-, nano- and microparticulate organic matter and bacterioplankton biomass in the oligotrophic Cretan Sea (NE Mediterranean). *Prog. Ocea nogr.* **2000**, *46*, 279–310. [CrossRef]

20. Navarro, J.M.; Thompson, R.J. Seas onal fluctuations in the size spectra, biochemical-composition and nutritive-value of the seston available to a suspension-feeding bivalve ina asub-arctic environment. *Mar. Ecol . Prog . Ser .* **1995**, *125*, 95–106. [CrossRef]

21. Kang, J.J.; Joo, H.; Lee, J.H.; Lee, J.H.; Lee, H.W.; Lee, D.; Kang, C.K.; Yun, M.S.; Lee, S.H. Comp arison of biochemical compositions of phytoplankton during spring and fall seasons in the northern East/Japan Sea. *Deep Sea Res. Part II* **2017**, *143*, 73–81. [CrossRef]

22. Lee, S.H.; Whitledge, T.E.; Kang, S. Rece nt carbon and nitrogen uptake rates of phytoplankton in Bering Strait and the Chukchi Sea. *Cont. Shel f Res.* **2007**, *27*, 2231–2249. [CrossRef]

23. Lee, J.H.; Kim, H.C.; Lee, T.; Lee, W.C.; Kang, J.J.; Jo, N.; Lee, D.; Kim, K.; Min, J.; Kang, S.; et al. Mont hly variations in the intracellular nutrient pools of phytoplankton in Jaran Bay, Korea. *J. Coas t. Res .* **2018**, *85*, 331–335. [CrossRef]

24. Welschmeyer, N.A. Fluo rometric analysis of chlorophyll a in the presence of chlorophyll b and phaeopigments. *Limnol. Ocea nogr.* **1994**, *39*, 1985–1992. [CrossRef]

25. Dubois, M.; Gilles, K.; Hamilton, J.K.; Rebers, P.A.; Smith, F. Colo rmetric method for determination of sugars and related substances. *Anal. Chem .* **1994**, *28*, 350–356. [CrossRef]

26. Stevens, J. *Applied Multivariate Statistics for the Social Science*; Lawrence Erlbaum: New York, NY, USA, 1986; p. 515.

27. Camdevyren, H.; Demyr, N.; Kanik, A.; Keskyn, S. Use principal component scores in multiple linear regression models for prediction of *Chlorophyll-a* in reservoirs. *Ecol. Mode l.* **2005**, *181*, 581–589. [CrossRef]

28. Wang, Y.S.; Lou, Z.P.; Sun, C.C.; Wu, M.L.; Han, S.H. Mult ivariate statistical analysis of water quality and phytoplankton characteristics in Daya Bay, China, from 1999 to 2002. *Oceanologia* **2006**, *48*, 193–211.

29. Deutsch, C.; Sarmiento, J.L.; Sigman, D.M.; Gruber, N.; Dunne, J.P. Spat ial coupling of nitrogen inputs and losses in the ocean. *Nature* **2007**, *445*, 163–167. [CrossRef]

30. Brun, P.; Vogt, M.; Payne, M.R.; Gruber, N.; O'Brien, C.J.; Buitenhuis, E.T.; Le Quere, C.; Leblanc, K.; Luo, Y.W. Ecol ogical niches of open ocean phytoplankton taxa. *Limnol. Ocea nogr.* **2015**, *60*, 1020–1038. [CrossRef]

31. Pedhazur, E. *Multiple regression in behavioral science*; Holt Rinehart & Winston: Fort Worth, TX, USA, 1982; p. 135.

32. Lee, Y.S.; Kang, C.K.; Kwon, K.Y.; Kim, S.Y. Orga nic and inorganic matter increase related to eutrophication in Gamak Bay, South Korea. *J. Envi ron. Biol* . **2009**, *30*, 373–380.

33. Lee, Y.S.; Kang, C.K. Caus es of COD increases in Gwangyang Bay, South Korea. *J. Envi ron. Moni t.* **2010**, *12*, 1537–1546. [CrossRef] [PubMed]

34. Lee, Y.S.; Lim, W.A.; Jung, C.S.; Park, J. Spat ial distributions and monthly variations of water quality in coastal seawater of Tongyeong, Korea. *J. Kore an Soc. Mar . Envi ron. Eng* . **2011**, *14*, 154–162. [CrossRef]

35. Jeong, D.H.; Shin, H.H.; Jung, S.W.; Lim, D.I. Vari ations and characters of water quality during flood and dry seasons in the Eastern coast of South Sea, Korea. *Korean J. Envi ron. Biol* . **2013**, *31*, 19–36. [CrossRef]

36. Park, J.S.; Yoon, Y.H.; Oh, S.J. Vari ational characteristics of phytoplankton community in the mouth parts of Gamak Bay, Southern Korea. *Korean J. Envi ron. Biol* . **2009**, *27*, 205–215.

37. Lee, J.H.; Kang, J.J.; Jang, H.K.; Jo, N.; Lee, D.; Yun, M.S.; Lee, S.H. Majo r controlling factors for spatio-temporal variations in the macromolecular composition and primary production by phytoplankton in Garolim and Asan bays in the Yellow Sea. *Reg. Stud . Mar . Sci* . **2020**, *36*, 101269. [CrossRef]

38. Desortová, B. Rela tionship between chlorophyll-a concentration and phytoplankton biomass in several reservoirs in Czechoslovakia. *Hydrobiology* **1981**, *66*, 153–169. [CrossRef]

39. Behrenfeld, M.J.; Boss, E.; Siegel, D.A.; Shea, D.M. Carb on-based ocean productivity and phytoplankton physiology from space. *Glob. Biog eochem. Cycl es* **2005**, *19*. [CrossRef]

40. Behrenfeld, M.J.; Boss, E. Beam attenuation and chlorophyll concentration as alternative optical indices of phytoplankton biomass. *J. Mar . Res* . **2006**, *64*, 431–451. [CrossRef]

41. Kruskopf, M.; Flynn, K.J. Chlo rophyll content and fluorescence responses cannot be used to gauge reliably phytoplankton biomass, nutrient status or growth rate. *New Phytol.* **2006**, *169*, 525–536. [CrossRef]

42. Kim, T.W.; Lee, K.; Lee, C.K.; Jeong, H.D.; Suh, Y.S.; Lim, W.A.; Kim, K.Y.; Jeong, H.J. Inte rannual nutrient dynamics in Korean coastal waters. *Harmful Algae* **2013**, *30*, 15–27. [CrossRef]

43. Son, Y.B.; Ryu, J.H.; Noh, J.H.; Ju, S.J.; Kim, S.-H. Clim atological variability of satellite-derived sea surface temperature and chlorophyll in the South Sea of Korea and East China Sea. *Ocean Polar Res.* **2012**, *34*, 201–218. [CrossRef]

Publisher's Note: MDPI stays neutral with regard to jurisdictional claims in published maps and institutional affiliations.

Article

Spatial Patterns of Macromolecular Composition of Phytoplankton in the Arctic Ocean

Keyseok Choe [1,2], Misun Yun [3,*], Sanghoon Park [1], Eunjin Yang [4], Jinyoung Jung [4], Jaejoong Kang [1], Naeun Jo [1], Jaehong Kim [1], Jaesoon Kim [1] and Sang Heon Lee [1,*]

[1] Department of Oceanography, College of Natural Science, Pusan National University, Busan 46241, Korea; keyseok.choe@gmail.com (K.C.); mossinp@pusan.ac.kr (S.P.); jaejung@pusan.ac.kr (J.K.); nadan213@pusan.ac.kr (N.J.); king9527@naver.com (J.K.); jaesoonkim1123@naver.com (J.K.)

[2] National Marine Biodiversity Institute of Korea, Seocheon 33662, Korea

[3] College of Marine and Environmental Sciences, Tianjin University of Science and Technology, Tianjin 300457, China

[4] Korea Polar Research Institute, Incheon 21990, Korea; ejyang@kopri.re.kr (E.Y.); jinyoungjung@kopri.re.kr (J.J.)

[*] Correspondence: misunyun@tust.edu.cn (M.Y.); sanglee@pusan.ac.kr (S.L.)

Abstract: The macromolecular concentrations and compositions of phytoplankton are crucial for the growth or nutritional structure of higher trophic levels through the food web in the ecosystem. To understand variations in macromolecular contents of phytoplankton, we investigated the macromolecular components of phytoplankton and analyzed their spatial pattern on the Chukchi Shelf and the Canada Basin. The carbohydrate (CHO) concentrations on the Chukchi Shelf and the Canada Basin were 50.4–480.8 µg L^{-1} and 35.2–90.1 µg L^{-1}, whereas the lipids (LIP) concentrations were 23.7–330.5 µg L^{-1} and 11.7–65.6 µg L^{-1}, respectively. The protein (PRT) concentrations were 25.3–258.5 µg L^{-1} on the Chukchi Shelf and 2.4–35.1 µg L^{-1} in the Canada Basin. CHO were the predominant macromolecules, accounting for 42.6% on the Chukchi Shelf and 60.5% in the Canada Basin. LIP and PRT contributed to 29.7% and 27.7% of total macromolecular composition on the Chukchi Shelf and 30.8% and 8.7% in the Canada Basin, respectively. Low PRT concentration and composition in the Canada Basin might be a result from the severe nutrient-deficient conditions during phytoplankton growth. The calculated food material concentrations were 307.8 and 98.9 µg L^{-1}, and the average calorie contents of phytoplankton were 1.9 and 0.6 kcal m^{-3} for the Chukchi Shelf and the Canada Basin, respectively, which indicates the phytoplankton on the Chukchi Shelf could provide the large quantity of food material and high calories to the higher trophic levels. Overall, our results highlight that the biochemical compositions of phytoplankton are considerably different in the regions of the Arctic Ocean. More studies on the changes in the biochemical compositions of phytoplankton are still required under future environmental changes.

Keywords: macromolecules; phytoplankton; Chukchi Shelf; Canada Basin; food material

Citation: Choe, K.; Yun, M.; Park, S.; Yang, E.; Jung, J.; Kang, J.; Jo, N.; Kim, J.; Kim, J.; Lee, S.H. Spatial Patterns of Macromolecular Composition of Phytoplankton in the Arctic Ocean. *Water* **2021**, *13*, 2495. https://doi.org/10.3390/w13182495

Academic Editor: Arantza Iriarte

Received: 4 August 2021
Accepted: 9 September 2021
Published: 11 September 2021

Publisher's Note: MDPI stays neutral with regard to jurisdictional claims in published maps and institutional affiliations.

1. Introduction

The Arctic Ocean is one of the most affected geographical locations in the world due to global climate change. In the Arctic, there has been a rapid decline of sea ice for several decades [1,2], which can be visualized by the downward trend of sea ice range through continuous satellite observations over the past decade [3–5]. In the last 30 years, the density of sea ice has decreased by about 9% every 10 years, and the sea ice thickness has also decreased [6]. With the disappearance of sea ice, various physico-chemical processes are likely to be altered [7–9].

Recent and rapid changes in the marine environment in the Arctic Ocean have been revealed to have significant impacts on the phytoplankton community [10–12]. For example, Kahru et al. [13] revealed that early phytoplankton blooms were caused by a decrease in sea ice in the Arctic Ocean. According to Ardyna et al. [14], the lengthening of the open water season in the Arctic Ocean was correlated with the increasing occurrence of the autumn bloom. In the areas of the Arctic Ocean where sea ice was absent, satellite observations have shown a significant increase in annual net primary production (NPP) [15–17]. Other than the quantitative changes of phytoplankton, the physiological conditions of phytoplankton appear to be affected by the recent environmental conditions [18–20].

In general, the organic matter produced by phytoplankton is composed of carbohydrates (hereafter, CHO), proteins (hereafter, PRT), and lipids (hereafter, LIP). The composition and synthesis of these major macromolecules of phytoplankton can provide important clues to the physiological status under the environment in which phytoplankton grow since they reflect the rapid adjustment of environmental conditions [21–23]. Furthermore, the relative amount of each macromolecular component in phytoplankton indicates the quality, or nutritional value, of phytoplankton as a food source [24]. The determination of the energy content for phytoplankton can be important since it could be transferred to the marine herbivores and the higher trophic levels and consequently determine the growth of higher trophic levels.

Previously, some studies on the macromolecular composition of phytoplankton were reported to understand their physiological conditions in the Polar Oceans [23,25–32]. The comparison of macromolecular compositions between phytoplankton and microzooplankton was conducted in the Arctic Ocean [27]. Kim et al. [28] revealed that the Antarctic phytoplankton indicated high protein composition under sustained high nutrient conditions, whereas the Arctic phytoplankton produced more lipids [19,20,27] or carbohydrates [30,31]. These studies mainly focused on the macromolecular compositions of or their vertical distributions within the euphotic zone or related environmental factors. Although the macromolecular composition of phytoplankton could be largely affected by environmental conditions in the regions, very little information is available on the spatial pattern of macromolecular concentration, composition, and nutritional value within the regions of the Arctic Ocean.

In this study, we examined the composition and concentration of the macromolecular pool (CHO, PRT, and LIP) of phytoplankton to understand the energy content of Arctic phytoplankton. In addition, we investigated how the spatial variation of the macromolecular composition is linked to physicochemical and biological parameters in different domains. Finally, the energy content of phytoplankton was calculated to estimate the nutritional value, which could be transferred to the organisms in the higher trophic levels in the Arctic ecosystem.

2. Materials and Methods

2.1. Research Area and Sampling

From 31 July to 23 August 2014, 21 survey stations on the Chukchi Shelf and the Canada Basin were occupied onboard R/V *Araon*. The samples for the macromolecular components were collected from 10 stations of shelf area (hereafter, Chukchi Shelf) and 11 stations of basin area (hereafter, Canada Basin) during the cruise period (Figure 1). The physical data were obtained through CTD (a Sea-Bird 911+) at each station, and seawater samples were obtained by using rosette samplers. A vertical irradiance profile (photosynthesis active irradiance (PAR), 400–700 nm) was obtained using an LI-COR underwater optical sensor mounted on a CTD/rosette sampler to determine the light depth.

2.2. Nutrients and Chlorophyll-a Analysis

Seawater samples for nutrient measurements were obtained from different photic depths (100%, 30%, and 1% depths of surface PAR) determined from the underwater PAR sensor. Samples were collected from all of the corresponding light depths, and the surface water collected was used for 100% light treatment. The nutrient concentrations were measured immediately using an automatic nutrient analyzer (SEAL, QuAAtro, Norderstedt, Germany).

Total chlorophyll-a concentrations (Whatman GF/F filter, ø = 24 mm) were analyzed using the method in [33]. The filters (Whatman GF/F filter, ø = 24 mm) were immediately frozen at −80 °C in each petri dish wrapped in aluminum foil until chlorophyll-a extraction at Pusan National University, South Korea. All the samples for chlorophyll-a concentrations were extracted with 90% acetone at −5 °C for 24 h, and the concentrations were measured using a fluorometer (Turner Designs, 10-AU, San Jose, CA, USA), which was calibrated before the analysis.

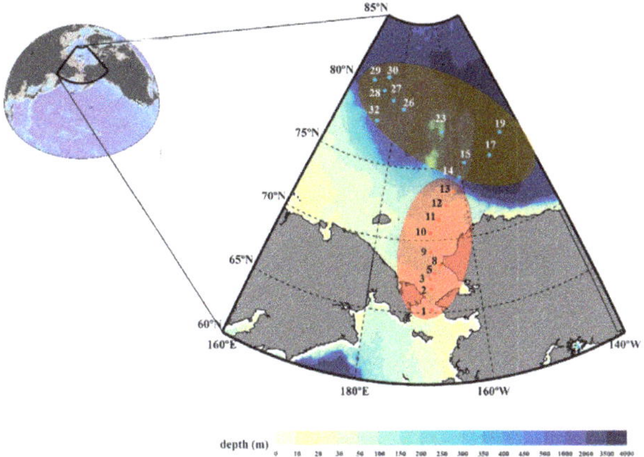

Figure 1. Map of the study area with functional regions highlighted in different colors. The red and blue colors indicate the sampling stations of the Chukchi Shelf and Canada Basin, respectively.

2.3. Macromolecular Concentration Analysis

Seawater samples (1L) were obtained from 100%, 30%, and 1% depths of surface PAR for the macromolecular concentration in the euphotic zone. The water sample was filtered through a 47-mm Whatman GF/F filter and then immediately stored at −80 °C until further analysis was performed at the home laboratory of Pusan National University. To extract CHO, PRT, and LIP, the phenol-sulfuric acid method [34], modified PRT method [35], and column method [36,37] were used, respectively. The detailed methods are available in [38].

2.4. Caloric Content Calculation

The Winberg [24] formula was used to calculate the calorie content (Kcal m^{-3}) of food material (FM; the sum of PRT, LIP, and CHO concentrations [39]).

$$\text{Calorie content (Kcal m}^{-3}) = \text{Kcal g FM}^{-1} \times \text{g FM m}^{-3}$$

3. Results

3.1. Temperature and Salinity Properties

Figure 2 shows the distributions of water temperature and salinity in the upper ocean. The range of water temperature and salinity of the entire study area varied from −1.8 to 9.7 °C and 26.7 to 32.4 psu, respectively. Regionally, the range of water temperature is from −1.4 to 9.7 °C (1.8 ± 3.4 °C) on the Chukchi Shelf and from −1.8 to 0.2 °C (−0.9 ± 0.5 °C) in the Canada Basin, showing a large difference between the Chukchi Shelf and the Canada Basin. The salinity ranges were 27.2–32.3 psu (30.1 ± 1.9 psu) on the Chukchi Shelf and 26.7–32.4 psu (30.1 ± 1.8 psu) in the Canada Basin, respectively. According to the distribution of salinity, a decreasing tendency was found to the north across the Chukchi Sea (Figure 2).

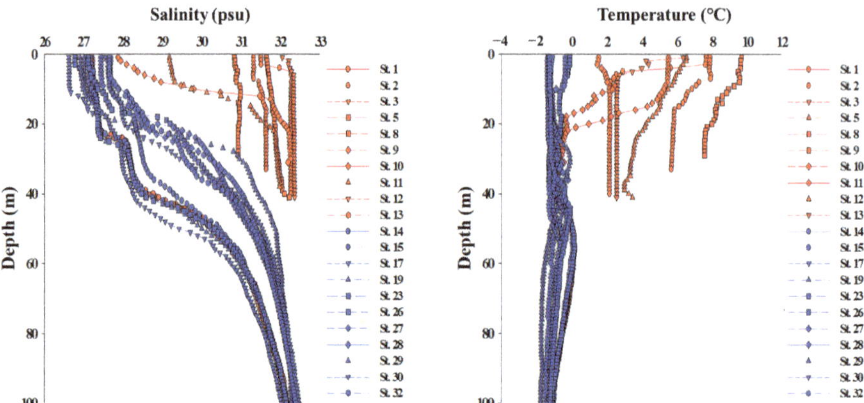

Figure 2. Vertical profiles of salinity (**left panel**) and temperature (**right panel**) in the stations of the study area.

3.2. Dissolved Inorganic Nutrients

The vertical patterns of dissolved inorganic nutrients showed spatial variations depending on the stations on the Chukchi Shelf, while they were relatively uniform in the Canada Basin (Figure 3). On the Chukchi Shelf, the concentration ranges of PO_4, $NO_2 + NO_3$, NH_4 and SiO_2 were 0.1–2.1, 0–14.3, 0–3.3, and 0.1–51.0 µM, respectively (Figure 3 upper panel). The average concentrations of PO_4, $NO_2 + NO_3$, NH_4, and SiO_2 are 0.7 ± 0.4, 1.7 ± 4.0, 0.5 ± 0.8, and 8.7 ± 10.1 µM, respectively. $NO_2 + NO_3$ was depleted at the upper layers on the Chukchi Shelf, and NH_4 showed low concentration rather than being depleted. In comparison, the concentration ranges of PO_4, NO_2+NO_3, NH_4, and SiO_2 in the Canada Basin were 0.5–1.6, 0–12.3, 0, and 1.6–31.5 µM, respectively (Figure 3 lower panel). The average concentrations of PO_4, $NO_2 + NO_3$, NH_4, and SiO_2 were 0.8 ± 0.3, 2.3 ± 3.8, 0 ± 0, and 7.3 ± 6.6 µM, respectively. Similar to what was observed on the Chukchi Shelf, NO_2+NO_3 was depleted at the surface of the Canada Basin. Furthermore, NH_4 was depleted in the entire water column.

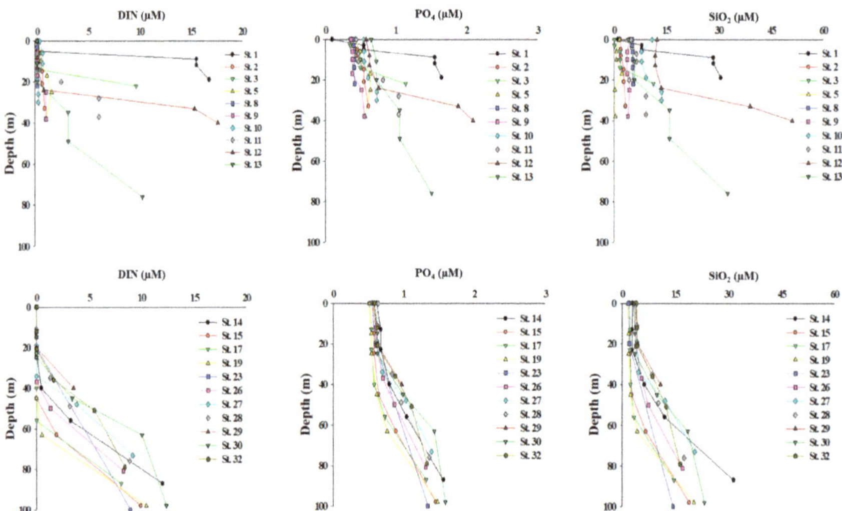

Figure 3. Vertical profiles of dissolved inorganic nutrients in stations of the Chukchi Shelf (**upper panel**) and the Canada Basin (**lower panel**).

3.3. Total Chlorophyll-a Concentration

During the cruise, the integrated chlorophyll-a concentration from the surface to 1% light depth in the entire study area was 66.3 ± 84.3 mg m^{-2}, with the total chlorophyll-a concentration showing a large regional variation as it ranged from 5.5 (station 27) to 376.2 mg m^{-2} (station 1) (Figure 4). The average concentration of chlorophyll-a on the Chukchi Shelf was 98.6 ± 104.3 mg m^{-2}, which was approximately three times that $(30.3 \pm 31.5$ mg m$^{-2})$ in the Canada Basin.

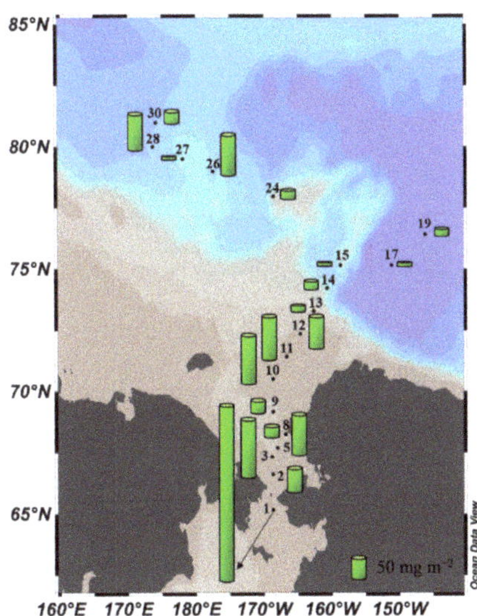

Figure 4. Spatial distribution of the chlorophyll-a concentration integrated from the surface to 1% light depth.

3.4. Vertical Distribution of Macromolecular Concentration and Composition on the Chukchi Shelf

Quantitative concentrations and relative ratios of CHO, PRT, and LIP on the Chukchi Shelf are summarized in Tables 1–3. The range of CHO, PRT, and LIP concentrations of phytoplankton was 48.2–409.3, 35.4–123.3, and 25.7–326.7 µg L^{-1}, respectively, at the surface of the Chukchi Shelf (Table 1). The average CHO concentration at each station was 114.7 ± 106.8 µg L^{-1} with a 43.6% contribution, being the dominant macromolecule found in phytoplankton. LIP (86.4 ± 93.1 µg L^{-1}) and PRT (60.2 ± 26.5 µg L^{-1}) contributed to 29.3% and 27.1% of the total compositions, respectively. The range of FM concentration was 117.6–859.3 µg L^{-1} (261.3 ± 221.2 µg L^{-1}). At 30% light depth (Table 2), the range of CHO concentration was determined to be 43.8–919.4 µg L^{-1} (167.3 ± 267.9 µg L^{-1}), in which the macromolecule accounted for 42.4% of the total composition. The range and contribution of LIP concentration were 19.2–546.4 µg L^{-1} (116.5 ± 161.3 µg L^{-1}) and 30.1%, respectively, and the range and contribution of PRT concentration were 27.2–586.8 µg L^{-1} (107.6 ± 170.8 µg L^{-1}) and 27.5%, respectively. Both the average concentration and contribution of LIP and PRT tended to increase compared to those at the surface layer. The range of FM concentration was 90.1–2052.6 µg L^{-1} (391.5 ± 597.0 µg L^{-1}), which was considerably higher than that at the surface. Compared to 30% light depth, the average composition of CHO concentration was slightly reduced to 41.7%, and its range was 34.9–259.0 µg L^{-1} (101.4 ± 64.0 µg L^{-1}) at 1% euphotic depth (Table 3). At this depth, the ranges of LIP and PRT concentrations were found to be 12.0–256.7 (90.5 ± 79.7 µg L^{-1}) and 10.0–185.8 µg L^{-1} (78.7 ± 52.8 µg L^{-1}). Consequently, the LIP contributed to 29.7% of the overall composition while PRT contributed to 28.6%, respectively. The CHO synthesis continued to decrease, whereas the PRT synthesis increased with depth. The range of FM concentration was 56.9–701.6 µg L^{-1} (270.6 ± 183.2 µg L^{-1}) at the 1% depth.

Table 1. Concentrations and compositions of macromolecular components (carbohydrates: CHO, proteins: PRT, and lipids: LIP), food material (FM), and calorie content of phytoplankton at 100% light depth on the Chukchi Shelf.

Station	Light Depth (%)	CHO (μg L^{-1})	PRT (μg L^{-1})	LIP (μg L^{-1})	FM (μg L^{-1})	CHO/FM (%)	PRT/FM (%)	LIP/FM (%)	Calorific Content (Kcal m^{-3})
1	100	409.3	123.3	326.7	859.3	47.6	14.3	38.0	5.5
2	100	80.1	54.3	135.9	270.3	29.6	20.1	50.3	1.9
3	100	139.1	77.6	123.9	340.6	40.8	22.8	36.4	2.2
5	100	62.0	67.5	43.0	172.6	35.9	39.1	24.9	1.0
8	100	113.6	70.0	53.8	237.4	47.8	29.5	22.6	1.4
9	100	65.6	38.2	29.5	133.3	49.2	28.7	22.1	0.8
10	100	74.0	40.4	26.9	141.3	52.4	28.6	19.1	0.8
11	100	48.2	43.6	25.7	117.6	41.0	37.1	21.9	0.7
12	100	82.3	51.5	40.1	173.9	47.4	29.6	23.0	1.0
13	100	73.1	35.4	58.1	166.7	43.9	21.2	34.9	1.0
Average		114.7	60.2	86.4	261.3	43.6	27.1	29.3	1.6
SD		106.8	26.5	93.1	221.2	6.9	7.6	10.1	1.4

Table 2. Concentrations and compositions of macromolecular components (carbohydrates: CHO, proteins: PRT, and lipids: LIP), food material (FM), and calorie content of phytoplankton at 30% light depth on the Chukchi Shelf.

Station	Light Depth (%)	CHO (μg L^{-1})	PRT (μg L^{-1})	LIP (μg L^{-1})	FM (μg L^{-1})	CHO/FM (%)	PRT/FM (%)	LIP/FM (%)	Calorific Content (Kcal m^{-3})
1	30	919.4	586.8	546.4	2052.6	44.8	28.6	26.6	12.2
2	30	88.9	62.5	54.2	205.7	43.2	30.4	26.3	1.2
3	30	199.7	126.2	215.4	541.3	36.9	23.3	39.8	3.6
5	30	78.9	62.2	76.2	217.3	36.3	28.6	35.1	1.4
8	30	96.8	67.2	45.0	209.1	46.3	32.1	21.5	1.2
9	30	53.7	40.4	22.8	116.8	45.9	34.6	19.5	0.7
10	30	59.6	36.5	28.3	124.4	47.9	29.3	22.8	0.7
11	30	43.8	27.2	19.2	90.1	48.6	30.1	21.3	0.5
12	30	70.2	37.2	74.4	181.8	38.6	20.4	40.9	1.2
13	30	62.3	30.4	83.4	176.1	35.4	17.3	47.4	1.2
Average		167.3	107.6	116.5	391.5	42.4	27.5	30.1	2.4
SD		267.9	170.8	161.3	597.0	5.1	5.4	9.9	3.5

Table 3. Concentrations and compositions of macromolecular components (carbohydrates: CHO, proteins: PRT, and lipids: LIP), food material (FM), and calorie content of phytoplankton at 1% light depth on the Chukchi Shelf.

Station	Light Depth (%)	CHO (μg L^{-1})	PRT (μg L^{-1})	LIP (μg L^{-1})	FM (μg L^{-1})	CHO/FM (%)	PRT/FM (%)	LIP/FM (%)	Calorific Content (Kcal m^{-3})
1	1	113.6	65.4	118.4	297.4	38.2	22.0	39.8	2.0
2	1	73.8	74.3	158.3	306.5	24.1	24.3	51.7	2.2
3	1	92.0	144.0	157.9	394.0	23.4	36.6	40.1	2.7
5	1	65.9	57.9	36.1	159.9	41.2	36.2	22.6	0.9
8	1	77.3	60.8	36.7	174.7	44.2	34.8	21.0	1.0
9	1	85.0	30.4	18.8	134.2	63.4	22.6	14.0	0.7
10	1	259.0	185.8	256.7	701.6	36.9	26.5	36.6	4.5
11	1	59.1	52.9	52.6	164.6	35.9	32.1	31.9	1.0
12	1	152.9	105.1	57.7	315.7	48.4	33.3	18.3	1.8
13	1	34.9	10.0	12.0	56.9	61.3	17.6	21.1	0.3
Average		101.4	78.7	90.5	270.6	41.7	28.6	29.7	1.7
SD		64.0	52.8	79.7	183.2	13.4	6.8	12.1	1.2

3.5. Vertical Distribution of Macromolecular Concentration and Composition in the Canada Basin

The range and contribution of CHO concentration of phytoplankton at the surface layer in the Canada Basin were dominant, being 28.8–177.2 µg L^{-1} (70.9 ± 42.3 µg L^{-1}) and 61.9%, respectively (Table 4). LIP concentration contributed to 29.7% of the total composition with a concentration range of 10.9–81.0 µg L^{-1} (34.2 ± 26.3 µg L^{-1}), and the range of PRT concentration was 3.2–13.6 µg L^{-1} (8.0 ± 3.2 µg L^{-1}) with a contribution of 8.4% at the surface layer. The range of FM concentration was 55.5–250.2 µg L^{-1} (113.1 ± 51.5 µg L^{-1}). At 30% euphotic depth (Table 5), CHO contributed to 57.6% of the total macromolecular composition and had a concentration range of 35.0–77.1 µg L^{-1} (48.2 ± 13.1 µg L^{-1}). Compared to the surface layer, the concentration of CHO was significantly reduced, and its contribution was somewhat decreased. On the other hand, the contribution of LIP concentration was 32.7%, which was increased compared to that at the surface, with a range of 10.4–58.7 µg L^{-1}. However, the average concentration of LIP was reduced to 29.3 µg L^{-1} (±18.7 µg L^{-1}) and smaller compared to that at the surface. The range and content of PRT concentration were 2.9–15.4 µg L^{-1} (8.4 ± 3.8 µg L^{-1}) and 9.8%, respectively, and it was identified to be greater than what was found at the surface. The range of FM concentration was between 56.0 and 110.0 µg L^{-1} (85.9 ± 19.1 µg L^{-1}). At 1% light depth (Table 6), both the range (33.3–92.9 µg L^{-1}) and content (62.2%) of CHO concentrations increased. The range and contribution of LIP concentration were determined to be 9.0–92.5 µg L^{-1} (32.2 ± 28.6 µg L^{-1}) and 29.9%, respectively. A decrease in the range and contribution of PRT concentration was observed. The range of PRT concentration was 0.7–83.8 µg L^{-1} (11.4 ± 24.1 µg L^{-1}) with a total contribution of 7.9% and the range of FM concentration was 46.8–269.2 µg L^{-1} (97.5 ± 63.1 µg L^{-1}) at 1% light depth in the Canada Basin.

Table 4. Concentrations and compositions of macromolecular components (CHO, PRT, and LIP), food material (FM), and calorie content of phytoplankton at 100% light depth in the Canada Basin.

Station	Light Depth (%)	CHO (µg L^{-1})	PRT (µg L^{-1})	LIP (µg L^{-1})	FM (µg L^{-1})	CHO/FM (%)	PRT/FM (%)	LIP/FM (%)	Calorific Content (Kcal m^{-3})
14	100	51.1	13.6	14.4	79.1	64.6	17.2	18.2	0.4
15	100	177.2	3.6	69.5	250.2	70.8	1.4	27.8	1.4
17	100	28.8	3.2	66.1	98.1	29.3	3.3	67.4	0.8
19	100	55.7	10.0	81.0	146.7	38.0	6.8	55.2	1.1
23	100	50.1	6.4	24.0	80.5	62.3	8.0	29.7	0.5
26	100	98.7	4.7	17.4	120.8	81.7	3.9	14.4	0.6
27	100	83.2	9.8	17.0	110.0	75.6	8.9	15.5	0.6
28	100	35.9	8.7	10.9	55.5	64.7	15.7	19.6	0.3
29	100	93.1	9.4	12.7	115.2	80.8	8.2	11.1	0.6
30	100	69.7	7.3	18.5	95.4	73.0	7.6	19.4	0.5
32	100	37.1	10.9	45.1	93.1	39.9	11.7	48.4	0.6
Average		70.9	8.0	34.2	113.1	61.9	8.4	29.7	0.7
SD		42.3	3.2	26.3	51.5	18.1	4.9	18.8	0.3

Table 5. Concentrations and compositions of macromolecular components (CHO, PRT, and LIP), food material (FM), and calorie content of phytoplankton at 30% light depth in the Canada Basin.

Station	Light Depth (%)	CHO (µg L^{-1})	PRT (µg L^{-1})	LIP (µg L^{-1})	FM (µg L^{-1})	CHO/FM (%)	PRT/FM (%)	LIP/FM (%)	Calorific Content (Kcal m^{-3})
14	30	36.4	6.8	55.4	98.6	37.0	6.9	56.1	0.7
15	30	47.3	2.9	14.6	64.8	73.0	4.4	22.6	0.3
17	30	35.0	5.0	58.7	98.8	35.5	5.1	59.5	0.7
19	30	54.7	15.4	12.6	82.7	66.2	18.6	15.3	0.4
23	30	39.9	5.7	10.4	56.0	71.2	10.2	18.6	0.3

Table 5. *Cont.*

Station	Light Depth (%)	CHO (μg L^{-1})	PRT (μg L^{-1})	LIP (μg L^{-1})	FM (μg L^{-1})	CHO/FM (%)	PRT/FM (%)	LIP/FM (%)	Calorific Content (Kcal m^{-3})
26	30	38.7	4.7	32.6	76.0	50.9	6.2	42.9	0.5
27	30	60.7	10.5	28.2	99.4	61.1	10.6	28.3	0.6
28	30	36.6	7.6	14.8	59.0	62.0	12.9	25.0	0.3
29	30	77.1	12.0	12.5	101.7	75.9	11.8	12.3	0.5
30	30	56.9	11.6	29.3	97.8	58.2	11.9	29.9	0.6
32	30	46.6	9.8	53.6	110.0	42.4	8.9	48.7	0.8
Average		48.2	8.4	29.3	85.9	57.6	9.8	32.7	0.5
SD		13.1	3.8	18.7	19.1	14.3	4.1	16.5	0.2

Table 6. Concentrations and compositions of macromolecular components (CHO, PRT, and LIP), food material (FM), and calorie content of phytoplankton at 1% light depth in the Canada Basin.

Station	Light Depth (%)	CHO (μg L^{-1})	PRT (μg L^{-1})	LIP (μg L^{-1})	FM (μg L^{-1})	CHO/FM (%)	PRT/FM (%)	LIP/FM (%)	Calorific Content (Kcal m^{-3})
14	1	38.7	3.6	24.4	66.6	58.1	5.4	36.6	0.4
15	1	45.9	0.7	18.2	64.8	70.8	1.1	28.1	0.4
17	1	41.8	4.6	72.1	118.5	35.2	3.9	60.8	0.9
19	1	47.7	8.2	9.0	64.9	73.4	12.7	13.9	0.3
23	1	70.2	4.6	57.0	131.8	53.3	3.5	43.2	0.9
26	1	81.0	2.9	22.6	106.5	76.1	2.7	21.2	0.6
27	1	33.3	2.5	10.9	46.8	71.3	5.4	23.3	0.3
28	1	55.3	6.5	9.4	71.2	77.6	9.2	13.2	0.4
29	1	92.9	83.8	92.5	269.2	34.5	31.1	34.3	1.7
30	1	39.1	3.6	10.1	52.9	74.0	6.9	19.2	0.3
32	1	47.3	4.4	28.0	79.7	59.4	5.5	35.1	0.5
Average		53.9	11.4	32.2	97.5	62.2	7.9	29.9	0.6
SD		19.2	24.1	28.6	63.1	15.7	8.3	14.1	0.4

3.6. Temperature, Salinity, Nutrients, and Macromolecular Concentration along the Shelf/Basin Gradient

The transect from station 10 to station 15 was run to understand the variability of environmental conditions and macromolecular concentrations along the shelf/basin gradient (Figure 5). Due to sea ice, the water column was in a freezing condition toward the basin. Based on the salinity and temperature distributions, station 12 was determined as the sea-ice edge. The dissolved inorganic nutrients at the depths around the sea-ice edge were distinctly high and decreased toward the basin. The chlorophyll-a distribution gradually decreased toward the basin. The chlorophyll-a concentration at the shelf was lowest at the surface and increased with depth.

The highest CHO concentration was observed at station 10 (>200 μg L^{-1}). Similarly, the LIP was highest at station 10. The upper layer of the basin showed a high LIP concentration. The distribution of PRT concentration was also similar to those of CHO or LIP, but PRT concentration at the ice-covered stations of the basin (from station 13 to station 15) was distinctly low. Interestingly, CHO and PRT concentrations were two or three times lower at the stations of the basin compared to the shelf area.

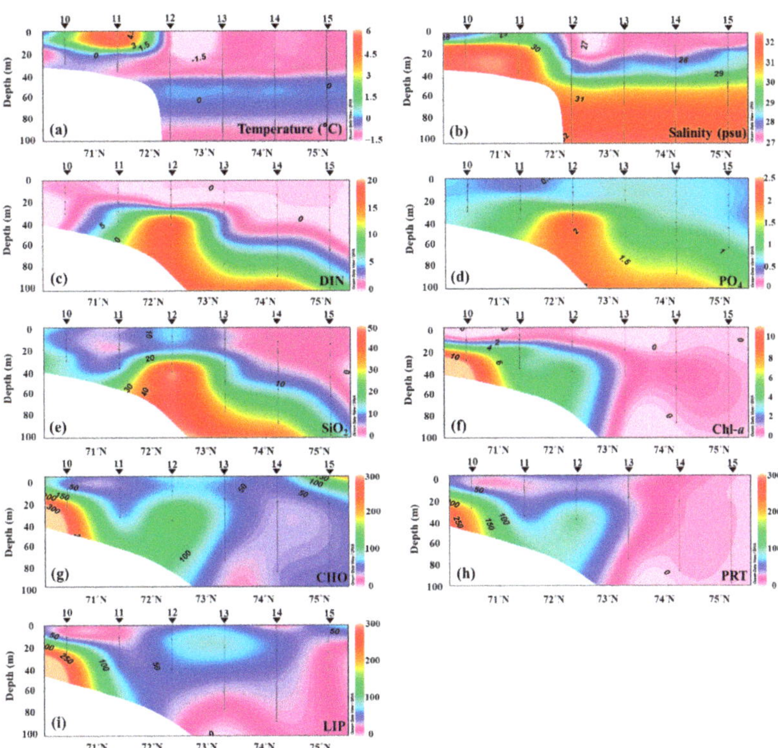

Figure 5. The temperature (**a**), salinity (**b**), DIN concentration (**c**), PO$_4$ concentration (**d**), SiO$_2$ concentration (**e**), chlorophyll-a concentration (**f**), CHO concentration (**g**), PRT concentration (**h**), and LIP concentration (**i**) distributions in the vertical section from the Chukchi shelf to Canada Basin in the Arctic Ocean.

3.7. Spatial Distribution of the Macromolecular Composition

No statistically significant difference in the relative percentage of each macromolecular component was found among the different light depths (*t*-test, $p > 0.05$). Thus, each component was averaged from three different light depths. Figure 6 shows a spatial distribution of the macromolecular composition over the euphotic zone during this study period. The contribution of CHO was the lowest at station 2 and highest at station 15. The LIP component accounted for 18.5–62.6% of the total macromolecular composition among the stations. The PRT contributed to 2.3–34.6% of the total macromolecular composition. Over the entire study area, CHO was identified as the biggest contributor (52.0%) to the overall average composition of phytoplankton, which was followed by LIP (30.3%) and PRT (17.8%). Regionally, the CHO contributed to 42.6% and 60.5% of the total macromolecular compositions for the Chukchi Shelf and the Canada Basin, respectively. The LIP contributions were 29.7% on the Chukchi Shelf and 30.8% in the Canada Basin, respectively. The average PRT composition was significantly low (*t*-test, $p < 0.05$) in the Canada Basin (8.7% of total) compared to the Chukchi Shelf (27.7% of total) (Figure 6).

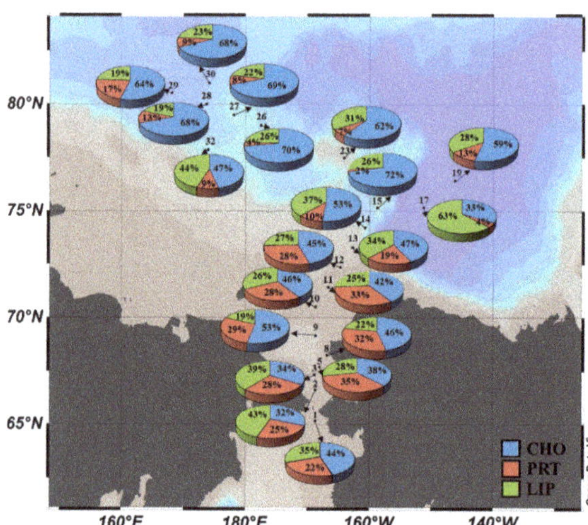

Figure 6. Spatial distribution of the macromolecular compositions of phytoplankton during the 2014 expedition.

4. Discussion

4.1. Major Controlling Factors for the Spatial Variation in Macromolecular Composition

In this study, the major macromolecule contributing to the overall average composition of phytoplankton was determined to be CHO on the Chukchi Shelf and the Canada Basin (Figure 7). It is interesting compared to those previously reported from the other regions of the polar oceans. For example, Yun et al. [27] observed a higher rate of LIP (58%) compared to CHO or PRT in the phytoplankton in the northern Chukchi Sea. In the Antarctic Ocean, Fabiano et al. [40,41] reported a high contribution, of 50% or more, of PRT to the total FM. Kim et al. [28] also found the high contribution of PRT (67%) to the macromolecular composition of phytoplankton in the Amundsen Sea due to the high concentrations of nitrate + nitrite. In general, the composition of PRT in the phytoplankton increases under nitrogen saturation conditions [40,42]. When the nitrogen or phosphorus is limited, triglycerides, which are energy stores, increase and are converted from PRT metabolism to LIP or CHO metabolisms [23,43,44]. Since CHO and LIP are not nitrogen-derived substrates, the accumulation of these storage compounds can be a reaction mechanism under nitrogen-deficient conditions [45]. In particular, LIP acts as a secondary storage material for the survival of long-term nitrogen conditions due to the fat synthetase system [46]. Thus, photosynthesis products are converted from CHO to LIP synthesis in a nitrogen-depleted environmental condition for an extended period [45,46], even though preferred accumulation in CHO or LIP compounds as a reservoir appears to be specific to species [45]. Some oily diatom species assimilate LIP as a major storage component under nitrogen or silicon restrictions [47,48]. Indeed, Ahn et al. [31] observed a sharp increase in LIP concentration with an increase in micro-phytoplankton in the Arctic Ocean. However, Harrison et al. [49] and Wear et al. [50] reported that diatoms have a relatively constant LIP under nitrogen deficiency while rapidly increasing CHO content and decreasing PRT. Therefore, the high CHO and moderate LIP compositions in the present study might have been due to the deficient nutrient conditions, which is consistent with previous findings [49,51], while PRT production predominates under nitrogen-rich conditions [40,42]. Consequently, it implies that major inorganic nutrients, especially nitrogen, are important controlling factors for the macromolecular composition of phytoplankton.

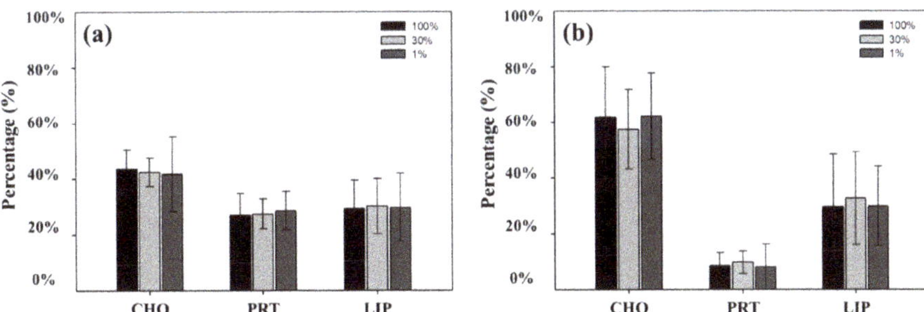

Figure 7. Contribution of each macromolecular component at three light depths on the Chukchi Shelf (**a**) and in the Canada Basin (**b**).

During the cruise, the phytoplankton in the Canada Basin were observed to have a significantly lower PRT (8.7%) compared to that of the Chukchi Shelf. This might be related to different conditions of nutrient limitation. Normally, nutrient deficiency can be indicated following as; nitrogen limit condition with N/P ratio with < 10 and Si/N ratio > 1, phosphorus restriction under N/P ratio > 22 and Si/P ratio > 22, and silicon limitation in Si/N ratio < 1 and Si/P ratio < 10 [52]. In terms of macromolecular composition, values with a higher PRT/CHO ratio (>1) are observed in the areas with high productivity or nitrogen-rich blooms [40], while the lower ratios (<1) are in nitrogen deficiency conditions [39]. Thus, low N/P ratios, PRT/CHO ratios, and PRT/LIP ratios indicate that nitrogen is particularly limited in the environment. In this study, the average molar ratio of $(NO_3 + NO_2 + NH_4):PO_4$ and $SiO_2:(NO_3 + NO_2 + NH_4)$ in the euphotic zone was 3.0:1 and 3.9:1 on the Chukchi Shelf, respectively, and 3.4:1 and 2.8:1 in the Canada Basin, respectively. Overall, the ratio of N/P was significantly lower than the Redfield ratio [53]. The PRT/CHO ratio and PRT/LIP ratio of the Canada Basin are 0.16 and 0.33, respectively, which are significantly lower (*t*-test, $p < 0.05$) than the PRT/CHO ratio (0.68) and PRT/LIP ratio (1) of the Chukchi Shelf. These results elucidate that the deficiency of nitrogen in the study area was present during the cruise, and it was especially noticeable that the nitrogen utilization by the phytoplankton in the Canada Basin was severely restricted. Therefore, the substantially low PRT composition of phytoplankton in the Canada Basin could be caused by the result of severe nitrogen deficiency during phytoplankton growth.

According to Suárez et al. [54], the light condition could act as an important factor for determining different macromolecular compositions of phytoplankton. The macromolecular composition of phytoplankton can vary depending on the amount of light [55,56]. For example, an increase in PRT is observed with a decrease in light intensity because of the lower illuminance saturation level of PRT in comparison to other macromolecules [54,57]. In contrast, the productions of CHO and LIP as storage materials can be observed under an excessive energy supply condition [54,58]. Suárez et al. [54] also reported that lower irradiance was more relevant in PRT synthesis than in LIP synthesis. However, no distinct pattern of macromolecular compositions was observed among three different light depths in this study (mentioned in Section 3.7), although the relatively higher protein concentrations were observed at deeper depths (30% and 1% light levels) than at the surface. In addition, the stations in the ice-covered Canada Basin showed significantly low PRT composition, even though it was thought to be a low light condition (Figure 5). Thus, we could conclude that the light condition might be insignificant in controlling the macromolecular composition during this study period.

4.2. The Implication of Macromolecular Composition as Energy Content Aspect

In this study, the concentration ranges of CHO, LIP, and PRT were substantially higher on the Chukchi Shelf than in the Canada Basin (Table 7). As a result, the average FMs were 307.8 µg L^{-1} on the Chukchi Shelf and 98.9 µg L^{-1} in the Canada Basin, respectively. The average calorie content of phytoplankton for the Chukchi Shelf and the Canada Basin during the cruise was 1.9 kcal m^{-3} and 0.6 kcal m^{-3}, respectively (Table 7). The overall FM concentration and calorie content of phytoplankton were three times higher on the Chukchi Shelf than in the Canada Basin, which implies that the phytoplankton on the Chukchi Shelf could provide higher FM and calories to the upper trophic levels in the Arctic ecosystem.

According to previous studies, the average calorie contents were 1.0 and 1.2 kcal m^{-3} in the northern Chukchi Sea of the Arctic Ocean [59,60]. Fabiano et al. [41] reported the calorie content of 1.6 kcal m^{-3} in the Ross Sea of the Antarctic Ocean. Recently, Kim et al. [32] observed the different calorie contents of phytoplankton between the two different periods in the Ross Sea of the Antarctic Ocean, indicating 1.3 kcal m^{-3} during the ice-free period and 0.6 kcal m^{-3} during the ice-covered season. Although Kim et al. [29] reported the exceptionally high-calorie content in the productive polynyas of the Amundsen Sea, the calorie content from our study is in a similar range with the previous studies in the polar oceans (Table 7). According to Kim et al. [32], the PRT concentration during the ice-free period in the Ross Sea of the Antarctic Ocean was 20 times increased than that during the ice-covered period, even though CHO or LIP concentrations showed a slight increase (Table 7). If it can be applied in the Arctic Ocean, the PRT concentration than other components might be largely increased under a decrease in sea ice conditions in the Arctic Ocean. Subsequently, the PRT composition that predominates as sea ice decreases could lead to a potential change in the calorie content from an energy point of view. In particular, the macromolecular composition or calorie content of the phytoplankton in ice-covered regions, such as the Canada Basin, might be anticipated to be changed. Consequently, the rich protein-containing FM might be transferred to the upper trophic levels under ongoing and future sea-ice decrease conditions. Thus, the potential effects of the different macromolecular compositions of phytoplankton on the upper trophic levels need to be further evaluated. Above all, in terms of sea ice change, the variability of macromolecular concentration, composition, and calorie content could be important in the Arctic Ocean under ongoing environmental changes.

Table 7. Comparison of carbohydrates (CHO), proteins (PRT), lipids (LIP), food material (FM) concentrations, and calorie content of phytoplankton at different regions of the Polar Oceans. Given were range or mean values.

Region	Season (Period)	CHO (μg L^{-1})	PRT (μg L^{-1})	LIP (μg L^{-1})	FM (μg L^{-1})	Caloric Content (kcal m^{-3})	References
Northern Chukchi Sea	30 July–19 August 2011	21.8–146.7	0.7–86.3	50.2–105.0	149.2 ± 36.5	1.0 ± 0.2	Kim et al. (2015) [59]
Northern Chukchi Sea	1 August–10 September 2012	15.9–88.0	9.2–183.1	37.0–147.4	156.4	1.2 ± 0.2	Yun et al. (2015) [27]
Chukchi Sea	7–24 August 2017	29.9–406.4	9.7–573.8	5.4–169.1	180.5 ± 195.3	-	Kim et al. (2020) [61]
Laptev and East Siberian Seas	21 August–22 September 2013	29–161	22–132	15–71	-	-	Ahn et al. (2019) [30]
Northern Kara Sea	18 August–30 September 2015	45.9–67.7	22.0–50. 8	15.4–44.0	-	-	Ahn et al. (2020) [31]
Laptev Sea		44.4–72.2	9.8–22.0	20.3–37.1	-	-	
Western East Siberian Sea		55.7–115. 5	1.7–30. 5	24.7–67.6	-	-	
Amundsen Sea	11 February–14 March 2012	2.8–216.0	5.9–396.2	13.2–36.9	219.4 ± 151.1	-	Kim et al. (2016) [28]
Amundsen Sea	31 December 2013–10 January 2014	89.3–991.1	69.9–360.5	25.4–199.3	671.5 ± 311.8	3.7 ± 1.6	Kim et al. (2018) [29]
Ross Sea	25 November 1989–7 January 1990	18–279	18–650	2–94	294.4 ± 228.1	-	Fabiano et al. (1993) [40]
Ross Sea (Terra Nova Bay)	austral summer	32–444	108–632	3–64	374.3	1.6 ± 1.3	Fabiano et al. (1996) [41]
Ross Sea (Terra Nova Bay)	February 2015 (ice-free period)	142.9 ± 55.9	143.6 ± 80.5	100.3 ± 59.1	386.9 ± 194.2	2.3 ± 1.2	Kim et al. (2021) [32]
	April–October 2015 (ice-covered period)	89.0 ± 23.0	7.4 ± 7.8	23.7 ± 4.6	121.1 ± 24.6	0.6 ± 0.1	
Chukchi Shelf	31 July–24 August 2014	50.4–480.8	25.3–258.5	23.7–330.5	307.8 ± 284.0	1.9 ± 1.8	This study
Canada Basin		35.2–90.1	2.4–35.1	11.7–65.6	98.9 ± 26.6	0.6 ± 0.2	This study

5. Conclusions

This study reported the spatial distributions of macromolecular concentrations, compositions, and energy contents of phytoplankton on the Chukchi Shelf and in the Canada Basin. CHO was the major macromolecular component of phytoplankton in the study area, accounting for 41.5% on the Chukchi Shelf and 58.4% in the Canada Basin. The LIP was moderate in both regions. Interestingly, the PRO composition was significantly different between the two regions, showing a low contribution in the Canada Basin (8.7%) and a relatively high contribution on the Chukchi Shelf (27.7%). Severe nutrient-deficient conditions for phytoplankton growth appear to be a major reason for the low PRT composition of phytoplankton in the Canada Basin. In terms of FM concentration and calorie content, a large quantity of high-calorie content food is available in the productive Chukchi Shelf compared to the Canada Basin.

Under the ongoing changes in Arctic environments, the concentration or composition of macromolecules of phytoplankton would be expected to change significantly. Since the different concentrations and compositions determine phytoplankton energy content and consequently regulate the growth and/or nutritional structure of upper trophic levels in the Arctic ecosystem, it is needed to monitor the variability of macromolecular concentration and compositions of phytoplankton. In particular, the macromolecular measurements of phytoplankton with different ocean conditions should be required to better understand how phytoplankton's energy content and its transfer to higher trophic levels are different in the regions of the Arctic Ocean. Recently, using satellite data, Roy et al. [62] tried to estimate the concentrations of CHO, PRT, and LIP and energy values of phytoplankton in the world's oceans. In situ measurement data in various oceans would be useful for improving the global scale estimation based on satellite-derived data.

Author Contributions: Conceptualization, K.C., E.Y., J.J. and S.L.; methodology, K.C., M.Y. and S.L.; validation, K.C., M.Y. and S.L.; formal analysis, K.C., S.P., J.K. (Jaejoong Kang), N.J., J.K. (Jaehong Kim), J.K. (Jaesoon Kim) and S.L.; investigation, K.C., E.Y. and J.J.; writing—original draft preparation, K.C.; writing—review and editing, M.Y. and S.L.; visualization, S.P., J.K. (Jaejoong Kang), N.J., J.K. (Jaehong Kim) and J.K. (Jaesoon Kim); supervision, S.L. All authors have read and agreed to the published version of the manuscript.

Funding: This research was a part of the project titled "Korea-Arctic Ocean Warming and Responses of Ecosystem (K-AWARE, KOPRI, 1525011760)", funded by the Ministry of Oceans and Fisheries, Korea.

Institutional Review Board Statement: Not applicable.

Informed Consent Statement: Not applicable.

Data Availability Statement: Not applicable.

Acknowledgments: We thank the captain and crew of the *ARAON* for their outstanding assistance during the cruise.

Conflicts of Interest: The authors declare no conflict of interest.

References

1. Comiso, J.C.; Parkinson, C.; Gersten, R.; Stock, L. Accelerated decline in the Arctic sea ice cover. *Geophys. Res. Lett.* **2008**, *35*, 1–6. [CrossRef]
2. Stroeve, J.; Notz, D. Insights on past and future sea-ice evolution from combining observations and models. *Glob. Planet. Change* **2015**, *135*, 154–196. [CrossRef]
3. Stroeve, J.; Holland, M.M.; Meier, W.; Scambos, T.; Serreze, M. Arctic sea ice decline: Faster than forecast. *Geophys. Res. Lett.* **2007**, *34*, 1–5. [CrossRef]
4. Stroeve, J.C.; Serreze, M.C.; Holland, M.M.; Kay, J.E.; Malanik, J.; Barrett, A.P. The Arctic's rapidly shrinking sea ice cover: A research synthesis. *Clim. Change* **2011**, *110*, 1005–1027. [CrossRef]
5. Stroeve, J.C.; Markus, T.; Boisvert, L.; Miller, J.; Barrett, A. Changes in Arctic melt season and implications for sea ice loss. *Geophys. Res. Lett.* **2014**, *41*, 1216–1225. [CrossRef]
6. Perovich, D.K.; Richter-Menge, J.A. Loss of sea ice in the Arctic. *Annu. Rev. Mar. Sci.* **2009**, *1*, 417–441. [CrossRef]

7.   Tremblay, J.-É.; Simpson, K.; Martin, J.; Miller, L.; Gratton, Y.; Barber, D.; Price, N. Vertical stability and the annual dynamics of nutrients and chlorophyll fluorescence in the coastal, southeast Beaufort Sea. *J. Geophys. Res. Space Phys.* **2008**, *113*, 113. [CrossRef]

8.   Codispoti, L.; Flagg, C.; Swift, J.H. Hydrographic conditions during the 2004 SBI process experiments. *Deep Sea Res.* **2009**, *56*, 1144–1163. [CrossRef]

9.   Lee, S.H.; Stockwell, D.; Whitledge, T.E. Uptake rates of dissolved inorganic carbon and nitrogen by under-ice phytoplankton in the Canada Basin in summer 2005. *Polar Biol.* **2010**, *33*, 1027–1036. [CrossRef]

10.   Li, W.K.W.; McLaughlin, F.A.; Lovejoy, C.; Carmack, E.C. Smallest algae thrive as the Arctic Ocean freshens. *Science* **2009**, *326*, 539. [CrossRef] [PubMed]

11.   Lee, S.H.; Kim, B.K.; Yun, M.S.; Joo, H.; Yang, E.J.; Kim, Y.N.; Shin, H.C.; Lee, S. Spatial distribution of phytoplankton productivity in the Amundsen Sea, Antarctica. *Polar Biol.* **2012**, *35*, 1721–1733. [CrossRef]

12.   Grebmeier, J.M.; Maslowski, W. (Eds.) *The Pacific Arctic Region: Ecosystem Status and Trends in a Rapidly Changing Environment*; Springer: Dordrecht, Germany, 2014.

13.   Kahru, M.; Brotas, V.; Manzano-Sarabia, M.; Mitchell, B.G. Are phytoplankton blooms occurring earlier in the Arctic? *Glob. Chang. Biol.* **2010**, *17*, 1733–1739. [CrossRef]

14.   Ardyna, M.; Babin, M.; Devred, E.; Forest, A.; Gosselin, M.; Raimbault, P.; Tremblay, J.-É. Shelf-basin gradients shape ecological phytoplankton niches and community composition in the coastal Arctic Ocean (Beaufort Sea). *Limnol. Oceanogr.* **2017**, *62*, 2113–2132. [CrossRef]

15.   Arrigo, K.R.; van Dijken, G. Secular trends in Arctic Ocean net primary production. *J. Geophys. Res. Space Phys.* **2011**, *116*, 1–15. [CrossRef]

16.   Arrigo, K.R.; van Dijken, G. Continued increases in Arctic Ocean primary production. *Prog. Oceanogr.* **2015**, *136*, 60–70. [CrossRef]

17.   Kahru, M.; Lee, Z.; Mitchell, B.G.; Nevison, C.D. Effects of sea ice cover on satellite-detected primary production in the Arctic Ocean. *Biol. Lett.* **2016**, *12*, 20160223. [CrossRef]

18.   Joo, H.; Lee, J.; Kang, C.-K.; An, S.; Kang, S.-H.; Lim, J.-H.; Joo, H.M.; Lee, S.H. Macromolecular production of phytoplankton in the northern Bering Sea, 2007. *Polar Biol.* **2013**, *37*, 391–401. [CrossRef]

19.   Yun, M.S.; Joo, H.T.; Park, J.W.; Kang, J.J.; Kang, S.-H.; Lee, S.H. Lipid-rich and protein-poor carbon allocation patterns of phytoplankton in the northern Chukchi Sea, 2011. *Cont. Shelf Res.* **2018**, *158*, 26–32. [CrossRef]

20.   Yun, M.S.; Joo, H.M.; Kang, J.J.; Park, J.W.; Lee, J.H.; Kang, S.-H.; Sun, J.; Lee, S.H. Potential implications of changing photosynthetic end-products of phytoplankton caused by sea ice conditions in the Northern Chukchi sea. *Front. Microbiol.* **2019**, *10*, 2274. [CrossRef]

21.   Laws, E.A. Photosynthetic quotients, new production and net community production in the open ocean. *Deep Sea Res. Part A Oceanogr. Res. Pap.* **1991**, *38*, 143–167. [CrossRef]

22.   Parrish, C.; McKenzie, C.; Macdonald, B.; Hatfield, E. Seasonal studies of seston lipids in relation to microplankton species composition and scallop growth in South Broad Cove, Newfoundland. *Mar. Ecol. Prog. Ser.* **1995**, *129*, 151–164. [CrossRef]

23.   Smith, R.; Gosselin, M.; Kattner, G.; Legendre, L.; Pesant, S. Biosynthesis of macromolecular and lipid classes by phytoplankton in the Northeast Water Polynya. *Mar. Ecol. Prog. Ser.* **1997**, *147*, 231–242. [CrossRef]

24.   Winberg, G.G. *Symbols, Units and Conversion Factors in Study of Fresh Waters Productivity*; International Biological Programme Control Office: London, UK, 1971; p. 23.

25.   Smith, R.E.H.; Clement, P.; Head, E. Biosynthesis and photosynthate allocation patterns of arctic ice algae. *Limnol. Oceanogr.* **1989**, *34*, 591–605. [CrossRef]

26.   Mock, T.; Gradinger, R. Changes in photosynthetic carbon allocation in algal assemblages of Arctic sea ice with decreasing nutrient concentrations and irradiance. *Mar. Ecol. Prog. Ser.* **2000**, *202*, 1–11. [CrossRef]

27.   Yun, M.S.; Lee, D.B.; Kim, B.K.; Kang, J.J.; Lee, J.H.; Yang, E.J.; Park, W.G.; Chung, K.H.; Lee, S.H. Comparison of phytoplankton macromolecular compositions and zooplankton proximate compositions in the northern Chukchi Sea. *Deep Sea Res.* **2015**, *120*, 82–90. [CrossRef]

28.   Kim, B.K.; Lee, J.H.; Joo, H.; Song, H.J.; Yang, E.J.; Lee, S.H. Macromolecular compositions of phytoplankton in the Amundsen Sea, Antarctica. *Deep Sea Res.* **2016**, *123*, 42–49. [CrossRef]

29.   Kim, B.K.; Lee, S.; Ha, S.-Y.; Jung, J.; Kim, T.W.; Yang, E.J.; Jo, N.; Lim, Y.J.; Park, J. Vertical distributions of macromolecular composition of particulate organic matter in the water column of the Amundsen Sea Polynya during the summer in 2014. *J. Geophys. Res. Oceans* **2018**, *123*, 1393–1405. [CrossRef]

30.   Ahn, S.H.; Whitledge, T.E.; Stockwell, D.A.; Lee, J.H.; Lee, H.W.; Lee, S.H. The biochemical composition of phytoplankton in the Laptev and East Siberian seas during the summer of 2013. *Polar Biol.* **2018**, *42*, 133–148. [CrossRef]

31.   Ahn, S.H.; Kim, K.; Jo, N.; Kang, J.J.; Lee, J.H.; Whitledge, T.E.; Stockwell, D.A.; Lee, H.W.; Lee, S.H. Fluvial influence on the biochemical composition of particulate organic matter in the Laptev and Western East Siberian seas during 2015. *Mar. Environ. Res.* **2020**, *155*, 104873. [CrossRef]

32.   Kim, K.; Park, J.; Jo, N.; Park, S.; Yoo, H.; Kim, J.; Lee, S.H. Monthly variation in the macromolecular composition of phytoplankton communities at Jang Bogo Station, Terra Nova Bay, Ross Sea. *Front. Microbiol.* **2021**, *12*, 618999. [CrossRef]

33.   Welschmeyer, A.N. Fluorometric analysis of chlorophyll a in the presence of chlorophyll b and pheopigments. *Limnol. Oceanogr.* **1994**, *39*, 1985–1992. [CrossRef]

34. Dubois, M.; Gilles, K.A.; Hamilton, J.K.; Rebers, P.A.; Smith, F. Colorimetric method for determination of sugars and related substances. *Anal. Chem.* **1956**, *28*, 350–356. [CrossRef]
35. Lowry, O.H.; Rosebrough, N.J.; Farr, A.L.; Randall, R.J. Protein measurement with the Folin phenol reagent. *J. Biol. Chem.* **1951**, *193*, 265–275. [CrossRef]
36. Bligh, E.G.; Dyer, W.J. A rapid method of total lipid extraction and purification. *Can. J. Biochem. Physiol.* **1959**, *37*, 911–917. [CrossRef] [PubMed]
37. Marsh, J.B.; Weinstein, D.B. Simple charring method for determination of lipids. *J. Lipid Res.* **1966**, *7*, 574–576. [CrossRef]
38. Bhavya, P.S.; Kim, B.K.; Jo, N.; Kim, K.; Kang, J.J.; Lee, J.H.; Lee, D.; Lee, J.H.; Joo, H.; Ahn, S.H.; et al. A review on the macromolecular compositions of phytoplankton and the implications for aquatic biogeochemistry. *Ocean Sci. J.* **2018**, *54*, 1–14. [CrossRef]
39. Danovaro, R.; Dell'Anno, A.; Pusceddu, A.; Marrale, D.; Della Croce, N.; Fabiano, M.; Tselepides, A. Biochemical composition of pico-, nano- and micro-particulate organic matter and bacterioplankton biomass in the oligotrophic Cretan Sea (NE Mediterranean). *Prog. Oceanogr.* **2000**, *46*, 279–310. [CrossRef]
40. Fabiano, M.; Povero, P.; Danovaro, R. Distribution and composition of particulate organic matter in the Ross Sea (Antarctica). *Polar Biol.* **1993**, *13*, 525–533. [CrossRef]
41. Fabiano, M.; Povero, P.; Danovaro, R. Particulate organic matter composition in Terra Nova Bay (Ross Sea, Antarctica) during summer 1990. *Antarct. Sci.* **1996**, *8*, 7–13. [CrossRef]
42. Lee, S.H.; Kim, H.-J.; Whitledge, T.E. High incorporation of carbon into proteins by the phytoplankton of the Bering Strait and Chukchi Sea. *Cont. Shelf Res.* **2009**, *29*, 1689–1696. [CrossRef]
43. Lombardi, A.; Wangersky, P. Influence of phosphorus and silicon on lipia class production by the marine diatom *Chaetoceros gracilis* grown in turbidostat cage cultures. *Mar. Ecol. Prog. Ser.* **1991**, *77*, 39–47. [CrossRef]
44. Takagi, M.; Watanabe, K.; Yamaberi, K.; Yoshida, T. Limited feeding of potassium nitrate for intracellular lipid and triglyceride accumulation of *Nannochloris* sp. UTEX LB1999. *Appl. Microbiol. Biotechnol.* **2000**, *54*, 112–117. [CrossRef]
45. Hu, Q. Environmental effects on cell composition. In *Handbook of Microalgal Culture: Applied Phycology and Biotechnology*; John Wiley & Sons: Hoboken, NJ, USA, 2007.
46. Fogg, G.E.; Thake, B. *Algal Cultures and Phytoplankton Ecology*; University of Wisconsin Press: Madison, WI, USA, 1987.
47. Moll, K.; Gardner, R.; Eustance, E.; Gerlach, R.; Peyton, B. Combining multiple nutrient stresses and bicarbonate addition to promote lipid accumulation in the diatom RGd-1. *Algal Res.* **2014**, *5*, 7–15. [CrossRef]
48. Sayanova, O.; Mimouni, V.; Ulmann, L.; Morant-Manceau, A.; Pasquet, V.; Schoefs, B.; Napier, J.A. Modulation of lipid biosynthesis by stress in diatoms. *Philos. Trans. R. Soc. B Biol. Sci.* **2017**, *372*, 1–14. [CrossRef] [PubMed]
49. Harrison, P.J.; Thompson, P.; Calderwood, G.S. Effects of nutrient and light limitation on the biochemical composition of phytoplankton. *Environ. Boil. Fishes* **1990**, *2*, 45–56. [CrossRef]
50. Wear, E.K.; Carlson, C.A.; Windecker, L.A.; Brzezinski, M.A. Roles of diatom nutrient stress and species identity in determining the short- and long-term bioavailability of diatom exudates to bacterioplankton. *Mar. Chem.* **2015**, *177*, 335–348. [CrossRef]
51. Shifrin, N.S.; Chisholm, S.W. Phytoplankton lipids: Interspecific differences and effects of nitrate, silicate and light-dark cycles 1. *J. Phycol.* **1981**, *17*, 374–384. [CrossRef]
52. Dortch, Q.; Whitledge, T.E. Does nitrogen or silicon limit phytoplankton production in the Mississippi River plume and nearby regions? *Cont. Shelf Res.* **1992**, *12*, 1293–1309. [CrossRef]
53. Redfield, A.C.; Ketchum, B.H.; Richards, F.A. The influence of organisms on the composition of seawater. In *The Composition of Seawater: Comparative and Descriptive Oceanography*; Hill, M.N., Ed.; The Sea: Ideas and Observations on Progress in the Study of the Seas; Interscience Publishers: New York, NY, USA, 1963; Volume 2, pp. 26–77.
54. Suárez, I.; Marañón, E. Photosynthate allocation in a temperate sea over an annual cycle: The relationship between protein synthesis and phytoplankton physiological state. *J. Sea Res.* **2003**, *50*, 285–299. [CrossRef]
55. Lancelot, C.; Mathot, S. Biochemical fractionation of primary production by phytoplankton in Belgian coastal waters during short- and long-term incubations with 14C-bicarbonate. *Mar. Biol.* **1985**, *86*, 227–232. [CrossRef]
56. Boëchat, I.G.; Giani, A. Seasonality affects diel cycles of seston biochemical composition in a tropical reservoir. *J. Plankton Res.* **2008**, *30*, 1417–1430. [CrossRef]
57. Morris, L.; Skea, W. Products of photosynthesis in natural populations of marine phytoplankton from the Gulf of Maine. *Mar. Biol.* **1978**, *47*, 303–312. [CrossRef]
58. Foy, R.H.; Smith, R.V. The role of carbohydrate accumulation in the growth of planktonic *Oscillatoria* species. *Br. Phycol. J.* **1980**, *15*, 139–150. [CrossRef]
59. Kim, B.K.; Lee, J.H.; Yun, M.S.; Joo, H.; Song, H.J.; Yang, E.J.; Chung, K.H.; Kang, S.-H.; Lee, S.H. High lipid composition of particulate organic matter in the northern Chukchi Sea, 2011. *Deep Sea Res.* **2015**, *120*, 72–81. [CrossRef]
60. Yun, M.S.; Whitledge, T.E.; Stockwell, D.; Son, S.H.; Lee, J.H.; Park, J.W.; Lee, D.B.; Lee, S.H. Primary production in the Chukchi Sea with potential effects of freshwater content. *Biogeosciences* **2016**, *13*, 737–749. [CrossRef]
61. Kim, B.K.; Jung, J.; Lee, Y.; Cho, K.H.; Gal, J.K.; Kang, S.H.; Ha, S.Y. Characteristics of the Biochemical Composition and Bioavailability of Phytoplankton-Derived Particulate Organic Matter in the Chukchi Sea, Arctic. *Water* **2020**, *12*, 2355.
62. Roy, S. Distributions of phytoplankton carbohydrate, protein and lipid in the world oceans from satellite ocean colour. *ISME J.* **2018**, *12*, 1457–1472. [CrossRef]

Article

In Situ Rates of Carbon and Nitrogen Uptake by Phytoplankton and the Contribution of Picophytoplankton in Kongsfjorden, Svalbard

Bo Kyung Kim, Hyoung Min Joo, Jinyoung Jung, Boyeon Lee and Sun-Yong Ha *

Division of Polar Ocean Sciences, Korea Polar Research Institute, 26 Songdomirae-ro, Yeonsu-gu, Incheon 21990, Korea; bkkim@kopri.re.kr (B.K.K.); hmjoo77@kopri.re.kr (H.M.J.); jinyoungjung@kopri.re.kr (J.J.); boyeon@kopri.re.kr (B.L.)
* Correspondence: syha@kopri.re.kr

Received: 20 August 2020; Accepted: 15 October 2020; Published: 17 October 2020

Abstract: Rapid climate warming and the associated melting of glaciers in high-latitude open fjord systems can have a significant impact on biogeochemical cycles. In this study, the uptake rates of carbon and nitrogen (nitrate and ammonium) of total phytoplankton and picophytoplankton (<2 μm) were measured in Kongsfjorden in early May 2017 using the dual stable isotope technique. The daily uptake rates of total carbon and nitrogen ranged from 0.3 to 1.1 g C m^{-2} day^{-1}, with a mean of 0.7 ± 0.3 g C m^{-2} day^{-1}, and 0.13 to 0.17 g N m^{-2} day^{-1}, with a mean of 0.16 ± 0.02 g N m^{-2} day^{-1}. Microphytoplankton (20–200 μm) accounted for 68.1% of the total chlorophyll a (chl-a) concentration, while picophytoplankton (<2 μm) accounted for 19.6% of the total chl-a, with a high contribution to the carbon uptake rate (42.9%) due to its higher particulate organic carbon-to-chl-a ratio. The contributions of picophytoplankton to the total nitrogen uptake rates were 47.1 ± 10.6% for nitrate and 74.0 ± 16.7% for ammonium. Our results indicated that picophytoplankton preferred regenerated nitrogen, such as ammonium, for growth and pointed to the importance of the role played by picophytoplankton in the local carbon uptake rate during the early springtime in 2017. Although the phytoplankton community, in terms of biovolume, in all samples was dominated by diatoms and *Phaeocystis* sp., a higher proportion of nano- and picophytoplankton chl-a (mean ± SD = 71.3 ± 16.4%) was observed in the relatively cold and turbid surface water in the inner fjord. Phytoplankton production (carbon uptake) decreased towards the inner fjord, while nitrogen uptake increased. The contrast in carbon and nitrogen uptake is likely caused by the gradient in glacial meltwater which affects both the light regime and nutrient availability. Therefore, global warming-enhanced glacier melting might support lower primary production (carbon fixation) with higher degrees of regeneration processes in fjord systems.

Keywords: phytoplankton productivity; carbon and nitrogen; stable isotopes; Kongsfjorden; Svalbard

---

1. Introduction

Marine phytoplankton play a critical role in global carbon and nitrogen cycling [1,2]. They uptake carbon dioxide from the atmosphere through photosynthesis and transform it into organic matter as well as uptaking nitrogen and other nutrients (e.g., phosphorous, silicate, trace metals), and they contribute approximately 50% of global primary production (carbon fixation) [3]. In marine systems, production depends on allochthonous inputs of nutrients (new production based on nitrate), while the remainder is mainly driven by the remineralization of nutrients (regenerated production; based on ammonium and urea) [4]. This distinction provides useful information for the proportion of the organic carbon exported out of the euphotic zone or into higher trophic levels because new production is assumed to be equivalent to the fraction of total production under steady-state conditions [5].

Climate change is expected to considerably influence the upper ocean in marine ecosystems, such as through rising temperatures and declining ice cover, altering the structure of phytoplankton communities, magnitudes of carbon, nitrogen, and export production, and the efficiency of the biological carbon pump [6–10]. Particularly, changes are more pronounced in the Arctic [11,12]. For example, patterns of increasing picophytoplankton (<2 μm) biomass and a decline in larger cells (2–20 μm) were found in the Canada Basin due to strong vertical stratification with a lower nitrate supply [11]. Although smaller phytoplankton is more competitive under low nutrient concentrations, picophytoplankton-dominated systems do not provide large carbon exports to the deep sea (carbon sequestration). Because picophytoplankton have a high surface-area-to-volume ratio and are more resistant (they do sink, form aggregates and such). As a result, picophytoplankton is considered a critical component of carbon and nitrogen cycling in high-latitude marine ecosystems and likely more important in warmer freshened ocean [11,13,14].

Our study area, Kongsfjorden, is located in the open fjords in western Spitsbergen and geographically belongs to the Arctic. The hydrological condition of this area is characterized by Atlantic water (T > 3 °C, S > 34.9), glacial meltwater, and Arctic water (T = 0.5–2 °C, S = 34.7–34.9) ([15–17] and references therein). The mixing process of these water masses influences the spatial and temporal variation in the food web [18,19]. Generally, phytoplankton biomass in Kongsfjorden begins to increase in early spring (March) and then decreases dramatically in late summer and early autumn [16,20]. The progress of the spring bloom is controlled by light and grazing pressure [21], and the primary production of phytoplankton can influence zooplankton production [22,23]. Therefore, knowing the local primary production and the contribution of the picophytoplankton fraction to the total primary production in the Arctic Ocean is useful for assessing the potential impacts caused by environmental changes, such as an intense ice melt season [12,24,25]. A number of studies have reported the composition of coupled pico- and nanoplankton communities (including heterotrophic bacteria) in water masses and/or phytoplankton bloom dynamics in Kongsfjorden [20,26,27]. Unfortunately, only a few in situ measurements of primary production have been conducted in Kongsfjorden [16,20,21,28]. Hence, the purposes of this study were (i) to investigate spatial variation in the phytoplankton community structure and the rates of carbon and nitrogen uptake and (ii) to estimate the contribution of picophytoplankton to the total biomass and carbon and nitrogen uptake rates during the spring period in Kongsfjorden, Svalbard.

2. Materials and Methods

2.1. Study Area and Sampling

Kongsfjorden is a fjord in western Spitsbergen and is located on the Svalbard archipelago. All of the samples except for the carbon and nitrogen uptake rates were collected at a total of ten stations in Kongsfjorden from 4–8 May 2017 onboard the Teisten (Figure 1 and Table 1). Our sampling stations were divided into three zones based on species distribution, substrate and the overriding environmental gradient described by Hop et al. [16]: the inner zone (St. 1, St. 2, and St. 3), the transition zone (St. 4, St. 5, and St. 6), and the middle zone (St. 7, St. 8, St. 9, and St. 10) (Table 1). At each station, we used a conductivity-temperature-depth (CTD) system with a turbidity sensor and Niskin bottles to obtain the water temperature, turbidity, and water samples, respectively. The water samples for the nutrients, chl-a concentrations, and taxonomy were obtained using a Niskin sampler from the surface to a depth of up to 100 m (Table 1). The characteristics of each station are summarized in Table 1. To estimate the daytime uptake rate, daily solar irradiance (W m^{-2}) data were obtained from the Baseline Surface Radiation Network (BSRN) for the sampling period. The daily solar irradiance was calculated using a 1 h resolution. One percent of the surface irradiance was defined as the euphotic zone, and different light levels were determined (100, 50, 30, 12, 5, and 1%) by a Secchi disc using the vertical attenuation coefficient (Kd = 1.7/Secchi depth) from Poole and Atkins [29]. The euphotic depth was measured only at experiment stations (St. 3, St. 5, St. 7, St. 9, and St. 10).

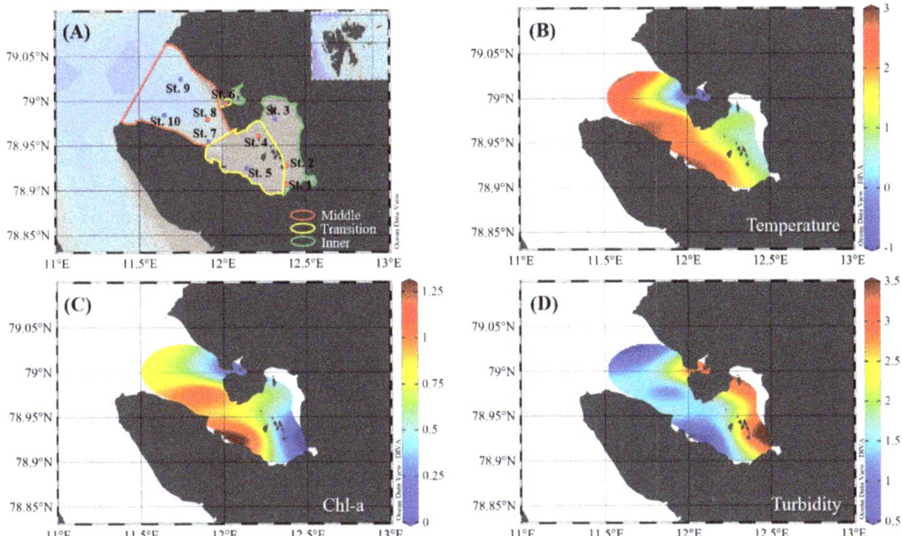

Figure 1. (**A**) Sampling stations in Kongsfjorden. The blue dots indicate productivity stations. Surface distribution of (**B**) water temperature (°C), (**C**) chlorophyll a (chl-a) (mg·m^{-3}), and (**D**) turbidity (FTU) in Kongsfjorden (Ocean Data View (ODV) version 5.1.0) (AWI, Bremerhaven, Germany, Schlitzer, R.).

Table 1. Sample location with corresponding sampling depth for nutrient, chl-a, andphytoplankton taxonomy samples in Kongsfjorden in 2017.

Station	Latitude (°N)	Longitude (°E)	Date (Day, Month Year)	Bottom Depth (m)	Euphotic Depth (m)	Sampling Depth (m)	Zone
St. 1	78.91	12.39	4 May 2017	63	-	0, 10, 30, 50	Inner
St. 2	78.93	12.39	4 May 2017	47	-	0, 10, 20, 40	Inner
St. 3	78.98	12.32	7 May 2017	55	16	0, 4, 16, 40	Inner
St. 4	78.96	12.22	7 May 2017	14	-	0, 5, 10	Transition
St. 5	78.93	12.15	4 May 2017	110	22	0, 6, 22, 80	Transition
St. 6	79	12.02	7 May 2017	63	-	0, 10, 20, 50	Transition
St. 7	78.96	11.92	7 May 2017	357	27	0, 7, 27, 80	Middle
St. 8	78.98	11.91	8 May 2017	226	-	0, 10, 40, 100	Middle
St. 9	79.03	11.75	8 May 2017	211	27	0, 7, 27, 50, 100	Middle
St. 10	78.99	11.65	8 May 2017	287	19	0, 5, 19, 50, 100	Middle

2.2. Major Inorganic Nutrient Analysis

The seawater samples for nutrients were collected in 250 mL acid-cleaned polythene bottles directly from CTD spigots without the use of a tube and filtered through a precombusted (450 °C, 4 h) glass fiber filter (GF/F, 47 mm in diameter, Whatman, Marlborough, MA, USA). The filtered seawater samples were preserved at approximately −20 °C until analysis in our laboratory. The major inorganic nutrients (nitrite and nitrate, ammonium, phosphate, and silicate) were measured using a 4-channel continuous AutoAnalyzer (QuAAtro, SEAL Analytical, Southampton, UK) with the settings recommended by the manufacturer's manual. Standard curves were run with each batch of samples using freshly prepared standards that spanned the range of concentrations in the samples.

2.3. Chlorophyll a, Identification and Counts of Phytoplankton

Water samples (1 L) for measuring the total chl-a concentrations of phytoplankton were filtered using a 0.7 μm pore-size Whatman GF/F (47 mm) at all stations during the cruise. The phytoplankton in this study were divided into 3 categories: micro- (>20 μm), nano- (2.0–20 μm), and picophytoplankton

(0.7–2.0 μm). For the size-fractionated chl-a concentrations, samples (1 L) were passed sequentially through 20 and 2 μm Nucleopore filters (47 mm, Whatman, UK) and 0.7 μm GF/F filters (47 mm). The chl-a concentrations were determined using a Trilogy fluorometer (Turner Designs, USA) after a 24 h extraction in 90% acetone at 4 °C [30].

To identify and count phytoplankton, aliquots of 125 mL were preserved with glutaraldehyde (final concentration: 1%). Phytoplankton net tows (20 μm mesh) were used to gather additional samples, which were preserved with glutaraldehyde (final concentration: 2%). Sample volumes of 50 to 100 mL were filtered through Gelman GN-6 Metricel filters (0.45 μm pore size, 25 mm diameter; Gelman Sciences, Inc., Port Washington, NY, USA). The filters were mounted on microscopic slides in a water-soluble embedding medium (HPMA, 2-hydroxypropyl methacrylate) in our laboratory. The HPMA slides were used for identification and estimation of the cell concentrations. Identification by light microscopy is time consuming and requires a high level of taxonomic skill but is still the most reliable method of microalgal identification [31]. Therefore, at least 300 cells were identified from each sample with an optical microscope (BX53TR-32FB3F0 microscope, Olympus, Inc., Tokyo, Japan) using a combination of light and epifluorescence microscopy at 400× magnification for microphytoplankton and at 1000× magnification for pico- and nanophytoplankton [32]. The genera were identified using the keys provided by Tomas [33]. The cell counts were converted into cell concentrations according to the method of Kang et al. [34].

2.4. Carbon and Nitrogen Uptake Experiments

The carbon and nitrogen uptake rates of phytoplankton were measured using ^{13}C-^{15}N dual isotope tracer techniques [14] at five in situ carbon and nitrogen uptake experiment stations (St. 3, St. 5, St. 7, St. 9, and St. 10; hereafter productivity stations). The samples were transferred into 1 L polycarbonate bottles wrapped with different neutral density screen films (LEE Filters, UK; Garneau et al. [35]) to simulate different light levels (100, 50, 30, 12, 5 and 1%) and then were enriched with ^{13}C (H^{13}CO$_3$) and ^{15}N (K^{15}NO$_3$ or ^{15}NH$_4$Cl) isotopes. The ^{13}C and ^{15}N enrichments made up approximately 5–10% of the total inorganic carbon and nitrogenous nutrients in the ambient water [4,36]. For phytoplankton incubation, the bottles containing isotopes were placed in incubators, surrounded by ambient seawater. After incubation (4–5 h), the samples were transferred into the laboratory and filtered onto precombusted (450 °C, 4 h) 25-mm GF/F (Whatman, 0.7 μm pore) filters to assess the carbon and nitrogen uptake rates of total phytoplankton. For the carbon and nitrogen uptake rates of small-sized cell (picophytoplankton) separations, the incubated samples were first filtered through 2-μm Nucleopore filters (47 mm) and then onto the GF/F filters (25 mm). The particulate organic carbon (POC) and nitrogen (PON) samples for the total phytoplankton and picophytoplankton were also obtained through the abovementioned filtration process. The filters were immediately stored at −80 °C until further analysis. The POC and PON concentrations and carbon and nitrogen isotope concentration were determined using a Finnigan Delta + XL mass spectrometer in the stable isotope laboratory at the University of Alaska Fairbanks, USA, after HCl fuming overnight to remove the carbonate. The carbon and nitrogen uptake rates were calculated according to Hama et al. [37] and Dugdale and Goering [4], respectively (Equation (1)).

$$\text{Carbon (or nitrogen) uptake rate} = \frac{(a_{is} - a_{ns})}{(a_{ic} - a_{ns})} \times \frac{P(t)}{t} \tag{1}$$

where a_{is} is the ^{13}C atom% of the POC in the incubated sample, a_{ns} is the ^{13}C atom% of the POC in the natural sample, P is the POC, a_{ic} is the total dissolved inorganic carbon, and t is the incubation time period. Similarly, the nitrogen (nitrate or ammonium) uptake rates were calculated using the same equation, where P denotes the particulate organic nitrogen, and a_{is} and a_{ns} are the ^{15}N atom% of PON in the incubated and natural samples, respectively. a_{ic} is the ambient dissolved inorganic nitrogen (nitrate or ammonium). The specific carbon (or nitrogen) uptake rates (photosynthetic rate; h^{-1}) are defined as the nutrients taken up per unit of POC (PON) and per unit of time, and the absolute carbon (nitrogen) uptake rates (expressed in mg C (or N)·m^{-3}·h^{-1}) are the product of the specific carbon (nitrogen) uptake rate and the POC (or PON). The uptake rate of the large-size fraction (>2 μm; micro- and nanophytoplankton) was calculated by subtracting the picophytoplankton value from the total phytoplankton uptake rate. The water column-integral chl-a concentrations and hourly carbon and nitrogen uptake rates at each station were estimated using trapezoidal integrations of volumetric values from 100 to 1% light levels [38].

2.5. Relative Preference Index and Turnover Time of Picophytoplankton

To estimate the utilization of a nitrogen compound, the relative preference index (RPI) value and turnover times were calculated using the equations of McCarthy et al. [39] (Equation (2)) and Gu and Alexander [40] (Equation (3)), respectively. For example, the RPI and turnover time for nitrate were calculated as follows:

$$\text{RPI}_{\text{nitrate}} = \frac{[\text{nitrate uptake rate}]/[\text{nitrogen uptake rate}]}{[\text{NO}_3^-]/[\text{DIN}]} \tag{2}$$

$$\text{Turnover time}_{\text{nitrate}} = \frac{[\text{NO}_3^-]}{[\text{nitrate uptake rate}]} \tag{3}$$

where the $\text{RPI}_{\text{nitrate}}$ of picophytoplankton is the nitrogen uptake rate in the combined nitrate and ammonium uptake rates for picophytoplankton. NO_3^- and DIN are the ambient nitrate concentration and dissolved inorganic nitrogen concentration (nitrate + nitrite + ammonium), respectively. The turnover time for the nitrogen substrate in picophytoplankton was calculated by assuming the consumption of the substrate.

2.6. Statistical Analysis

A *t* test was used to determine whether there was a significant difference between the two independent groups. Correlations between the investigated variables were examined parametrically using Pearson's correlation. Statistical analysis was performed with a significance threshold of $p < 0.05$.

3. Results

3.1. Physical and Nutrient Properties in Kongsfjorden

The stations experienced a gradient of influence from a primarily oceanographic influence to a primarily glacier influence towards the inner zone of the fjord. Water temperature ranged from −0.7 to 2.6 °C with an average of 2.0 °C (standard deviation (SD) = ± 0.8 °C) from the surface to depth of 100-m deep. The middle zone (mean ± SD = 2.4 ± 0.1 °C) of the fjord was composed of relatively warmer water than the inner zone (mean ± SD = 0.8 ± 0.7 °C). In particular, the cold water temperatures (<0 °C) observed in the surface water at St. 6 and in 30–50 m of St. 1, respectively, were related to the extent of input from the melting of land-based ice during the sampling period because sea ice was rare in the fjord (Figure 1B). Overall, warm water has been intruding into the fjord, and meltwater flows out the fjord in Kongsfjorden.

As observed for water temperature, there was a clear difference in nutrient concentration between the inner and remaining parts of the fjord. The phosphate and nitrite + nitrate concentrations were 0.2–0.6 μM and 1.5–8.1 μM from the surface to 100-m water depth, respectively (Figure 2). The silicate concentration ranged from 3.4 to 5.0 μM with a mean of 4.5 μM (SD = ± 0.3 μM) (Figure 2). Most of the concentrations of nitrite and nitrate, phosphate, and silicate gradually increased with depth in the inner zone and were nearly constant throughout the water in the transition and middle zones. However, ammonium did not show a specific trend and ranged from 0.6 to 1.9 μM with a mean of 1.0 μM (SD = ± 0.3 μM) (Figure 2).

During the survey period, the measured daily solar irradiance ranged from 24 to 543 W m^{-2}, with a mean of 247.8 ± 146.4 W m^{-2}; the variability and trend of the daily solar irradiance remained consistent except for on 4 May. At the productivity stations, the vertical light attenuation coefficient (Kd) ranged from 0.17 to 0.28 m^{-1} (mean ± SD = 0.22 ± 0.05 m^{-1}), leading to a shift in depth from 16 m at St. 3 to 27 m at St. 7 and St. 9, with an average depth of approximately 22.2 m (SD = ± 4.9 m) (Table 1). The spatial distribution of turbidity (FTU) at the sampling stations is shown in Figure 1D. The turbidity ranged from 0.3 to 6.1 FTU, with an average of 1.1 ± 1.0 FTU at a 100-m depth. In near-glacier stations (St. 1, St. 2, St. 3, St. 4, and St. 6), turbidity exceeds 1.7 FTU at the surface and then decreases with a depth of ~20 m, whereas in the rest of stations (except St. 7; no vertical data) is low (mean of 0.5 FTU) throughout the water column (within 100-m depth) (data not shown). Therefore, the turbidity of the surface water was used to analyze the amount of inorganic particles from glacial runoff in this study.

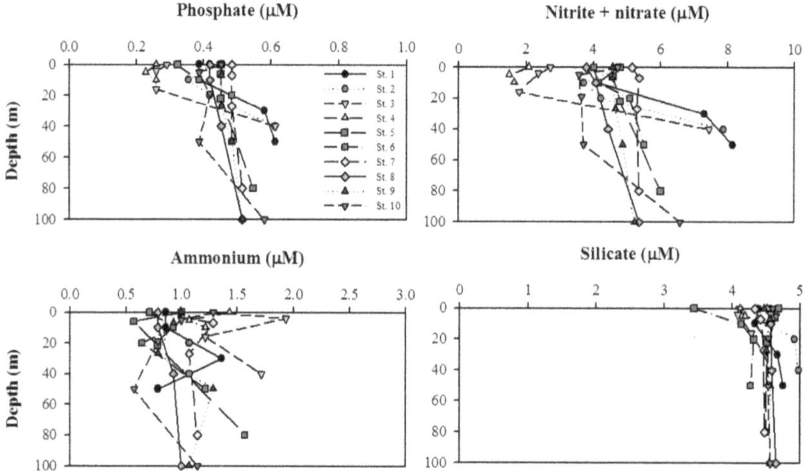

Figure 2. Vertical patterns of phosphate (μM), nitrite + nitrate (μM), ammonium (μM), and silicate (μM) concentrations from the surface to the 100-m water depth at sampling stations.

3.2. Chlorophyll a, Size Distribution, and Community of Phytoplankton

At a surface to depth of 100 m, the concentration of chl-a varied by an order of magnitude from high chl-a (0.09 to 1.52 mg m^{-3}, with a mean of 0.80 ± 0.33 mg m^{-3}). The phytoplankton community was dominated by microphytoplankton (20–200 μm), which accounted for 63.3% (SD = ± 16.0%) of the total chl-a concentration, followed by pico- (0.7–2.0 μm; mean ± SD = 20.5 ± 8.8%) and nanophytoplankton (2.0–20 μm; mean ± SD = 16.2 ± 8.8%) in Kongsfjorden (Table 2). The surface chl-a increased from 0.3 mg m^{-3} in the inner part to 0.9 mg m^{-3} in the middle part, which was accompanied by a relative increase in the micro- and decrease in nano- and picophytoplankton fractions (28.7 to 78.6%, 42.2 to 6.9%, and 29.1 to 14.5% from the inner to middle, respectively). In contrast, the other depths in Kongsfjorden showed that microphytoplankton were dominant, with a mean value of 65.3 ± 10.8%. The total chl-a concentration was integrated from the surface to a 1% light level and ranged from 14.6 to 26.1 mg chl-a m^{-2}, with a mean of 19.7 mg chl-a m^{-2} (SD = ± 4.9 mg chl-a m^{-2}) at the productivity stations (Figure 3). Among the productivity stations, St. 3 had the lowest microphytoplankton composition (<40%) at the surface (100% light level), while microphytoplankton was dominant throughout the euphotic zone.

The number of phytoplankton species ranged from 7 to 27 species, including unidentified micro-, nano-, and picophytoplankton within 100 m, and the cell abundance ranged from 1.9 × 10^5 to 3.9 × 10^6 cells L^{-1}, with a mean of 1.6 × 10^6 ± 8.8 × 10^5 cells L^{-1} (data not shown). Diatoms (Bacillariophyceae) and Prymnesiophyceae (only *Phaeocystis* sp. were identified) dominated the phytoplankton community in the samples (Figure 4). The most abundant species were *Phaeocystis* sp. and the diatom *Fragilariopsis cylindrus,* with maximum cell concentrations of 3.1 × 10^6 and 2.9 × 10^5 cells L^{-1}, respectively. At most stations, the diatoms and *Phaeocystis* sp. contributed approximately 88.3 and 77.9% of the total abundance and biovolume of phytoplankton, respectively (Figure 4). Notably, the unidentified picophytoplankton contributed up to 40.1% of the total phytoplankton abundance at St. 6 (Figure 4). The proportions of the total biovolume showed that the diatoms (mean ± SD = 51.2 ± 4.7%) were higher than those of *Phaeocystis* sp. (mean ± SD = 33.1 ± 7.3%) in the inner part of fjord, whereas the phytoplankton community was characterized by the prevalence of *Phaeocystis* sp. (mean ± SD = 64.8 ± 4.8%) in the middle part of fjord (Figure 4). However, in abundance, the *Phaeocystis* sp. was higher than diatoms at all stations of fjord. At the productivity stations, the diatom (Bacillariophyceae) was positively correlated between N/P (DIN to phosphate) and Si/P (silicate to phosphate) molar ratios (r = 0.59, p < 0.05 and r = 0.69, p < 0.001, n = 14, respectively), and a negative correlation existed between diatom abundance and temperature (r = −0.92, n = 14, p < 0.001) within the euphotic zone, while no relationship was found in *Pheaocystis* sp. (Prymneosiophyceae).

Table 2. The relative contribution of chl-a in the different size fractions from surface to 100-m depth.

Station	Depth (m)	Microphytoplankton (%)	Nanophytoplankton (%)	Picophytoplankton (%)
St. 1	0	38.1	37.6	24.2
	10	54.0	26.0	20.0
	30	71.6	13.5	14.9
	50	80.0	11.2	8.8
St. 2	0	9.7	53.5	36.8
	10	48.3	21.3	30.3
	20	79.2	5.7	15.1
	40	73.8	12.7	13.5
St. 3	0	38.2	35.4	26.4
	4	52.7	28.6	18.8
	16	61.1	21.9	17.0
	40	57.0	38.7	4.3
St. 4	0	64.1	19.3	16.6
	5	75.5	10.8	13.7
	10	73.6	14.8	11.6
St. 5	0	78.8	6.5	14.7
	6	82.1	7.3	10.6
	22	81.5	5.4	13.1
	80	57.8	9.8	32.4
St. 6	0	29.5	48.8	21.7
	10	55.1	37.0	7.9
	20	71.5	9.9	18.6
	50	53.1	24.2	22.7
St. 7	0	80.1	7.8	12.1
	7	68.1	9.5	22.4
	27	69.6	7.1	23.3
	80	52.7	11.2	36.0
St. 8	0	86.5	4.4	9.1
	10	57.0	9.6	33.4
	40	47.3	10.3	42.3
	100	60.5	13.0	26.5
St. 9	0	81.4	4.2	14.5
	7	61.1	9.5	29.4
	27	75.1	6.2	18.7
	50	74.4	7.5	18.1
	100	74.3	6.5	19.2
St. 10	0	66.6	11.2	22.2
	5	68.6	10.5	20.9
	19	56.3	13.5	30.3
	50	53.8	13.3	32.9
	100	75.9	7.7	16.4

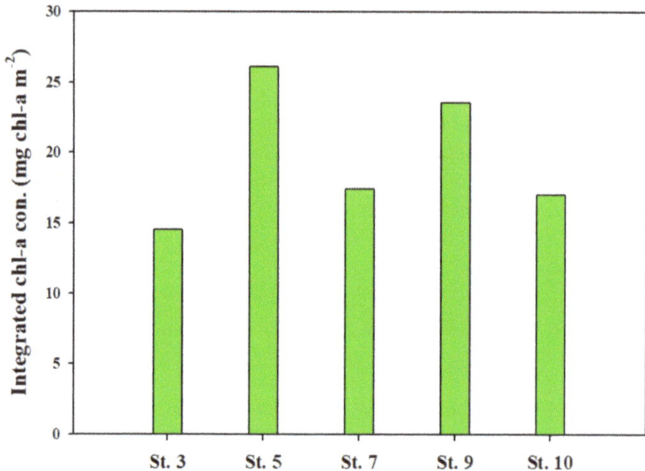

Figure 3. The distribution of chl-a concentrations (mg chl-a m^{-2}) integrated from the surface to 1% light depth at productivity stations.

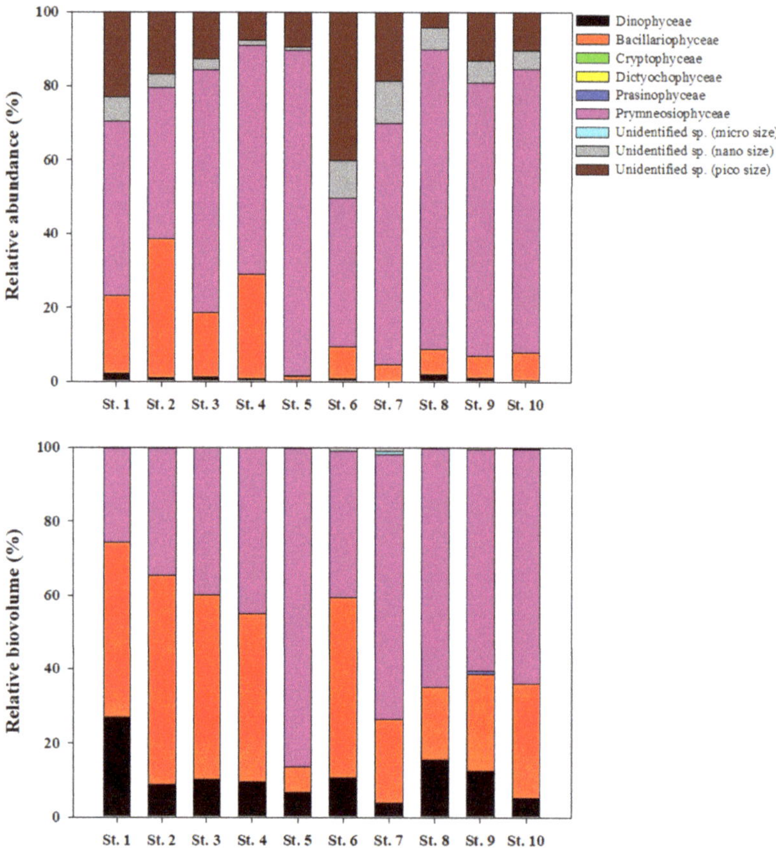

Figure 4. Relative abundance and biovolume of the phytoplankton taxa.

3.3. Carbon Uptake Rates of Phytoplankton

The total absolute and specific carbon uptake rates of phytoplankton in Kongsfjorden are shown in Figure 5. The absolute rate varied from 0.2 to 3.2 mg C m^{-3} h^{-1} (with a mean of 1.4 ± 0.9 mg C m^{-3} h^{-1}) and the specific rate varied from 0.0006 and 0.0105 h^{-1} (mean ± SD = 0.0046 ± 0.0030 h^{-1}). The total absolute carbon uptake rate was lowest at the surface or a 1% light level. Likewise, the highest specific carbon uptake rate was also observed in the absolute uptake rate profile (Figure 5).

The average total hourly carbon uptake rate integrated from 100 to 1% was 30.5 mg C m^{-2} h^{-1} (SD = ± 12.5 mg C m^{-2} h^{-1}) in this study (Figure 6). The lowest uptake rate was 11.5 mg C m^{-2} h^{-1} at St. 3, whereas the highest uptake rate was 46.1 mg C m^{-2} h^{-1} at St. 9. The range of hourly carbon uptake rates of picophytoplankton (<2 μm) ranged from 5.4 (St. 3) to 15.5 (St. 7) mg C m^{-2} h^{-1}, with a mean value of 12.7 mg C m^{-2} h^{-1} (SD = ± 4.3 mg C m^{-2} h^{-1}) (Figure 6). The picophytoplankton contributed 42.9% (± 5.9%) of the total carbon production. The highest contribution of the large size fraction was found at St. 9 (Figure 6).

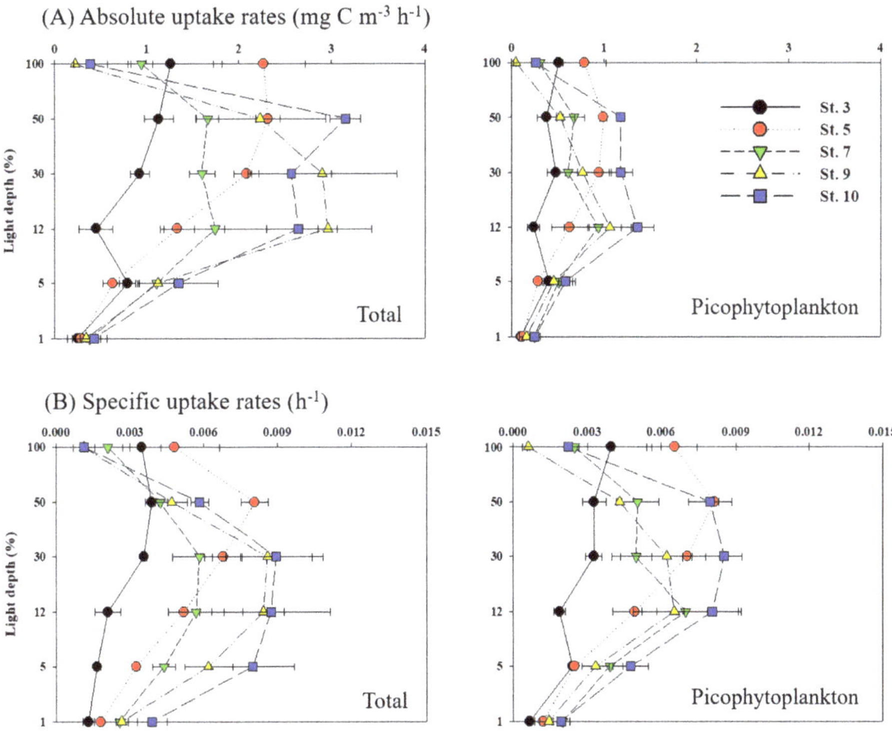

Figure 5. The vertical pattern of (**A**) absolute and (**B**) specific carbon uptake rates for total and picophytoplankton at each light depth (100, 50, 30, 12, 5, and 1% light depth).

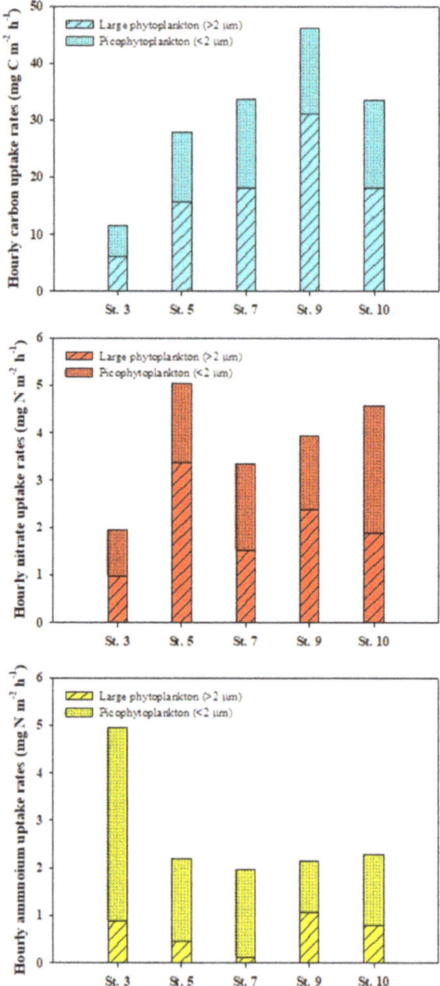

Figure 6. Hourly carbon, nitrate and ammonium uptake rates of phytoplankton integrated from the surface to 1% light depth at the productivity stations.

3.4. Nitrogen Uptake Rates of Phytoplankton

The absolute total nitrate uptake rates (new production) ranged between 0.003 and 0.450 mg N m^{-3} h^{-1} (mean ± SD = 0.127 ± 0.107 mg N m^{-3} h^{-1}). The absolute total ammonium uptake rates ranged between 0.013 and 0.423 mg N m^{-3} h^{-1} (mean ± SD = 0.120 ± 0.092 mg N m^{-3} h^{-1}). The specific total nitrate and ammonium uptake rates ranged between 0.0002 and 0.0112 h^{-1} (mean ± SD = 0.0039 ± 0.0027 h^{-1}) and 0.0005–0.0125 h^{-1} (mean ± SD = 0.0037 ± 0.0027 h^{-1}), respectively (Figure 7). The absolute total uptake rate generally showed a similar pattern to that of the specific total uptake rate (Figure 7). The temperature and ammonium concentrations showed a significant negative and positive correlation with the absolute and specific ammonium uptake rates, respectively ($p < 0.05$) (Table 3). In contrast, there was no significant difference between the nitrate concentration and the nitrate uptake rate ($p > 0.05$). The specific uptake rates of nitrate and ammonium were identical, thus suggesting that the utilization of nitrate and ammonium are similar for phytoplankton growth.

(A) Absolute uptake rates (mg N m⁻³ h⁻¹)

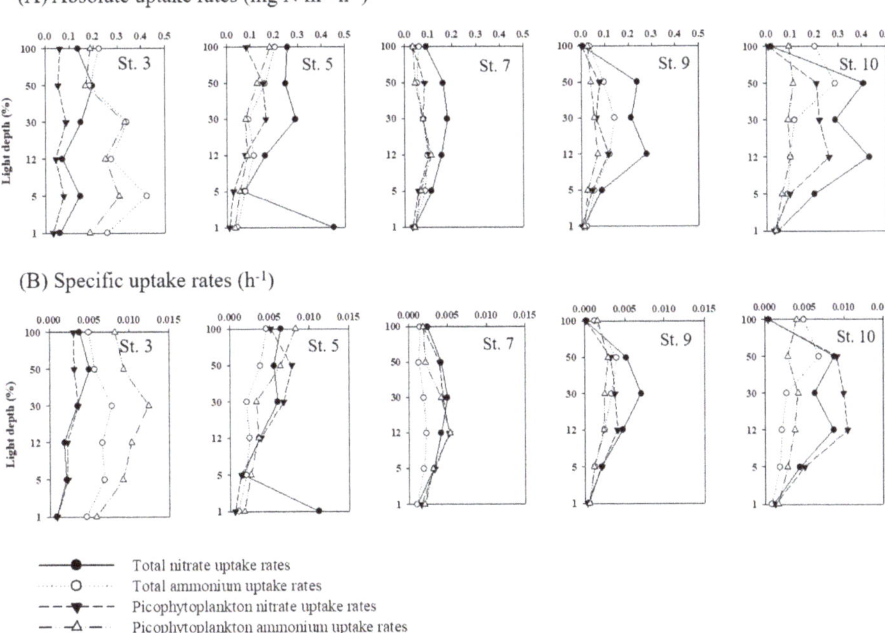

(B) Specific uptake rates (h⁻¹)

Figure 7. The vertical pattern of (**A**) absolute and (**B**) specific nitrogen (nitrate and ammonium) uptake rates for total and picophytoplankton at each light depth (100, 50, 30, 12, 5, and 1% light depth).

The total hourly nitrate and ammonium uptake rates integrated from 100 to 1% light levels were 2.0–5.0 mg N m⁻² h⁻¹ and 2.0–5.0 mg N m⁻² h⁻¹, respectively, at the productivity stations (Figure 6). The average nitrate uptake rate was 3.8 mg N m⁻² h⁻¹ (SD = ± 1.2 mg N m⁻² h⁻¹), which was higher than 2.7 mg N m⁻² h⁻¹ (SD = ± 1.3 mg N m⁻² h⁻¹) for ammonium, but the rates were not significantly different (*t* test, $p > 0.05$). The total nitrogen (nitrate and ammonium) uptake rates of the total phytoplankton and picophytoplankton ranged from 5.3 to 7.2 mg N m⁻² h⁻¹ (mean ± SD = 6.5 ± 0.8 mg N m⁻² h⁻¹) and 2.6 to 5.0 mg N m⁻² h⁻¹ (mean ± SD = 3.8 ± 0.9 mg N m⁻² h⁻¹), respectively (Figure 6). The contributions of picophytoplankton to the total nitrogen uptake rates were 47.1 ± 10.6% for nitrate and 74.0 ± 16.7% for ammonium (Figure 6).

Table 3. Pearson's correlation analysis of specific and absolute total carbon and nitrogen uptake rates ($n = 15$). Significant correlations are indicated as follow: * $p < 0.05$, ** $p < 0.01$.

	Temperature	Phosphate	Nitrite + Nitrate	Silicate	Ammonium	Specific Carbon	Absolute Carbon	Specific Nitrate	Absolute Nitrate	Specific Ammonium	Absolute Ammonium
Temperature	1										
Phosphate	0.868 **	1									
Nitrite + Nitrate	0.797 **	0.974 **	1								
Silicate	0.795 **	0.706 **	0.598 *	1							
Ammonium	−0.664 **	−0.630 *	−0.503	−0.651 **	1						
Specific Carbon	0.200	0.140	0.140	0.353	−0.130	1					
Absolute Carbon	0.165	0.101	0.102	0.326	−0.109	0.998 **	1				
Specific Nitrate	0.232	0.116	0.161	0.248	−0.187	0.567 *	0.566 *	1			
Absolute Nitrate	0.234	0.069	0.102	0.272	−0.209	0.525 *	0.531 *	0.979 **	1		
Specific Ammonium	−0.698 **	−0.716 **	−0.621 *	−0.469	0.795 **	0.183	0.207	0.026	0.026	1	
Absolute Ammonium	−0.732 **	−0.784 **	−0.701 **	−0.501	0.768 **	0.141	0.167	0.006	0.015	0.987 **	1

4. Discussion

4.1. Environmental and Phytoplankton Community Characteristics of Early Springtime in Kongsfjorden

In Kongsfjorden, the bloom timing and dominant species varied from April to the end of May in relation to the environmental factors (i.e., sea ice, glacier and Atlantic water inflow events) [16,41,42]. During the spring bloom, the phytoplankton community was dominated by diatoms, with high integrated chl-a concentrations (up to 35 mg chl-a m^{-2}) and nutrient-depleted conditions (<1.0 μM for silicate and <0.2 μM for nitrate) at depths of less than 50 m, while the dominant species shifted from *Fragilariopsis* spp. and diatoms in April and mid-May to *Phaeocystis pouchetii* colonies after mid-May with the low nutrient concentrations (level of spring bloom) [41]. This finding was consistent with that of von Quillfeldt [43], who found that diatoms dominated the early phytoplankton spring bloom.

In our study, the mean integrated chl-a value (ca. 20 mg chl-a m^{-2}) in the euphotic zone was similar to that on 8 May (after peak bloom; 22 mg chl-a m^{-2} for 20 m) in Hodal et al.'s study [41] and higher than the winter values (6–7 mg chl-a m^{-2}) in Eilertsen et al.'s study [21]. The observed values for nitrate and nitrite (mean ± SD = 4.2 ± 1.1 μM) were approximately half of the winter values (10–12 μM; Hop et al. [16]). The silicate (mean ± SD = 4.5 ± 0.2 μM) concentration was similar to that of the maximum level (4.8 μM) of early springtime (9 April to 12 May 2008) [44]. In addition, the N/P molar ratio (mean ± SD = 12.8 ± 1.4) at depths of less than 50 m was slightly lower than the Redfield ratio (16; Redfield [45]) and higher than that reported by Iversen and Seuthe [20] for the upper 50 m of the water column (3.2) during the spring bloom in Kongsfjorden. Although the Si/P molar ratio (mean ± SD = 11.1 ± 2.7) was less than the Redfield ratio (15:1) during the sampling period, this finding is not an indication of severe nutrient deficiency for phytoplankton growth because the proportion of microphytoplankton was approximately 68.1% of the total chl-a concentration. It is generally considered that communities dominated by large cells are responsible for phytoplankton biomass accumulation in nutrient-replete conditions, while small cells are typical of oligotrophic systems [46].

Previous studies indicated that in spring blooms, diatoms were the most important components of the phytoplankton communities in the fjord ([16,41] and references therein). However, the major contributor to the phytoplankton assemblage was *Phaeocystis* sp., which accounted for more than 40% of the total phytoplankton abundance in our study. The lower presence of the diatom compared to *P. pouchetii* and generally very low values of chl-a (<1 mg m^{-3}) relative to the enhanced inflow of warm Atlantic water during the spring bloom were features observed in Kongsfjorden [41,42]. In fact, the waters of the middle zone (St. 7, St. 8, St. 9, and St. 10) at depths of less than 100 m were composed of warmer water (>2 °C), which was coincidental with a higher abundance of *Phaeocystis* sp. than the inner zone (St. 1, St. 2, and St. 3; mean ± SD = 0.9 ± 0.7 °C) in our study. The relative abundance of diatoms showed an inverse relationship with temperature. These relationships reflected the spatial distribution and taxonomic composition of phytoplankton communities and water masses. An advection of the haptophyte *P. pouchetii* into Kongsfjorden was suggested by van De Poll et al. [47]. In addition, the unidentified sp. (pico size; <2 μm) was quite important in terms of abundance in Kongsfjorden (~40%) at St. 6, which is near the glacier. A higher proportion of nano- and picophytoplankton (mean ± SD = 71.3 ± 16.4%) with relatively cold surface water was observed at the inner fjord, suggesting that glacial influence may have a predominant effect on phytoplankton size classes. The observed size classes of phytoplankton were in agreement with Piquet et al. [44], who conducted sampling near the Kongsfjorden glacier during early spring. Hence, we inferred that the sampling period was considered the early phase of the spring bloom and not a post-bloom state in Kongsfjorden based on chl-a nutrient data. The phytoplankton community was affected by glacial meltwater input and influenced by Atlantic waters (presumably as a result of warmer water).

4.2. Total Carbon and Nitrogen Uptake Rates of Phytoplankton

Generally, the carbon and nitrogen uptake rates reflected the light conditions, biomass, ambient nutrient concentration, size class, composition, and physiological state of the phytoplankton and their bloom timing [48–50]. The integrated carbon uptake rates of phytoplankton and chl-a decreased towards the inner part of the fjord and were even completely suppressed at the stations closest to the glacier (St. 3), which may have been caused by glacial sediment. The melting glacier releases particles into the fjord, leading to a lower euphotic zone and highly unfavorable conditions for phytoplankton growth [17,47,51,52]. In our study, due to their location and the nearest tidewater glacier in Kongsfjorden, the inner stations showed high turbidity (Figure 1D). This finding was consistent with that of Piquet et al. [44]; the relatively high Kd values near the Kongsfjorden glacier were caused by enriched sediment particles as a result of the early onset of glacial meltwater during early spring. Van de Poll et al. [53] also demonstrated that reduced light penetration into the water column associated with glacial sediments limits the buildup of depth-integrated chl-a in the inner Kongsfjorden. In this sense, at St. 3 in the observation area, the euphotic depth (16 m) was the lowest, which suggests that increases in meltwater influx led to increased turbidity and reduced light penetration into the water, resulting in a lower carbon uptake rate.

More specifically, phytoplankton were less active in shallow water than in deep water within the euphotic zone. As shown in Figure 5, the maximum total carbon uptake rates were relatively high between 10 and 30 m, where the chl-a concentrations were highest. The specific carbon uptake of the total phytoplankton, except for the phytoplankton biomass (particulate organic matter; POC), also showed a similar trend. This result mirrored the findings of Hodal et al. [41] in Kongsfjorden; they reported that a high primary production and low biomass of phytoplankton were observed at 5 or 10 m and at the surface, respectively, due to photoinhibition during mid-May. Notably, at St. 3 (which had high turbidity), the maximum total carbon uptake rate was observed in surface water, which coincided with the high picophytoplankton uptake rate and relatively low contribution of microphytoplankton (38.2%). There is a positive relationship between the contribution of picophytoplankton chl-a and turbidity in surface data ($r = 0.75$, $n = 10$, $p < 0.05$). From these results, it has been suggested that smaller phytoplankton have a competitive advantage over microphytoplankton under turbidity as a strategic response to light. This potential mechanism aligned with the finding that absorption efficiency was higher in smaller cells than in larger cells due to their reduced chromophore self-shading [54–56]. Unlike the carbon uptake rate, the absolute concentration of inorganic nutrients, except for ammonium, was constant within the euphotic zone in this study. As a result, there was no statistical significance between the nutrient parameters and carbon uptake rate ($p > 0.05$) (Table 3). Likewise, they did not show any relationship between species composition. In this sense, the high turbidity, which represented low light conditions across transects, was the primary factor influencing light attenuation among stations in our study.

In this study, the daily uptake rates were calculated based on the daylight duration (24 h). The mean carbon uptake rate (0.7 g C m^{-2} day^{-1}) was up to two times lower than that reviewed by Hop et al. [16] (0.8–1.4 g·C·m^{-2} day^{-1}) during July 1996 and much higher than that of Eilertsen et al. [21] in the transitional zone of Kongsfjorden (0.03 g C·m^{-2} day^{-1} for 26 July 1979 and 0.17 g C m^{-2} day^{-1} for 31 July 1980) (Table 4). In May, Hodal et al. [41] observed the carbon uptake rate of total phytoplankton during the peak bloom in 2002 (1 May; 1.5 g C m^{-2} day^{-1} for 30 m), and their mean value (0.5 g·C·m^{-2} day^{-1} from 1–22 May) was slightly lower than the values reported in our data. Similarly, Iversen and Sethue [20] and van de Poll et al. [53] observed low values (0.5 g C·m^{-2}·day^{-1}) at the end of May and in June, respectively. The variability in these data reflected large spatial and temporal abundance and the physiological state of phytoplankton and their bloom timing. Previous studies have reported that the carbon uptake rate ranges from 0.01 to 1.5 g C m^{-2} day^{-1} and is very variable in Kongsfjorden; the highest values are observed in April or May and then diminish [16,20,21,41].

Table 4. Investigations of daily carbon uptake rate in Kongsfjorden.

Sampling Period	Daily Carbon Uptake Rate (g C m^{-2} day^{-1})	Reference	Remarks
July 1979 July 1980	0.128 0.024	Eilertsen et al. [21]	Daily mean production rate 76% of noon; calculated by Hop et al. [16]
July 1996	0.8–1.4	Hop et al. [16] and references therein	Night production 32% of daytime production; calculated by Hop et al. [16]
1–22 May 2002	0.466	Hodal et al. [41]	Integrated to 30 m
23–27 July 2002	0.088	Piwosz et al. [28]	Integrated to 30 m, daily production was calculated based on 24 h
18 March 2006 25 April 2006 30 May 2006 4 July 2006 16 September 2006	0.004 0.405 0.445 0.155 0.080	Iversen and Seuthe [20]	Integrated to 50 m
Jun 2015	0.528	van De Poll et al. [53]	Integrated to euphotic zone (defined as the depth interval down to 0.1% irradiance)
4–8 May 2017	0.733	This study	Integrated to 1% light depth

The assimilation index (integrated carbon uptake rates to integrated chl-a concentrations) provides a useful indicator of the physiological state of phytoplankton [38,41,57]. It is mainly dependent on light conditions, nutrient supply, and/or phytoplankton species composition [58]. In Kongsfjorden, 14–36 mg·C (mg chl-a)$^{-1}$ day^{-1} was documented, demonstrating good growth conditions [41]. The value that we calculated for the assimilation index ranged from 19.0 to 47.4 (mg C (mg chl-a)$^{-1}$ day^{-1}), with a mean of 37.1 mg·C (mg chl-a)$^{-1}$ day^{-1} (SD = ± 13.7 mg C (mg chl-a)$^{-1}$ day^{-1}); the integrated value (16~27 m) was higher than that of a marginal ice zone during a bloom in the northern Barents Sea (Hodal and Kristiansen [38]; 3–10 mg C (mg chl-a)$^{-1}$ day^{-1}) and the values of Kongsfjorden during May (mean ± SD = 22.6 ± 8.3 mg C (mg chl-a)$^{-1}$ day^{-1}, integrated to 30 m (Hodal et al. [41]). The result of the assimilation index indicated a phytoplankton community that is not in a poor physiological condition. Despite the good growth conditions, the difference between the daily carbon uptake rate (0.7 g C m^{-2} day^{-1}) in this study and that of Hodal et al. [41] (0.5 g C m^{-2} day^{-1}) could be an indication of a difference between sampling periods. Hodal et al. [41] conducted measurements at one point station and covered peak- and post-bloom values, unlike our study.

In this study, the uptake rates of total nitrogen (nitrate and ammonium) were on average 0.16 ± 0.02 g·N·m^{-2}·day^{-1}, which was similar to that of the non-polynya area of the Amundsen Sea, Antarctica (0.2 g N·m^{-2} day^{-1}; Lee et al. [59]), and slightly lower than those of the Northeast Atlantic fjord, Faroe Islands (May–September; 0.27 g N·m^{-2}·day^{-1}; Gaard et al. [60]). Compared to the rates previously reported in the Arctic Ocean [61–64], our values were relatively high. Lee and Whitledge [61] found that the average total nitrogen uptake rates were 0.03 g N m^{-2} day^{-1} (SD = 0.02 g N·m^{-2}·day^{-1}) in the Canada Basin, whereas Lee et al. [62] obtained various values (ranging from 0.02 to 0.5 g·N·m^{-2}·day^{-1}) from the Chukchi Sea during summer. In contrast, Yun et al. [63] found relatively low average rates (0.005 g·N·m^{-2}·day^{-1}) in the Canada Basin. The lower uptake rates in the Arctic Ocean compared to in the fjord area and the Southern Ocean may be due to the lower nitrogen availability because the Arctic Ocean is a nitrogen-limited system [65].

Overall, the daily total nitrate uptake rates (mean ± SD = 0.09 ± 0.03 g N m^{-2} day^{-1}) were slightly higher than the ammonium uptake rates (mean ± SD = 0.07 ± 0.03 g N m^{-2} day^{-1}), which implies that the ecosystem has potentially near-equal contributions from new and regenerated production to the total productivity within the euphotic zone. Generally, strong new production is observed in the exponential growth phase of phytoplankton, whereas regenerated production appears post-bloom due to the nutrients released by grazing and microbial processes (bacterial degradation and viral lysis) [4]. Considering that zooplankton abundance is low in spring in Kongsfjorden [66,67], the phytoplankton were not in the exponential growth phase, at least during our observation period. However, St. 3 had a somewhat unexpectedly high depth-integrated ammonium uptake rate with the lowest euphotic depth. In particular, the absolute and specific ammonium uptake rates coincided with peaks in ambient ammonium concentration and lower temperature in our study. These results suggest that the nitrogen uptake rate of phytoplankton was more strongly correlated with nutrients (with ammonium being of particular importance) than with light conditions. According to Halbach et al. [68], tidewater glacier and subglacial discharge, Atlantic water, and nutrient release from the seafloor can have both direct and indirect effects on the nutrient dynamics in Kongsfjorden. In this sense, supply from glacial discharge and induced upwelling with ammonium released from the seafloor could be considered possible sources of the observed ammonium at St. 3.

4.3. Contribution of Picophytoplankton to Carbon and Nitrogen Uptake Rates

Although the importance of picophytoplankton (<2 μm) as a major primary producer is well established in oligotrophic [69] and high-latitude environments, little is known about the abundance and carbon uptake rates of the picophytoplankton response to environmental forcing in Kongsfjorden. Recently, the contribution of picophytoplankton to total primary production has been highly variable and was reported to be up to 90% of the carbon fixation in the Atlantic Ocean [70] and 60% of that in the Kara, Laptev, and East Siberian Seas [71]. Picophytoplankton abundance is high in the Arctic Ocean [72,73] and accounts for 60–90% of the chl-a concentration in the eastern Fram Strait [74].

In this study, the depth-integrated carbon uptake rates of the picophytoplankton in Kongsfjorden ranged from 0.1 to 0.4 g C m^{-2} day^{-1} and decreased from the middle zone (St. 7, St. 9, and St. 10) to the inner zone (St. 3). Microphytoplankton (>20 μm) accounted for >60% of the phytoplankton chl-a, whereas picophytoplankton (<2 μm) was a minor primary producer, contributing approximately 20% of the total chl-a. However, the contribution of picophytoplankton to the total carbon uptake rates ranged from 32.6 to 46.5%, with a mean of 42.9 ± 5.9% (Figure 6), which was higher than the contribution of chl-a. In the Arctic Ocean, the mean contribution values for this study were similar to those of Hodal and Kristiansen [38] (46% for <10 μm-sized cells) in the marginal ice zone of the northern Barents Sea during the spring bloom, whereas these values were lower than those of the western Canada Basin (64.0 ± 9.1%; Yun et al. [75]) and higher than those in the Chukchi Sea (31.7 ± 23.6%) for 0.7–5 μm-sized phytoplankton [14].

Despite the low contribution of chl-a, it had a relatively high contribution to production due to the higher particulate organic carbon (POC)-to-chl-a ratio of picophytoplankton (867.8 ± 218.8) than that of larger phytoplankton (>2 μm; 270.6 ± 37.6). This finding was consistent with Lee et al. [14]. These authors reported that chl-a as a proxy for total phytoplankton biomass may not be as effective as a proxy for biomass in picophytoplankton. Additionally, the carbon-to-chl-a ratio of phytoplankton is highly regulated in response to the nutrient, temperature, and irradiance conditions [26]. In our study, we could not find any trend between total carbon uptake rates and the contribution of picophytoplankton chl-a, whereas a small negative correlation (r = −0.60) was observed with the picophytoplankton contribution to POC. Therefore, if we consider carbon to be the basis for phytoplankton biomass, the contribution of picophytoplankton is expected to decrease the total primary production in Kongsfjorden.

The nitrate uptake rates of picophytoplankton ranged from 0.02 to 0.06 g N m^{-2} day^{-1}, with a mean of 0.04 g N m^{-2} day^{-1} (SD = ± 0.01 g N m^{-2} day^{-1}), whereas the ammonium uptake ranged from 0.03 to 0.10 g N m^{-2} day^{-1}, with a mean of 0.05 g N m^{-2} day^{-1} (SD = ± 0.03 g N m^{-2} day^{-1}). Overall, the contributions of picophytoplankton to the total nitrogen uptake rates were 47.1 ± 10.6% for nitrate and 74.0 ± 16.7% for ammonium (Figure 6). Consistent with this observation, the relative preference index (RPI) of picophytoplankton for ammonium and nitrate ranged from 1.3 to 7.0 with a mean of 2.9 (SD = ± 1.4) and 0.1 to 0.9 with a mean of 0.5 (SD = ± 0.3), respectively (Figure 8). The RPI and turnover time values can be valuable tools for estimating the utilization of a nitrogen compound. RPI > 1 indicates positive nutrient selection for phytoplankton nitrogen uptake, and RPI < 1 indicates the opposite. Assuming little change in nitrate and ammonium concentrations during spring and only use by picophytoplankton, the mean turnover time for nitrate and ammonium was 105.8 (SD = ± 197.0 days) and 9.1 days (±6.6 days), respectively (Figure 8). Therefore, the RPI and turnover time of picophytoplankton suggested a predominance of ammonium uptake. This result was consistent with those of several studies [14,48,49,76], which found that large cells used nitrate for growth, whereas small phytoplankton (<5 μm) preferred regenerated nitrogen, such as ammonium. As suggested below, the ambient ammonium concentration may also have been an inhibitor of nitrate uptake.

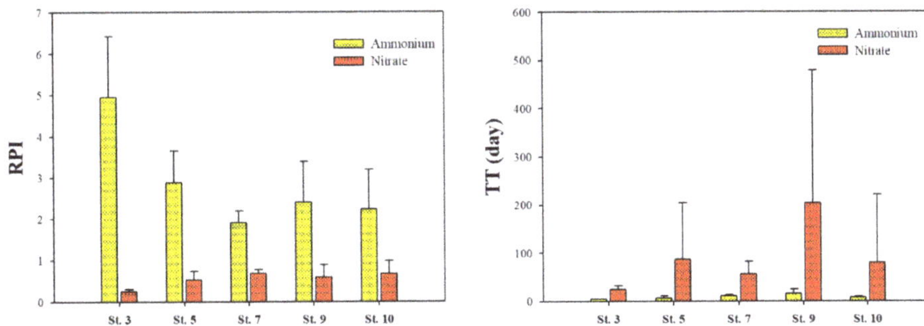

Figure 8. The relative preference index (RPI) and turnover time (TT; day) for picophytoplankton. An RPI > 1 indicates positive nutrient selection for phytoplankton nitrogen uptake.

The regulation and inhibition of the nitrate uptake rate of ammonium have been reported in many studies ([77–79] and references therein). More specifically, Goeyens et al. [77] found that nitrate uptake rates decrease when ammonium stocks exceed 1.7% of the total inorganic nitrogen in Southern Ocean marginal ice zones. Various threshold concentrations of inhibition have been reported in different studies, with a wide range of 0.1 to 2 μM [50,80–83]. In particular, the ammonium inhibition of nitrate uptake by picophytoplankton (<2 μm) was more pronounced than that of large (>2 μm) fractionated phytoplankton in the northeastern basin of the Atlantic [84]. In our study, the contribution of ammonium to the total inorganic nitrogen ranged from 11.1 to 45.0%, with a mean of 21.3% (±10.0%). Regardless of the phytoplankton cell size, there was no significant relationship between the ammonium contribution (%) and the absolute and specific nitrate uptake rates, which differed from the findings of Goeyens et al. [77]. However, as shown in Figure 6, the highest ammonium and lowest nitrate uptake rates of picophytoplankton at St. 3 in the fjord coincided with a relatively high concentration (approximately 1.5 μM) and contribution (39.2 ± 6.5%) of ammonium to the total inorganic nitrogen. Our results agreed with previous studies that reported that the ammonium uptake rate at >1 μM ammonium was higher than that at concentrations below 1 μM [50,77,79,82]. Hence, it was speculated that a major determinant of ammonium and nitrate uptake rates in this region may be the result of the ammonium concentration, suggesting that ammonium levels are capable of suppressing nitrate uptake.

However, our samples (<2 μm) contained not only phytoplankton but also microorganisms, such as bacteria, which can play a role in nitrogen uptake. In our study, the bacterial biomass within the euphotic zone was not quantified, but to verify this hypothesis, we compared the δ^{13}C (−27.1 ± 1.4% and −26.2 ± 0.5% in total and pico-POC, respectively) and the carbon-to-nitrogen molar ratio (C/N) values (9.2 ± 1.7 and 9.6 ± 1.4) of the POC in this study. The observed values were within the ranges that were previously published. According to earlier studies, generally, the δ^{13}C and C/N values of phytoplankton ranged between −27.5 and −25.6% and from 6 to 10, respectively ([85–88] and references therein). In addition, heterotrophic systems usually have low nutrient concentrations or nutrient-limited conditions. Based on the δ^{13}C, C/N and nutrient conditions, we assumed that POC was mainly derived from phytoplankton in this study [14,64,87–90]. However, Fouilland et al. [91] reported that heterotrophic bacteria accounted for a large portion of the total uptake of nitrogen (up to 78%) in the North Water polynya. Nitrate and ammonium uptake by heterotrophic bacteria accounted for 4–14% and 22–39% during the North Atlantic spring bloom, respectively (May 1989; Kirchman et al. [92]).

In Kongsfjorden, the heterotrophic bacteria accounted for 17% of the total POC in May [20]. Therefore, further investigations are needed to understand the regulation of nitrogen uptake by phytoplankton and bacteria in fjord ecosystems, as well as their interaction under nitrogen-limited conditions.

5. Summary and Conclusions

This present study investigated the carbon and nitrogen uptake rates of phytoplankton and the contribution of picophytoplankton to total production in Kongsfjorden. Our results suggest that low light availability by glacier melt inputs with high turbidity has an important negative effect on springtime primary production, restricting large phytoplankton productivity. The picophytoplankton contributed most to the uptake of ammonium (74.0 ± 16.7%) than nitrate (47.1 ± 10.6%) and carbon (42.9 ± 5.9%). This finding indicated that picophytoplankton was largely based on regenerated nutrients (ammonium as a nitrogen source) and a slightly lower assimilation of carbon than in large phytoplankton (>2 μm). During the study period, diatoms and *Phaeocystis* sp. were clearly the most important producers in Kongsfjorden, but if they were replaced by the nano- and picophytoplankton size classes, then the fjord ecosystem would become less productive, which would not sustain higher trophic levels of biomass (e.g., zooplankton).

Author Contributions: S.-Y.H. conceived of the study, participated in its design and helped to draft the manuscript. B.K.K. drafted the manuscript and performed the field and laboratory experiments. H.M.J. and J.J. carried out the analysis of the phytoplankton taxonomy, chl-a, and nutrients. B.L. critically reviewed the manuscript. All authors have read and agreed to the published version of the manuscript.

Funding: This study was supported by the Korea Polar Research Institute (KOPRI) and undertaken as part of "Carbon assimilation rate of sea ice ecosystem in the Kongsfjorden MIZ, Arctic (PE18170)" and "Adaptation and Assessment of coastal marine (benthic–pelagic) ecosystem impacted by rapid glacier retreat, Antarctica (PE20120)".

Conflicts of Interest: The authors declare no conflict of interest.

References

1. Azam, F. Oceanography: Microbial Control of Oceanic Carbon Flux: The Plot Thickens. *Science* **1998**, *280*, 694–696. [CrossRef]
2. Arrigo, K.R. Marine microorganisms and global nutrient cycles. *Nature* **2005**, *437*, 349–355. [CrossRef] [PubMed]
3. Falkowski, P.G.; Raven, J.A. *Aquatic Photosynthesis*, 2nd ed.; Princeton University Press: Princeton, NJ, USA, 2007.
4. Dugdale, R.C.; Goering, J.J. Uptake of new and regenerated forms of nitrogen in primary productivity. *Limnol. Oceanogr.* **1967**, *12*, 196–206. [CrossRef]
5. Eppley, R.W.; Peterson, B.J. Particulate organic matter flux and planktonic new production in the deep ocean. *Nature* **1979**, *282*, 677–680. [CrossRef]
6. Ducklow, H.; Steinberg, D.; Buesseler, K. Upper Ocean Carbon Export and the Biological Pump. *Oceanography* **2001**, *14*, 50–58. [CrossRef]
7. Wassmann, P.; Slagstad, D.; Riser, C.W.; Reigstad, M. Modelling the ecosystem dynamics of the Barents Sea including the marginal ice zone. *J. Mar. Syst.* **2006**, *59*, 1–24. [CrossRef]
8. Olli, K.; Wassmann, P.; Reigstad, M.; Ratkova, T.N.; Arashkevich, E.; Pasternak, A.; Matrai, P.A.; Knulst, J.; Tranvik, L.J.; Klais, R.; et al. The fate of production in the central Arctic Ocean—top–down regulation by zooplankton expatriates? *Prog. Oceanogr.* **2007**, *72*, 84–113. [CrossRef]
9. Richardson, T.L.; Jackson, G.A. Small Phytoplankton and Carbon Export from the Surface Ocean. *Science* **2007**, *315*, 838–840. [CrossRef]
10. Henson, S.A.; Le Moigne, F.A.C.; Giering, S.L.C. Drivers of Carbon Export Efficiency in the Global Ocean. *Glob. Biogeochem. Cycles* **2019**, *33*, 891–903. [CrossRef]
11. Li, W.K.W.; McLaughlin, F.A.; Lovejoy, C.; Carmack, E.C. Smallest Algae Thrive as the Arctic Ocean Freshens. *Science* **2009**, *326*, 539. [CrossRef]
12. Comiso, J.C.; Hall, D.K. Climate trends in the Arctic as observed from space. *Wiley Interdiscip. Rev. Clim. Chang.* **2014**, *5*, 389–409. [CrossRef] [PubMed]

13. Morán, X.A.G.; López-Urrutia, Á.; Calvo-Díaz, A.; Li, W.K.W. Increasing importance of small phytoplankton in a warmer ocean. *Glob. Chang. Biol.* **2010**, *16*, 1137–1144. [CrossRef]

14. Lee, S.H.; Yun, M.S.; Kim, B.K.; Joo, H.; Kang, S.-H.; Chang-Keun, K.; Whitledge, T.E. Contribution of small phytoplankton to total primary production in the Chukchi Sea. *Cont. Shelf Res.* **2013**, *68*, 43–50. [CrossRef]

15. Hopkins, T.S. The GIN Sea—A synthesis of its physical oceanography and literature review 1972–1985. *Earth Sci. Rev.* **1991**, *30*, 175–319. [CrossRef]

16. Hop, H.; Pearson, T.; Hegseth, E.N.; Kovacs, K.M.; Wiencke, C.; Kwasniewski, S.; Eiane, K.; Mehlum, F.; Gulliksen, B.; Wlodarska-Kowalczuk, M.; et al. The marine ecosystem of Kongsfjorden, Svalbard. *Polar Res.* **2002**, *21*, 167–208. [CrossRef]

17. Payne, C.M.; Roesler, C. Characterizing the influence of Atlantic water intrusion on water mass formation and phytoplankton distribution in Kongsfjorden, Svalbard. *Cont. Shelf Res.* **2019**, *191*, 104005. [CrossRef]

18. Svendsen, H.; Beszczynska-Møller, A.; Hagen, J.O.; Lefauconnier, B.; Tverberg, V.; Gerland, S.; Ørbæk, J.B.; Bischof, K.; Papucci, C.; Zajaczkowski, M.; et al. The physical environment of Kongsfjorden-Krossfjorden, an Arctic fjord system in Svalbard. *Polar Res.* **2002**, *21*, 133–166.

19. Stempniewicz, L.; Błachowiak-Samołyk, K.; Węsławski, J.M. Impact of climate change on zooplankton communities, seabird populations and arctic terrestrial ecosystem—A scenario. *Deep Sea Res. Part II* **2007**, *54*, 2934–2945. [CrossRef]

20. Iversen, K.R.; Seuthe, L. Seasonal microbial processes in a high-latitude fjord (Kongsfjorden, Svalbard): I. Heterotrophic bacteria, picoplankton and nanoflagellates. *Polar Biol.* **2011**, *34*, 731–749. [CrossRef]

21. Eilertsen, H.C.; Taasen, J.P.; Weslawski, J.M. Phytoplankton studies in the fjords of West Spitzbergen: Physical environment and production in spring and summer. *J. Plankton Res.* **1989**, *11*, 1245–1260. [CrossRef]

22. Richardson, A.J.; Schoeman, D.S. Climate impact on plankton ecosystems in the Northeast Atlantic. *Science* **2004**, *305*, 1609–1612. [CrossRef] [PubMed]

23. Trudnowska, E.; Basedow, S.L.; Blachowiak-Samolyk, K. Mid-summer mesozooplankton biomass, its size distribution, and estimated production within a glacial Arctic fjord (Hornsund, Svalbard). *J. Mar. Syst.* **2014**, *137*, 55–66. [CrossRef]

24. Stroeve, J.; Markus, T.; Meier, W.N.; Miller, J. Recent changes in the Arctic melt season. *Ann. Glaciol.* **2006**, *44*, 367–374. [CrossRef]

25. Serreze, M.C.; Holland, M.M.; Stroeve, J. Perspectives on the Arctic's Shrinking Sea-Ice Cover. *Science* **2007**, *315*, 1533–1536. [CrossRef] [PubMed]

26. Wang, G.Z.; Guo, C.Y.; Luo, W.; Cai, M.H.; He, J.F. The distribution of picoplankton and nanoplankton in Kongsfjorden, Svalbard during late summer 2006. *Polar Biol.* **2009**, *32*, 1233–1238. [CrossRef]

27. Sinha, R.K.; Krishnan, K.P.; Hatha, A.A.M.; Rahiman, M.; Thresyamma, D.D.; Kerkar, S. Diversity of retrievable heterotrophic bacteria in Kongsfjorden, an Arctic fjord. *Environ. Microbiol.* **2017**, *48*, 51–61. [CrossRef]

28. Piwosz, K.; Walkusz, W.; Hapter, R.; Wieczorek, P.; Hop, H.; Wiktor, J. Comparison of productivity and phytoplankton in a warm (Kongsfjorden) and a cold (Hornsund) Spitsbergen fjord in mid-summer 2002. *Polar Biol.* **2009**, *32*, 549–559. [CrossRef]

29. Poole, H.H.; Atkins, W.R.G. Photo-electric measurements of submarine illumination throughout the year. *J. Mar. Biol. Assoc. UK* **1929**, *16*, 297–324. [CrossRef]

30. Parsons, T.R.; Maita, Y.; Lalli, C.M. *A Manual of Chemical and Biological Methods for Seawater Analysis*; Pergamon Press: Oxford, UK, 1984; 173p.

31. Bérard-Therriault, L.; Poulin, M.; Bossé, L. *Guide D'identification du Phytoplancton Marin de L'estuaire et du Golfe du Saint-Laurent Incluant également Certains Protozoaires*; Conseil National de Researches du Canada: Ottawa, ON, Canada, 1999.

32. Booth, B.C. Estimating cell concentration and biomass of autotrophic plankton using microscopy. In *Handbook of Methods in Aquatic Microbial Ecology*; Kemp, P.F., Sherr, B.F., Sherr, E.B., Cole, J.J., Eds.; Lewis Publishers: Boca Raton, FL, USA, 1993; pp. 199–205.

33. Tomas, C.R. *Identifying Marine Phytoplankton*; Academic Press: San Diego, CA, USA, 1997.

34. Kang, S.H.; Fryxell, G.A.; Roelke, D.L. *Fragilariopsis cylindrus* compared with other species of the diatom family Bacillariaceae in Antarctic marginal ice-edge zones. *Nova Hedwig. Beih.* **1993**, *106*, 335–352.

35. Garneau, M.-È.; Gosseling, M.; Klein, B.; Tremblay, J.-È.; Fouilland, E. New and regenerated production during a late summer bloom in an Arctic polynya. *Mar. Ecol. Prog. Ser.* **2007**, *345*, 13–26. [CrossRef]

36. Dugdale, R.C.; Wilkerson, F.P. The use of ^{15}N to measure nitrogen uptake in eutrophic ocean; experimental considerations. *Limnol. Oceanogr.* **1986**, *31*, 637–689. [CrossRef]

37. Hama, T.; Miyazaki, T.; Ogawa, Y.; Iwakuma, T.; Takahashi, M. Measurement of Photosynthetic Production of a Marine Phytoplankton Population Using a STable 13C Isotope. *Mar. Biol.* **1983**, *73*, 31–36. [CrossRef]

38. Hodal, H.; Kristiansen, S. The importance of small-called phytoplankton in spring blooms at the marginal ice zone in the northern Barents Sea. *Deep Sea Res. Part II* **2008**, *55*, 2176–2185. [CrossRef]

39. McCarthy, J.J.; Taylor, W.R.; Taft, J.L. Nitrogenous nutrition of the plankton in the Chesapeake Bay. I. Nutrient availability and phytoplankton preferences. *Limnol. Oceanogr.* **1977**, *22*, 996–1011. [CrossRef]

40. Gu, B.; Alexander, V. Dissolved nitrogen uptake by a Cyanobacterial bloom (*Anabaena flos-aquae*) in a subarctic lake. *J. Appl. Environ. Microbiol.* **1993**, *59*, 422–430. [CrossRef] [PubMed]

41. Hodal, H.; Falk-Petersen, S.; Hop, H.; Kristiansen, S.; Reigstad, M. Spring bloom dynamics in Kongsfjorden, Svalbard: Nutrients, phytoplankton, protozoans and primary production. *Polar Biol.* **2012**, *35*, 191–203. [CrossRef]

42. Hegseth, E.N.; Tverberg, V. Effect of Atlantic water inflow on timing of the phytoplankton spring bloom in a high Arctic fjord (Kongsfjorden, Svalbard). *J. Mar. Syst.* **2013**, *113–114*, 94–105. [CrossRef]

43. von Quillfeldt, C.H. Common Diatom Species in Arctic Spring Blooms: Their Distribution and Abundance. *Bot. Mar.* **2000**, *43*, 499–516. [CrossRef]

44. Piquet, A.M.-T.; van de Poll, W.H.; Visser, R.J.W.; Wiencke, C.; Bolhuis, H.; Buma, A.G.J. Springtime phytoplankton dynamics in Arctic Krossfjorden and Kongsfjorden (Spitsbergen) as a function of glacier proximity. *Biogeosciences* **2014**, *11*, 2263–2279. [CrossRef]

45. Redifield, A.C. The biological control of chemical factors in the environment. *Am. Sci.* **1958**, *46*, 205–221.

46. Thingstad, T.F.; Sakshaug, E. Control of phytoplankton growth in nutrient recycling ecosystems. Theory and Terminology. *Mar. Ecol. Prog. Ser.* **1990**, *63*, 261–272. [CrossRef]

47. van De Poll, W.H.; Maat, D.S.; Fischer, P.; Rozema, P.D.; Daly, O.B.; Koppelle, S.; Visser, R.J.W.; Buma, A.G.J. Atlantic advection driven changes in glacial meltwater: Effects on phytoplankton chlorophyll a and taxonomic composition in Kongsfjorden, Spitsbergen. *Front. Mar. Sci.* **2016**, *3*, 200. [CrossRef]

48. Probyn, T.A. Nitrogen uptake by size-fractionated phytoplankton populations in the southern Benguela upwelling system. *Mar. Ecol. Prog. Ser.* **1985**, *22*, 249–258. [CrossRef]

49. Koike, I.; Holm-Hansen, O.; Biggs, D.C. Inorganic nitrogen metabolism by Antarctic phytoplankton with special reference to ammonium cycling. *Mar. Ecol. Prog. Ser.* **1986**, *30*, 105–116. [CrossRef]

50. Dugdale, R.C.; Wilkerson, F.P.; Hogue, V.E.; Marchi, A. The role of ammonium and nitrate in spring bloom development in San Francisco Bay. *Estuar. Coast. Shelf Sci.* **2007**, *73*, 17–29. [CrossRef]

51. Keck, A.; Wiktor, J.; Hapter, R.; Nilsen, R. Phytoplankton assemblages related to physical gradients in an arctic, glacier-fed fjord in summer. *ICES J. Mar. Sci.* **1999**, *56*, 203–214. [CrossRef]

52. Hop, H.; Falk-Petersen, S.; Svendsen, H.; Kwasniewski, S.; Pavlov, V.; Pavlova, O.; Søreide, J.E. Physical and biological characteristics of the pelagic system across Fram Strait to Kongsfjorden. *Prog. Oceanogr.* **2006**, *71*, 182–231. [CrossRef]

53. Van de Poll, W.H.; Kulk, G.; Rozema, P.D.; Brussaard, C.P.D.; Visser, R.J.W.; Buma, A.G.J. Contrasting glacial meltwater effects on post-bloom phytoplankton on temporal and spatial scales in Kongsfjorden, Spitsbergen. *Elem. Sci. Anthr.* **2018**, *6*, 50. [CrossRef]

54. Raven, J.A. The twelfth Tansley Lecture. Small is beautiful: The picophytoplankton. *Funct. Ecol.* **1998**, *12*, 503–513. [CrossRef]

55. Callieri, C. Picophytoplankton in freshwater ecosystems: The importance of small-sized phototrophs. *Freshw. Rev.* **2008**, *1*, 1–28. [CrossRef]

56. Finkel, Z.V.; Beardall, J.; Flynn, K.J.; Quigg, A.; Rees, T.A.V.; Raven, J.A. Phytoplankton in a changing world: Cell size and elemental stoichiometry. *J. Plankton Res.* **2010**, *32*, 119–137. [CrossRef]

57. Lourey, M.J.; Thompson, P.A.; McLaughlin, M.J.; Bonham, P.; Feng, M. Primary production and phytoplankton community structure during a winter shelf-scale phytoplankton bloom off Western Australia. *Mar. Biol.* **2013**, *160*, 335–369. [CrossRef]

58. Huot, Y.; Babin, M.; Bruyant, F.; Grob, C.; Twardowski, M.; Claustre, H. Relationship between photosynthetic parameters and different proxies of phytoplankton biomass in the subtropical ocean. *Biogeosciences* **2017**, *4*, 853–868. [CrossRef]

59. Lee, S.H.; Kim, B.K.; Yun, M.S.; Joo, H.; Yang, E.J.; Kim, Y.N.; Shin, H.C.; Lee, S. Spatial distribution of phytoplankton productivity in the Amundsen Sea, Antarctica. *Polar Biol.* **2012**, *35*, 1721–1733. [CrossRef]

60. Gaard, E.; Norði, Á.G.; Simonsen, K. Environmental effects on phytoplankton production in a Northeast Atlantic fjord, Faroe Islands. *J. Plankton Res.* **2011**, *33*, 947–959. [CrossRef]

61. Lee, S.H.; Whitledge, T.E. Primary and new production in the deep Canada Basin during summer 2002. *Polar Biol.* **2005**, *28*, 190–197. [CrossRef]

62. Lee, S.H.; Whitledge, T.E.; Kang, S.-H. Recent carbon and nitrogen uptake rates of phytoplankton in Bering Strait and the Chukchi Sea. *Cont. Shelf Res.* **2007**, *27*, 2231–2249. [CrossRef]

63. Yun, M.S.; Chung, K.H.; Zimmermann, S.; Zhao, J.; Joo, H.M.; Lee, S.H. Phytoplankton productivity and its response to higher light levels in the Canada Basin. *Polar Biol.* **2012**, *35*, 257–268. [CrossRef]

64. Lee, S.H.; Lee, J.H.; Lee, H.; Kang, J.J.; Lee, J.H.; Lee, D.; An, S.H.; Stockwell, D.A.; Whitledge, T.E. Field-obtained carbon and nitrogen uptake rates of phytoplankton in the Laptev and East Siberian seas. *Biogeosci. Discuss.* **2017**, 1–32. [CrossRef]

65. Harrison, W.G.; Cota, G.F. Primary production in polar waters: Relation to nutrient availability. *Polar Res.* **1991**, *10*, 87–104. [CrossRef]

66. Willis, K.J.; Cottier, F.R.; Kwasniewski, S. Impact of warm water advection on the winter zooplankton community in an Arctic fjord. *Polar Biol.* **2008**, *31*, 475–481. [CrossRef]

67. Walkusz, W.; Kwasniewski, S.; Falk-Petersen, S.; Hop, H.; Tverberg, V.; Wieczorek, P.; Weslawski, J.M. Seasonal and spatial changes in zooplankton composition in the glacially influenced Kongsfjorden, Svalbard. *Polar Res.* **2009**, *28*, 254–281. [CrossRef]

68. Halbach, L.; Vihtakari, M.; Duarte, P.; Everett, A.; Granskog, M.A.; Hop, H.; Kauko, H.M.; Kristiansen, S.; Myhre, P.I.; Pavlov, A.K.; et al. Tidewater Glaciers and Bedrock Characteristics Control the Phytoplankton Growth Environment in a Fjord in the Arctic. *Front. Mar. Sci.* **2019**, *6*, 254. [CrossRef]

69. Partensky, F.; Garczarek, L. *Prochlorococcus*: Advantages and limits of minimalism. *Annu. Rev. Mar. Sci.* **2010**, *2*, 305–331. [CrossRef]

70. Poulton, A.J.; Holligan, P.M.; Hickman, A.; Kim, Y.-N.; Adey, T.R.; Stinchcombe, M.C.; Holeton, C.; Root, S.; Woodward, E.M.S. Phytoplankton carbon fixation, chlorophyll-biomass and diagnostic pigments in the Atlantic Ocean. *Deep Sea Res. Part II* **2006**, *53*, 1593–1610. [CrossRef]

71. Bhavya, P.S.; Lee, J.H.; Lee, H.W.; Kang, J.J.; Lee, J.H.; Lee, D.; An, S.H.; Stockwell, D.A.; Whitledge, T.E.; Lee, S.H. First in situ estimations of small phytoplankton carbon and nitrogen uptake rates in the Kara, Laptev, and East Siberian seas. *Biogeosciences* **2018**, *15*, 5503–5517. [CrossRef]

72. Tremblay, G.; Belzile, C.; Gosselin, M.; Poulin, M.; Roy, S.; Tremblay, J.-É. Late summer phytoplankton distribution along a 3500 km transect in Canadian Arctic waters: Strong numerical dominance by picoeukaryotes. *Aquat. Microb. Ecol.* **2009**, *54*, 55–70. [CrossRef]

73. Ardyna, M.; Gosselin, M.; Michel, C.; Poulin, M.; Tremblay, J.-É. Environmental forcing of phytoplankton community structure and function in the Canadian High Arctic: Contrasting oligotrophic and eutrophic regions. *Mar. Ecol.* **2011**, *442*, 37–57. [CrossRef]

74. Metfies, K.; von Appen, W.-J.; Kilias, E.; Nicolaus, A.; Nöthig, E.-M. Biogeography and Photosynthetic Biomass of Arctic Marine Pico-Eukaroytes during Summer of the Record Sea Ice Minimum 2012. *PLoS ONE* **2016**, *11*, e0148512. [CrossRef]

75. Yun, M.S.; Kim, B.K.; Joo, H.T.; Yang, E.J.; Nishino, S.; Chung, K.H.; Kang, S.-H.; Lee, S.H. Regional productivity of phytoplankton in the Western Arctic Ocean during summer in 2010. *Deep Sea Res. Part II* **2015**, *120*, 61–71. [CrossRef]

76. Tremblay, J.-É.; Legendre, L.; Klein, B.; Therriault, J.-C. Size-differential uptake of nitrogen and carbon in a marginal sea (Gulf of St. Lawrence, Canada): Significance of diel periodicity and urea uptake. *Deep Sea Res. Part II* **2000**, *47*, 489–518. [CrossRef]

77. Goeyens, L.; Tréguer, P.; Baumann, M.E.M.; Baeyens, W.; Dehairs, F. The leading role of ammonium in the nitrogen uptake regime of Southern Ocean marginal ice zones. *J. Mar. Syst.* **1995**, *6*, 345–361. [CrossRef]

78. Xu, J.; Glibert, P.M.; Liu, H.; Yin, K.; Yuan, X.; Chen, M.; Harrision, P. Nitrogen sources and rates of phytoplankton uptake in different regions of Hong Kong Waters in summer. *Estuar. Coast.* **2012**, *35*, 559–571. [CrossRef]

79. Kim, B.K.; Joo, H.T.; Song, H.J.; Yang, E.J.; Lee, S.H.; Hahm, D.; Rhee, T.S.; Lee, S.H. Large seasonal variation in phytoplankton production in the Amundsen Sea. *Polar Biol.* **2015**, *38*, 319–331. [CrossRef]

80. MacIsaac, J.J.; Dugdale, R.C. The kinetics of nitrate and ammonia uptake by natural populations of marine phytoplankton. *Deep Sea Res.* **1969**, *16*, 45–57. [CrossRef]

81. Paasche, E.; Kristiansen, S. Nitrogen nutrition of the phytoplankton in the Oslofjord. *Estuar. Coast. Shelf Sci.* **1982**, *14*, 237–249. [CrossRef]

82. Dortch, Q. The interaction between ammonium and nitrate uptake in phytoplankton. *Mar. Ecol. Prog. Ser.* **1990**, *61*, 183–201. [CrossRef]

83. Cochlan, W.P.; Harrison, P.J. Uptake of nitrate, ammonium, and urea by nitrogen-starved cultures of *Micromonas pusilla* (Prasinophyceae): Transient responses. *J. Phycol.* **1991**, *27*, 673–679. [CrossRef]

84. L'Helguen, S.; Maguer, J.-F.; Caradec, J. Inhibition kinetics of nitrate uptake by ammonium in size-fractionated oceanic phytoplankton communities: Implications for new production and f-ratio estimates. *J. Plankton Res.* **2008**, *30*, 1179–1188. [CrossRef]

85. Wada, E.; Terazaki, M.; Kabaya, Y.; Nemoto, T. ^{15}N and ^{13}C abundances in the Antarctic Ocean with emphasis on the biogeochemical structure of the food web. *Deep Sea Res. Part A* **1987**, *34*, 829–841. [CrossRef]

86. Corbisier, T.N.; Petti, M.A.V.; Skowronski, R.S.P.; Brito, T.A.S. Trophic relationships in the nearshore zone of Martel Inlet (King George Island, Antarctica): δ^{13}C stable-isotope analysis. *Polar Biol.* **2004**, *27*, 75–82. [CrossRef]

87. Kim, B.K.; Lee, J.H.; Joo, H.T.; Song, H.J.; Yang, E.J.; Lee, S.H.; Lee, S.H. Macromolecular compositions of phytoplankton in the Amundsen Sea, Antarctica. *Deep Sea Res. Part II* **2016**, *123*, 42–49. [CrossRef]

88. Baumann, M.E.M.; Lancelot, C.; Brandini, F.P.; Sakshaug, E.; John, D.M. The taxonomic identity of the cosmopolitan prymnesiophyte *Phaeocystis*: A morphological and ecophysiological approach. *J. Mar. Syst.* **1994**, *5*, 5–22. [CrossRef]

89. Kendall, C.; Silva, S.R.; Kelly, V.J. Carbon and nitrogen isotopic compositions of particulate organic matter in four large river systems across the United States. *Hydrol. Process.* **2001**, *15*, 1301–1346. [CrossRef]

90. Schoemann, V.; Becquevort, S.; Stefels, J.; Rousseau, V.; Lancelot, C. *Phaeocystis* blooms in the global ocean and their controlling mechanisms: A review. *J. Sea Res.* **2005**, *53*, 43–66. [CrossRef]

91. Fouilland, E.; Gosselin, M.; Rivkin, R.B.; Vasseur, C.; Mostajir, B. Nitrogen uptake by heterotrophic bacteria and phytoplankton in Arctic surface waters. *J. Plankton Res.* **2007**, *29*, 369–376. [CrossRef]

92. Kirchman, D.L.; Ducklow, H.W.; McCarthy, J.J.; Garside, C. Biomass and nitrogen uptake by heterotrophic bacteria during the spring phytoplankton bloom in the North Atlantic Ocean. *Deep Sea Res. Part I* **1994**, *41*, 879–895. [CrossRef]

Publisher's Note: MDPI stays neutral with regard to jurisdictional claims in published maps and institutional affiliations.

Article

Physiological Changes and Elemental Ratio of *Scrippsiella trochoidea* and *Heterosigma akashiwo* in Different Growth Phase

Xiaofang Liu [1,†], Yang Liu [2,†], Md Abu Noman [3], Satheeswaran Thangaraj [3] and Jun Sun [3,*]

1   Research Centre for Indian Ocean Ecosystem, Tianjin University of Science and Technology, Tianjin 300457, China; liuxiaofang@mail.tust.edu.cn
2   Institute of Marine Science and Technology, Shandong University, Qingdao 266237, China; youngliu@mail.sdu.edu.cn
3   College of Marine Science and Technology, China University of Geosciences, Wuhan 430074, China; abu.noman.nstu@gmail.com (M.A.N.); satheeswaran1990@gmail.com (S.T.)
*   Correspondence: phytoplankton@163.com; Tel.: +86-22-6060-1116
†   Xiaofang Liu and Yang Liu contributed equally to this work.

Citation: Liu, X.; Liu, Y.; Noman, M.A.; Thangaraj, S.; Sun, J. Physiological Changes and Elemental Ratio of *Scrippsiella trochoidea* and *Heterosigma akashiwo* in Different Growth Phase. *Water* **2021**, *13*, 132. https://doi.org/10.3390/w13020132

Received: 22 October 2020
Accepted: 3 January 2021
Published: 8 January 2021

Abstract: The elemental ratios in phytoplankton are important for predicting biogeochemical cycles in the ocean. However, understanding how these elements vary among different phytoplankton taxa with physiological changes remains limited. In this paper, we determine the combined physiological–elemental ratio changes of two phytoplankton species, *Scrippsiella trochoidea* (Dinophyceae) and *Heterosigma akashiwo* (Raphidophyceae). Our results show that the cell growth period of *S. trochoidea* (26 days) was significantly shorter than that of *H. akashiwo* (32 days), with an average cell abundance of 1.21×10^4 cells·mL^{-1} in *S. trochoidea* and 1.53×10^5 cells·mL^{-1} in *H. akashiwo*. The average biovolume of *S. trochoidea* (9.71×10^3 μm^3) was higher than that of *H. akashiwo* (0.64×10^3 μm^3). The physiological states of the microalgae were assessed based on elemental ratios. The average ratios of particulate organic nitrogen (PON) to chlorophyll-a (Chl-*a*) and particulate organic carbon (POC) to Chl-*a* in *S. trochoidea* (57.32 and 168.16) were higher than those of *H. akashiwo* (9.46 and 68.86); however, the ratio of POC/PON of the two microalgae was nearly equal (6.33 and 6.17), indicating that POC/Chl-*a* may be lower when the cell is actively growing. The physiological variation, based on the POC/Chl-*a* ratio, in different phytoplankton taxa can be used to develop physiological models for phytoplankton, with implications for the marine biogeochemical cycle.

Keywords: *Scrippsiella trochoidea*; *Heterosigma akashiwo*; biovolume; chlorophyll-*a*; particulate organic nitrogen; particulate organic carbon

1. Introduction

Phytoplankton is the basis of the marine food web, accounting for more than 50% of the primary production [1]. The cell size, chlorophyll-*a* (Chl-*a*) content, and elemental composition of microalgae (i.e., organic carbon and organic nitrogen) are important indicators for assessing biomass, as they have strong influences on the food web structure and the biogeochemical cycling of carbon, as well as cell growth rates [2–4]. However, these indicators vary with physical environmental parameters (i.e., irradiance and nutrient availability) and the growth phase of microalgae [5,6]. Such variations in physiological changes and elemental ratios depend on how efficiently phytoplankton biomass is remineralized by bacteria, deep ocean export flux, or utilization by consumers [7,8]. In such cases, these variations may provide a mechanistic basis for modeling phytoplankton growth, elemental ratios, and quantifying the flow energy among organisms in an ecosystem [9].

Estimation of chlorophyll is an easy technique to assess the fundamental photosynthetic performance (mainly in photosystems II) of microalgae under various environmental conditions, which reflects the photosynthetic performance of the microalgae [10]. Therefore, it can be used as a tool to estimate phytoplankton productivity. The elemental composition of cells and the content of their constituent organic matters are two important physiological indicators for phytoplankton [11]. Particulate organic matters (POM) can influence processes such as transport, transformation, and removal of elements, which subsequently determine biological life processes and primary productivity. Thus, it is an important parameter for evaluating marine productivity [12]. Therefore, we can provide a preliminary view for evaluating the productivity of the sea based on the analysis of physiological parameters and elemental ratios.

Phytoplankton produces organic carbon by absorbing nutrients from seawater and, in turn, nutrients are returned to the seawater through remineralization. It is well-understood that studies on particulate organic carbon and nitrogen (POC/PON) have great significance for productivity modeling and the quantification of carbon in the ocean [13]. However, particulate organic nitrogen (PON) and particulate organic carbon (POC) in microalgae cells cannot be distinguished from zooplankton and non-living organic matter, so their contents are not directly available in the ocean [14,15]. Therefore, previous studies have commonly used chlorophyll content to estimate changes in intracellular carbon content, as it accurately reflects microalgae growth and photosynthetic rates [16,17]. Interestingly, the ratio between POC and Chl-*a* varies. Earlier studies have demonstrated that POC/Chl-*a* ratios are influenced by nitrogen concentrations [5]. This ratio decreases when phytoplankton are exposed to low light levels, as well as nutrient-rich zones [18,19]. Furthermore, the POC/Chl-*a* ratio has been found to be lower in an estuary, compared to the open ocean [5]. Moreover, Sathyendranath et al. studied the POC/Chl-*a* of phytoplankton using high-performance liquid chromatography (HPLC) and observed that the ratio ranged from 15 to 176 [20]. Most recent studies have examined POC/Chl-*a* and elemental content changes in phytoplankton with respect to the changes in coastal ecosystems, especially focusing on seasonal variation. However, there is scant information regarding the changes in carbon and nitrogen content during phytoplankton growth [21].

S. trochoidea (Dinophyceae) and *H. akashiwo* (Raphidophyceae) are two widely distributed harmful algal species in Chinese coastal waters. They have significant physiological differences (e.g., cell size and growth) due to variation in their taxonomic positions. Most studies have focused on growth rates and biomass under different environmental conditions, such as temperature or nutrients. For example, Qi Yu et al. studied the life history of *S. trochoidea* and found that the emergence of cysts and its reproduction rate were closely related to nitrogen concentration [22]. Furthermore, Xiao et al. pointed out a relationship between the cysts of *S. trochoidea* and vegetative cells in Daya Bay, showing that formation of cysts led to a reduction in vegetative cells [23]. Futhermore, some investigations have explained the effects of physicochemical methods for *S. trochoidea* under environmental conditions. These methods include heating, ultraviolet (UV) radiation, and others [24]. Similarly, a variation in pH also influenced the carbon assimilation and acquisition of *H. akashiwo*, along with physiological changes of growth rate [25]. Despite the fact that these studies have provided significant contributions, very few studies have examined the elemental ratios of these two important species in different stages of their growth phase [26,27].

The objective of this study was to explore the variations in POC and PON during the entire growth cycles of *S. trochoidea* and *H. akashiwo*. To accomplish this, we estimated the physiological changes (cell abundance, cell volume, Chl-*a*), element ratios (POC and PON), and inferred POC/Chl-*a* ratios from *S. trochoidea* and *H. akashiwo* during different life phases. We hope that our results provide comprehensive information on how elemental composition varies with the size difference of phytoplankton. This may be useful for marine biogeochemical modeling.

2. Material and Methods

2.1. Microalgal Species and Pre-Culture Conditions

S. trochoidea and *H. akashiwo* used in the present study were isolated, in 96-well microplates, from the coastal waters of Qinhuang Island, China. They were subsequently identified and maintained in our laboratory. Microalgal cells were cultured in f/2 medium [28] under the following culture conditions: 25 ± 1 °C; 12 h:12 h for light-dark ratio; and 200 ± 10 µmol photons m^{-2} s^{-1} for irradiance, which ensured that they were not limited by light intensity.

2.2. Determination of Microalgal Growth

Microalgae cells were collected in a sterile bench when they reached the exponential phase of growth. The collected microalgae cells were then centrifuged at 5000 rpm for 5 min and washed with sterile seawater for another 5 min for the same treatment [29]. In this way, we removed the interference of the original medium on the concentration of the new medium. The collected cells were then transferred into 2000 mL conical flasks containing 1500 mL of f/2 medium, again under the above-mentioned light conditions. The initial cell abundances of the two microalgal cell cultures were approximately 360 cells·mL^{-1} and 3000 cells·mL^{-1} at the beginning of the experiment, based on the consistent in vivo fluorescence values. Algal cells were then placed into an algal shaker with a speed of 180 rpm/min. Aliquots of approximately 100 µL of microalgae solution were transferred into a blood cell counting chamber and counted immediately using an inverted microscope (XS-213). The different growth phases of both of the considered algae were determined based on the in vivo fluorescence values and cell abundances.

2.3. Sample Analyses

Individual growth rates and doubling times of single species cultured in the laboratory were calculated based on the following equations [30]:

$$\text{Specific growth rate } (\mu, \text{d}^{-1}) = (\ln N_b - \ln N_a)/(t_b - t_a)$$

$$\text{Doubling time } (T_d, t) = \ln 2/\mu$$

where N_b and N_a denote the algal cell abundance (cells·mL^{-1}) at times t_b and t_a, respectively.

Cell volume was determined by constructing a geometric simulation graph [31]. Micrographic photos are presented in Figure 1. Chl-*a* was determined using a CE Turner Designs Fluorometer [32,33]. The PON and POC contents of the microalgae were analyzed using a Costech Elemental Analyzer (ECS4010, Milan, Italy).

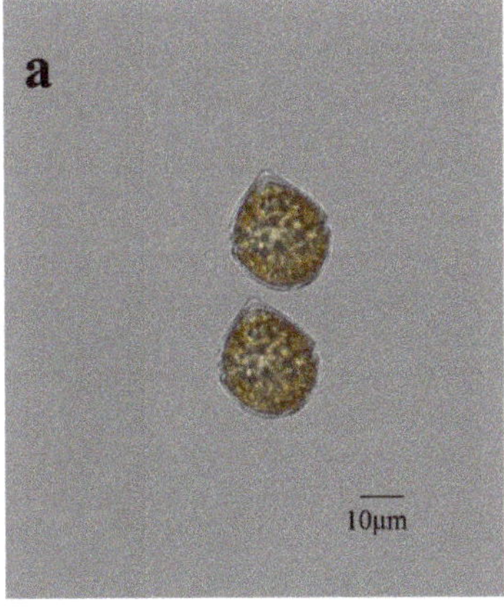

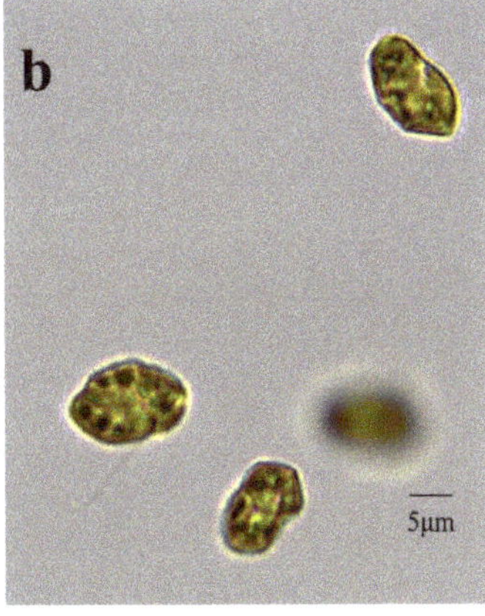

Figure 1. Light micrographic photos of two microalgae. (**a**) *S. trochoidea*; (**b**) *H. akashiwo*.

2.4. Date Analyses

All measured data are presented as mean $\pm$ standard deviation, plotted using the GraphPad Prism 8.0.2 software. Furthermore, all parameters (cell abundance, cell volume, Chl-*a*, and elements) between different species and growth phases were statistically analyzed using the SPSS software (version 21.0). Student's t-test was applied to test the differences between *S. trochoidea* and *H. akashiwo*. One-way analysis of variance (ANOVA) was used to analyze the difference of growth phases (e.g., lag, exponential, stationary, and degradation) for the two species. The levels of significance of the difference were set as $p < 0.05$ and $p < 0.01$.

3. Results

3.1. Growth Profiles of Scrippsiella trochoidea and Heterosigma akashiwo

During the experiment, the cell abundance of *S. trochoidea* significantly increased ($p < 0.01$, Figure 2a) from 360 cells·mL^{-1} to 2.60×10^4 cells·mL^{-1}. Similarly, the cell abundance of *H. akashiwo* also showed significant variation ($p < 0.01$, Figure 2b), from 3000 cells·mL^{-1} to 2.79×10^5 cells·mL^{-1}. However, the specific growth rates of *S. trochoidea* (0.86 ± 0.17 d^{-1}) and *H. akashiwo* (1.52 ± 0.38 d^{-1}) did not show any significant differences ($p > 0.05$). The doubling times of the two microalgae, *S. trochoidea* (0.80 d^{-1}) and *H. akashiwo* (0.45 d^{-1}), also had a statistically significant difference ($p < 0.05$). To simulate the growth curves of the two microalgae, the beta growth and decay model was applied in this study (Table 1). With this model, we estimated the environmental capacities of these two species as 2.19×10^4 cells·mL^{-1} and 2.59×10^5 cells·mL^{-1}, respectively (Table 1). The R^2 values of the two fitting equations were 0.87 and 0.97, respectively, illustrating good fitting results. The average cell volumes of *S. trochoidea* and *H. akashiwo* were 9.71×10^3 μm^3 and 0.64×10^3 μm^3, respectively. As shown in Figure 2c,d, a clear distinction was observed between the lag, exponential, and stationary phases ($p < 0.01$).

Table 1. The growth curve models of two microalgal species.

Species	Model	Formula	R-Squared	K ($10^4 \cdot$mL^{-1})	T (d)	T$_m$ (d)
Scrippsiella trochoidea	Beta growth then decay	$Y_t = K \times (1+ (T_m - t)/(T_m - T)) \times$	0.872	2.185	7.21	24.09
Heterosigma akashiwo		$(t/T_m)\hat{}(T_m/(T_m - T))$	0.970	25.88	9.55	16.60

Y$_t$, cell abundance during a given day; K, environmental capacity; T, time of maximum specific growth rate; T$_m$, time of maximum cell abundance. t, time of culture.

3.2. Chlorophyll-a Content of Two Microalgal Species

The Chl-*a* content of the two microalgae showed similar variation (Figure 2e). In both species, Chl-*a* values increased initially and then decreased. *S. trochoidea* reached a maximum of 0.43 mg·L^{-1} at day 16, whereas the higher value (1.25 mg·L^{-1}) for *H. akashiwo* was observed on day 15. Nevertheless, the Chl-*a* concentration of *H. akashiwo* was significantly higher than that of *S. trochoidea* ($p < 0.05$).

3.3. POC and PON Content in Two Microalgal Species

For *S. trochoidea* and *H. akashiwo*, the POC and PON per milliliter contents showed a similar trend, accumulating with the growth of the microalgae. The POC and PON contents of *S. trochoidea* reached their maximum range during the stationary phase, at 19.15 μg·mL^{-1} and 4.37 μg·mL^{-1} respectively. Surprisingly, a similar phenomenon occurred in *H. akashiwo*, with 85.42 μg·mL^{-1} for POC content and 10.33 μg·mL^{-1} for PON content (Figure 3a,b).

The POC content per cell of *S. trochoidea* decreased significantly during the first six days of the experiment, with a minimum value of 0.83×10^3 pg·cell^{-1} (exponential) and a maximum value of 3.78×10^3 pg·cell^{-1} (degradation) (Table 2). However, there was no significant difference between the exponential and stationary phase in terms of POC content per cell. Moreover, in terms of POC, the lag, exponential, and stationary phases were significantly lower than the degradation phase ($p < 0.05$) (Figure 3c). The lowest POC value per cell of *H. akashiwo* was 0.12×10^3 pg·cell^{-1} in the stationary phase, whereas the highest value was 0.50×10^3 pg·cell^{-1} in the degradation of incubation (Table 2). Therefore, it is obvious that the POC value per cell of *H. akashiwo* accumulated across the degradation phase. Moreover, although the POC content per cell of *H. akashiwo* in the first three phases (lag., exp., and stat.) declined gradually, the rate of decreases was very low.

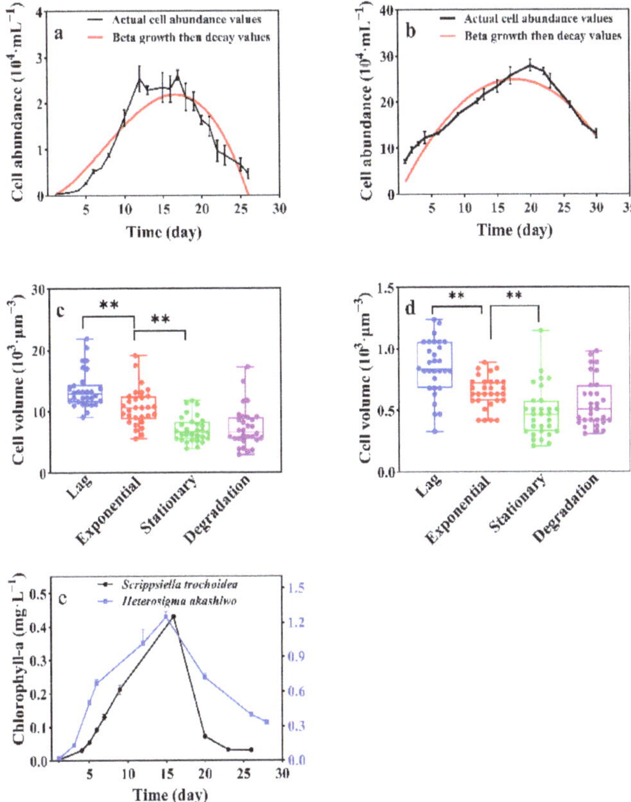

Figure 2. The growth process of *S. trochoidea* and *H. akashiwo*. (**a**) The growth curve of *S. trochoidea*; (**b**) the growth curve of *H. akashiwo*; (**c**) the cell volume of *S. trochoidea*; (**d**) the cell volume of *H. akashiwo*; (**e**) the Chl-*a* of two microalgal species; ** $p < 0.01$.

The PON content per cell of two species showed unusual differences during the experiment. The PON content per cell of *S. trochoidea* increased during the initial four days and remained relatively stable over the following 16 days, with an average value of 0.25×10^3 pg·cell^{-1} (Figure 3c). Similar to the POC content per cell of *H. akashiwo*, PON content accumulation occurred towards the end of the growth phase, with an average value of PON of 0.04×10^3 pg·cell^{-1} and a maximum of 56.66 pg·cell^{-1} (Table 2).

The PON and POC densities of the two species also showed a consistent trend. For example, the PON density in exponential and stationary phases was lower than that in the degradation phase. Interestingly, the POC density was significantly higher in the degradation phase than the other three phases ($p < 0.05$, Figure 3e,f). However, the average PON (0.03 pg·µm^{-3}) and POC (0.15 pg·µm^{-3}) density of *S. trochoidea* was lower than the PON (0.06 pg·µm^{-3}) and POC (0.40 pg·µm^{-3}) of *H. akashiwo* (Table 2).

The PON/Chl-*a* ratios of *S. trochoidea* and *H. akashiwo* had different trends throughout the growth stage, with average values of 57.32 and 9.46, respectively. The PON/Chl-*a* ratio for *S. trochoidea* showed a maximum value on day 16, then decreased rapidly in the following days (Figure 4a). The PON/Chl-*a* ratio of *H. akashiwo* changed very little during the first 20 days and rapidly increased to 25.80 on day 26 (Figure 4b). The POC/Chl-*a* ratio of both microalgal species showed similar trends throughout the growth phases. Moreover, the POC/Chl-*a* ratio of *S. trochoidea* was significantly higher than that of *H. akashiwo*. Both *S. trochoidea* and *H. akashiwo* showed an increasing trend in POC/PON during the degradation phase, which may be due to the PON loss rate being higher than that of POC (Figure 4c).

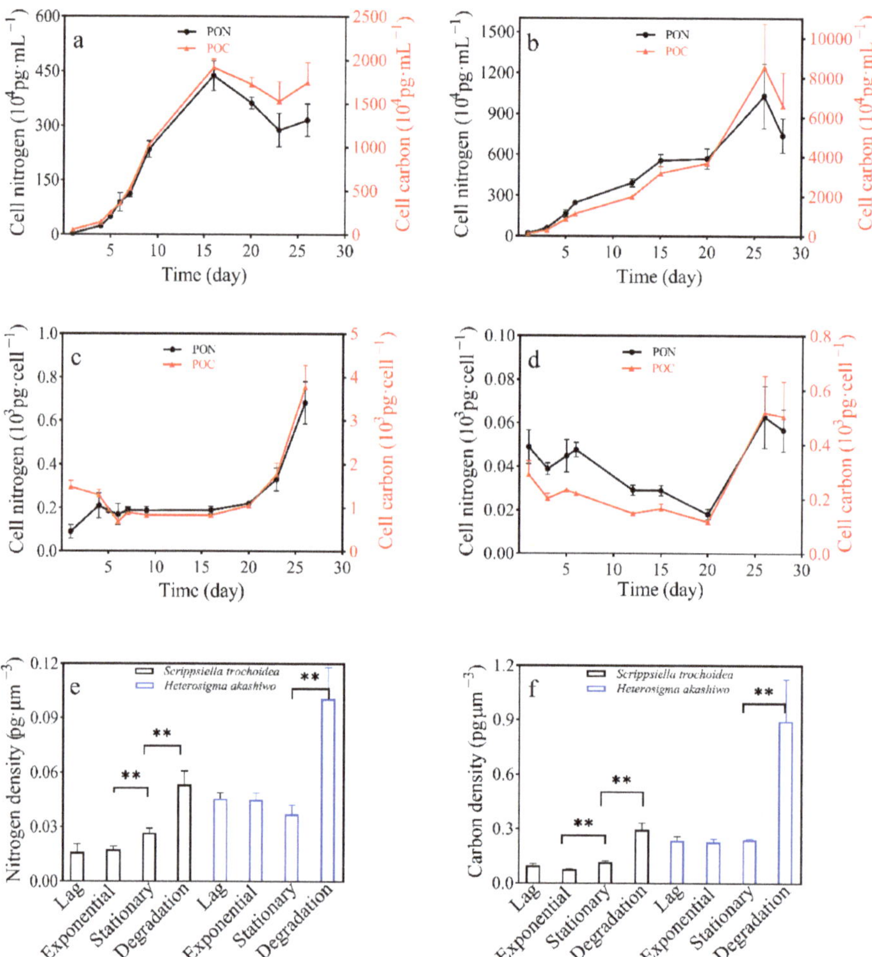

Figure 3. The organic cell composition of two species varied in different growth phases. (**a**) Cell nitrogen and carbon contents per unit milliliter of *S. trochoidea*; (**b**) cell nitrogen and carbon contents per unit milliliter of *H. akashiwo*; (**c**) nitrogen and carbon contents per unit cell of *S. trochoidea*; (**d**) nitrogen and carbon contents per unit cell of *H. akashiwo*; (**e**) cellular nitrogen density content of two species; (**f**) cellular carbon density content of two species; ** $p < 0.01$.

Table 2. Result of POC, PON quotas, and densities of two microalgal species in different growth phases.

		POC Quotas (Average ± SD) /pg·Cell^{-1}	PON Quotas (Average ± SD) /pg·Cell^{-1}	POC Densities (Average ± SD) /pg·μm^{-3}	PON Densities (Average ± SD) /pg·μm^{-3}
S. trochoidea	lag	1285.3 ± 126.55 [a]	210.21 ± 57.92 [a]	0.0947 ± 0.0093 [a]	0.0159 ± 0.0043 [a]
	exponential	825.98 ± 21.70 [b]	187.65 ± 18.50 [b]	0.0768 ± 0.0020 [a]	0.0175 ± 0.0017 [a]
	stationary	830.35 ± 41.80 [b]	189.60 ± 17.71 [b]	0.1168 ± 0.0058 [b]	0.0267 ± 0.0023 [b]
	degradation	3778.90 ± 497.56 [c]	683.53 ± 97.55 [c]	0.5101 ± 0.0671 [c]	0.0532 ± 0.0076 [c]
H. akashiwo	lag	202.18 ± 18.73 [a]	38.86 ± 2.73 [a]	0.2376 ± 0.0220 [a]	0.0457 ± 0.0032 [a]
	exponential	148.46 ± 10.00 [b]	29.20 ± 2.33 [b]	0.2298 ± 0.0155 [a]	0.0452 ± 0.0036 [a]
	stationary	116.92 ± 2.06 [b]	18.09 ± 2.36 [c]	0.2415 ± 0.0043 [a]	0.0374 ± 0.0049 [a]
	degradation	503.83 ± 129.22 [c]	56.66 ± 9.69 [a]	0.8958 ± 0.2297 [b]	0.1007 ± 0.0172 [b]

The different letter represents a significant difference in value within the same column. ($p < 0.05$).

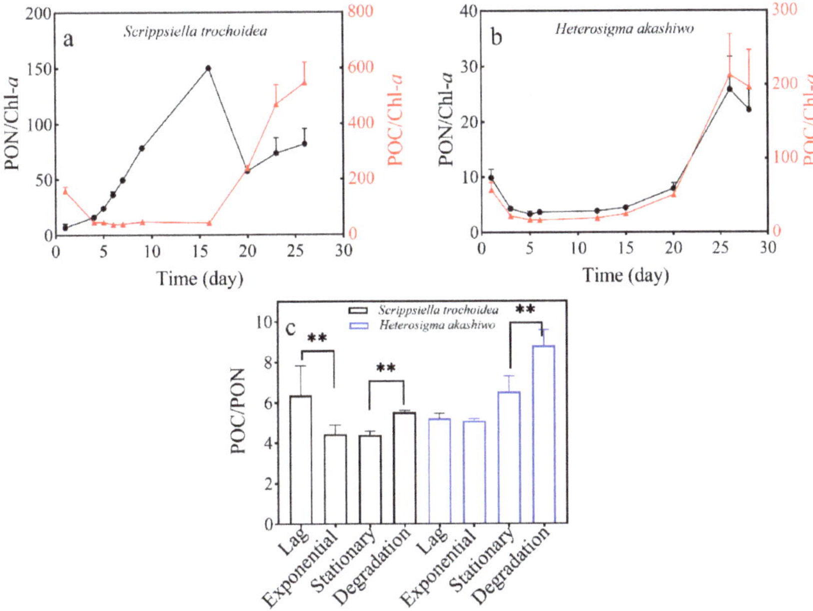

Figure 4. The change of ratios relationship for *S. trochoidea*, *H. akashiwo* in different growth phases; ** $p < 0.01$. (**a**) The ratio of PON/Chl-*a* and POC/Chl-*a* for *S. trochoidea*; (**b**) the ratio of PON/Chl-*a* and POC/Chl-*a* for *H. akashiwo*; (**c**) the ratio of POC/PON for two species.

4. Discussion

Our results showed a notable difference between the specific growth rate and cell abundance of two different phytoplankton taxa isolated from the Bohai sea. The specific growth rate results were relatively close to those of earlier reports by Xu et al. [34] and Ono et al. [35]. They noted that the maximum specific growth rate of *S. trochoidea*, under orthogonal test results, was 0.76 d^{-1} [34], while that of *H. akashiwo* did not exceed 2 d^{-1} under optimal conditions in the laboratory [35]. However, our results showed that the maximum growth rates of *S. trochoidea* and *H. akashiwo* were 0.863 d^{-1} and 1.524 d^{-1}, respectively. Interestingly, the highest abundances of *S. trochoidea* and *H. akashiwo* in this experiment were only 2.60 × 10^4 and 2.79 × 10^5 cells·mL^{-1}, respectively. Earlier studies have shown that, during "red tide", the cell abundances of *S. trochoidea* at Daya Bay and *H. akashiwo* at Dalian Bay reached 4.06 × 10^4 and 72 × 10^4 cells·mL^{-1}, respectively [23,36]. In these coastal ecosystems, the cell abundance is influenced by many environmental factors, such as the light intensity of surface water, temperature, and nutrients. Cell abundance reaches its maximum when these environmental factors are optimal [37]. Futhermore, Daya Bay and Dalian Bay have special geographical characteristics of semi-enclosed coastal inlets, which has led to increased nutrient loading by rapidly expanding mariculture and human population growth in these regions [37,38]. However, the nutrients were depleted by microalgae during batch culture in the lab. Therefore, the cell abundances of the two microalgae species in Daya Bay and Dalian Bay were higher than those under laboratory incubation. Compared with the larger-sized *S. trochoidea* in the same culture conditions, we observed that the smaller *H. akashiwo* presented high cell abundance values. A similar phenomenon was observed, by Marañón, for the relationship between cell abundance and cell size in the surface waters of a coastal site. It has been demonstrated that cell abundance decreased with increased cell volume in different microalgae species [2]. Past studies have shown that cell size can determine the nutrient diffusion and nutrient requirements of microalgae cells. Small cells can easily maintain their growth in low-nutrient conditions [39]. Therefore, the cell abundance under large biovolume was lower than that of smaller cells in similar conditions.

Microalgae can transform inorganic matter into organic matter through photosynthesis, which plays an important role in the ocean carbon sink. In addition to Chl-*a*, intracellular carbon and nitrogen content were also used as indicators to characterize the abundances of microalgae [23]. In this experiment, the maximum Chl-*a* values of *S. trochoidea* and *H. akashiwo* were preceded by the maximum cell abundance. A similar result has been found in the varied N/P ratios of microalgae growth by past studies. The trend could be explained by good adaptation of microalgae in new experiments. This adaptation, in advance, could supply cellular substances for further proliferation [40]. Furthermore, the Chl-*a* concentration is also an important parameter for characterizing algal blooms. As the two algae in this study were isolated from Qinhuangdao Island during blooming conditions, the Chl-*a* concentration of the natural population in the coastal water likely reached a maximum before the cell abundance. Therefore, the Chl-*a* concentration exceeding a certain threshold may serve as an indicator for microalgae blooms. POC and PON content per cell of two microalgae increased in the degradation phase. As observed previously, cell division slowed down in the stationary phase, and energy began to accumulate [21,41]. For microalgae in their natural environment, the accumulated POC remains in living and non-living organisms. Eventually, it is deposited on the seafloor, thus completing the biological carbon sink [1]. In this study, the carbon content per cell of both microalgae species was reduced at the beginning of culture. Microalgae have two fundamental mechanisms of physiological characteristics in the environment, including absolute and relative adaptation [42]. Culture studies have shown that microalgae have a relative adaptation period when nutrients are not limited in a new environment, which leads to an ability to synthesize organic matter slowly for algae at the beginning of the culture [40]. Interestingly, the POC content per unit cell of *S. trochoidea* was higher than that of *H. akashiwo*, while the POC and PON density content of *S. trochoidea* were lower than those of *H. akashiwo*. We hypothesize that such variations lead to the contributions of different species to algal blooms. According to a study by Finkel et al., small cells have more elemental stoichiometry density content, which has a larger side effect than larger individuals [39]. However, Emilio et al. concluded that blooms are dominated by large, rather than small, cells [2]. Due to the large size of the cell, it has higher metabolic rates than small individuals [43]. Some observations have implied that resource acquisition and assimilation rates can be maximized by microalgae with higher biovolume. Meanwhile, they can minimize loss and material transport rates in order to survive in pelagic ecosystems [44]. Compared with large individuals, small cells have a longer growth cycle and are more easily preyed upon by consumers [45]. Based on earlier studies regarding algal bloom occurrence in the Qingdao sea, we observed that the frequency of algal blooms in this area was more severe with respect to *S. trochoidea* compared to *H. akashiwo* [46].

Therefore, compared with *H. akashiwo*, we can further extend this explanation and conclude that *S. trochoidea* may dominate the algal bloom.

The amount of carbon in phytoplankton is influenced by the cell size, as well as environmental factors [47]. However, it is difficult to directly determine the carbon biomass of phytoplankton, as zooplankton and non-living organic matter may interfere in its measurement. Therefore, it is important to estimate the biomass of phytoplankton by using the ratio of carbon to Chl-*a* content [20]. Our results suggest that the POC/Chl-*a* values of microalgae increased in the degradation phase. This is due to the depletion of nutrients while the microalgae were growing. This led to a decrease in biological activity and the accumulation of carbon content. Several studies support our findings that low values of POC/Chl-*a* exist in active microalgae cells [48]. Interestingly, we found that the POC/Chl-*a* value of *S. trochoidea* was higher than that of *H. akashiwo*. In our results, the cell biovolume tended to decrease during the growth period. In eukaryotes, renewable components, such as membranes and nucleic acids, are increased with microalgal growth. This leads to a decline in other renewable and catalytic components of the cytoplasmic proportion, such as enzyme activity and the synthesis of organic matter. Therefore, compared to larger cells, small individuals transfer nutrients to biomass at a slow rate and have low POC/Chl-*a* values [49].

The POC/PON ratio of the phytoplankton community in seawater has been found to be 106:16 [50]. However, this ratio is not consistent between lab cultures and during the process of microalgae blooming in the ocean. Although the POC/PON values for the two microalgae decreased in the exponential phase, they were close to the Redfield ratio in the degradation phase. The chemometrics of deviation states have been associated with changes in the structural elements of microalgae. Different intracellular chemometrics have different properties, which reflect differences in their ability to store intracellular nutrients during the metabolic process [51]. Mature cells can increase their allocation of nutrients in the exponential phase. Under nutrient-restricted conditions, they resynthesize substances from other parts of the cell [52]. Therefore, we observed high values during the degradation phase.

5. Conclusions

Our findings demonstrated both physiological and elemental variations in two different phytoplankton, which can serve to strengthen efforts to mechanistically quantify and assess biogeochemical processes. The similar variation of the growth curve, biomass, and elements in this study may present a useful constraint for predicting variability of phytoplankton POC, PON, POC/Chl-*a*, and PON/Chl-*a* values in cellular trait-based models. We also demonstrated taxonomic variation in POC and PON accumulation mechanisms, highlighting the impacts that biochemical traits and the biogeography of Dinophyceae and Raphidophyceae may have on ocean element patterns, which may be valuable for marine biogeochemical modeling based on Dinophyceae and Raphidophyceae.

Author Contributions: Conceptualization, J.S.; methodology, J.S.; resources, J.S.; writing—original draft preparation, X.L.; writing—review and editing, J.S., Y.L.; visualization, Y.L.; supervision, J.S.; project administration, J.S.; funding acquisition, M.A.N.; S.T. helped to modify all syntax errors of the manuscript. All authors have read and agreed to the published version of the manuscript.

Funding: This research was financially supported by the National Nature Science Foundation of China (Grant No. 41876134, 41676112 and 41276124), the Tianjin 131 Innovation Team Program (20180314), and the Changjiang Scholar Program of Chinese Ministry of Education (T2014253) to Jun Sun.

Institutional Review Board Statement: Not appicable.

Informed Consent Statement: Not appicable.

Data Availability Statement: Data is contained within the article.

Acknowledgments: This research was financially supported by the National Key Research and Development Project of China (2019YFC1407805), the National Natural Science Foundation of China (41876134, 41676112 and 41276124), the Tianjin 131 Innovation Team Program (20180314), and the Changjiang Scholar Program of Chinese Ministry of Education (T2014253) to Jun Sun.

Conflicts of Interest: The authors state that they have no conflicts of interest.

References

1. Sun, J. Marine phytoplankton and biological carbon sink. *Acta Ecol. Sin.* **2011**, *31*, 5372–5378.
2. Marañón, E. Cell size as a key determinant of phytoplankton metabolism and community structure. *Ann. Rev. Mar.* **2014**, *7*, 241–264. [CrossRef]
3. Duarte, C.M.; Conley, D.J.; Carstensen, J.; Sánchez-Camacho, M. Return to neverland: Shifting baselines affect eutrophication restoration targets. *Estuaries Coast* **2009**, *2*, 29–36. [CrossRef]
4. Longhurst, A.R.; Harrison, W.G. The biological pump: Profiles of plankton production and consumption in the upper ocean. *Prog. Oceanogr.* **1989**, *22*, 47–123. [CrossRef]
5. Jakobsen, H.H.; Markager, S. Carbon-to-chlorophyll ratio for phytoplankton in temperate coastal waters: Seasonal patterns and relationship to nutrients. *Limnol. Oceanogr.* **2016**, *61*, 1853–1868. [CrossRef]
6. Vidoudez, C.; Pohnert, G. Comparative metabolomics of the diatom *Skeletonema marinoi* in different growth phases. *Metabolomics* **2012**, *8*, 654–669. [CrossRef]
7. Laws, E.A.; Falkowski, P.G.; Smith, W.O., Jr.; Ducklow, H.; McCarthy, J.J. Temperature effects on export production in the open ocean. *Glob. Biogeochem. Cycles* **2000**, *14*, 1231–1246. [CrossRef]
8. Armstrong, R.A.; Lee, C.; Hedges, J.I.; Honjo, S.; Wakeham, S.G. A new, mechanistic model for organic carbon fluxes in the ocean based on the quantitative association of POC with ballast minerals. *Deep Sea Res. Part II* **2001**, *49*, 219–236. [CrossRef]
9. Vrede, K.; Heldal, M.; Norland, S.; Bratbak, G. Elemental composition (C, N, P) and cell volume of exponentially growing and nutrient-limited bacterioplankton. *Appl. Environ. Microbiol.* **2002**, *68*, 2965–2971. [CrossRef]
10. Markou, G.; Depraetere, O.; Muylaert, K. Effect of ammonia on the photosynthetic activity of Arthrospira and Chlorella: A study on chlorophyll fluorescence and electron transport. *Algal Res.* **2016**, *16*, 449–457. [CrossRef]
11. Bopaiah, B.; Benner, R. Carbon, nitrogen, and carbohydrate fluxes during the production of particulate and dissolved organic matter by marine phytoplankton. *Limnol. Oceanogr.* **1997**, *42*, 506–518. [CrossRef]
12. Liang, C.J.; Wang, J.T.; Tan, L.J. Distribution of particulate organic carbon in Qingdao coastal waters in summer and winter. *Mar. Environ. Sci.* **2010**, *29*, 12–16. [CrossRef]
13. Mongin, M.; Nelson, D.M.; Pondaven, P.; Tréguer, P. Simulation of upper-ocean biogeochemistry with a flexible-composition phytoplankton model: C, N and Si cycling and Fe limitation in the Southern Ocean. *Deep Sea Res. Part II* **2006**, *53*, 601–619. [CrossRef]
14. Eppley, R.W.; Chavez, F.P.; Barber, R.T. Standing stocks of particulate carbon and nitrogen in the equatorial Pacific at 150 W. *J. Geophys. Res. C: Ocean.* **1992**, *97*, 655–661. [CrossRef]
15. Oubelkheir, K.; Claustre, H.; Sciandra, A.; Babin, M. Bio-optical and biogeochemical properties of different trophic regimes in oceanic waters. *Limnol. Oceanogr.* **2005**, *50*, 1795–1809. [CrossRef]
16. Behrenfeld, M.J.; Boss, E.; Siegel, D.A.; Shea, D.M. Carbon-based ocean productivity and phytoplankton physiology from space. *Glob. Biogeochem.* **2005**, *19*, GB1006. [CrossRef]
17. Platt, T.; Sathyendranath, S.; Forget, M.H.; White, G.N., III; Caverhill, C.; Bouman, H. Operational estimation of primary production at large geographical scales. *Remote Sens. Environ.* **2008**, *112*, 3437–3448. [CrossRef]
18. Terry, K.L.; Hirata, J.; Laws, E.A. Light-limited growth of two strains of the marine diatom *Phaeodactylum tricornutum* Bohlin-Chemical composition, carbon partitioning and the diel periodicity of physiological processes. *J. Exp. Mar. Biol. Ecol.* **1983**, *68*, 209–227. [CrossRef]
19. Markager, S.; Sand-Jensen, K. Implications of thallus thickness for growth-irradiance relationships of marine macroalgae. *Eur. J. Phycol.* **1996**, *31*, 79–87. [CrossRef]
20. Sathyendranath, S.; Stuart, V.; Nair, A.; Oka, K.; Nakane, T.; Bouman, H.; Platt, T. Carbon-to chlorophyll ratio and growth rate of phytoplankton in the sea. *Mar. Ecol. Prog. Ser.* **2009**, *383*, 73–84. [CrossRef]
21. Massie, T.M.; Blasius, B.; Weithoff, G.; Gaedke, U.; Fussmann, G.F. Cycles, phase synchronization, and entrainment in single-species phytoplankton populations. *Proc. Natl. Acad. Sci. USA* **2010**, *107*, 4236–4241. [CrossRef]
22. Qi, Y.Z.; Zheng, L.; Wang, R. The life cycle of *Scrippsiella trochoidea* and its physiol-ecological control. *Chin. J. Oceanol. Limnol* **1997**, *28*, 588–593.
23. Xiao, Y.Z. The relationship between *Scrippsiella trochoidea* red tide and cysts in the Daya Bay. *Mar. Sci.* **2001**, *25*, 50–54.
24. Wang, Z.H.; Liang, W.B.; Guo, X.; Liu, L. Inactivation of *Scrippsiella trochoidea* cysts by different physical and chemical methods: Application to the treatment of ballast water. *Mar. Pollut. Bull.* **2018**, *126*, 150–158. [CrossRef] [PubMed]
25. Liu, Y.; Lou, Y.D.; Wang, H.X.; Liu, Q. Effects of carbon source restricted on pH chang during growth of red tides algae. *Mar. Environ. Sci.* **2019**, *38*, 286–293.
26. Zhou, C.X.; Wang, F.X.; Yan, X.J. Effects of temperature, salinity and irradiance on the cell stability of *Heterosigma akashiwo* Hada. *Mar. Environ. Sci.* **2008**, *27*, 17–20. [CrossRef]

27. Qin, X.M.; Qin, P.Y.; Zou, J.Z. Effects of nitrogen and phosphorus on the growth of a red tide dinoflagellate *Scrippsiella trochoidea* (stein) loeblch III. *Chin. J. Oceanol. Limnol.* **1999**, *17*, 212–218. [CrossRef]
28. Guillard, R.R.L.; Hargraves, P.E. Stichochrysis immobilis is a diatom, not a chrysophyte. *Phycologia* **1993**, *32*, 234–236. [CrossRef]
29. Liu, Y.; Lv, J.P.; Feng, J.; Liu, Q.; Nan, F.R.; Xie, S.L. Treatment of real aquaculture wastewater from a fishery utilizing phytoremediation with microalgae. *J. Chem. Technol. Biotechnol.* **2019**, *94*, 900–901. [CrossRef]
30. Sun, J.; Ning, X.R. Marine phytoplankton specific growth rate. *Adv. Earth Sci.* **2005**, *20*, 939–945. [CrossRef]
31. Sun, J.; Liu, D.Y. Geometric models for calculating cell biovolume and surface area for phytoplankton. *J. Plankton Res.* **2003**, *25*, 1331–1346. [CrossRef]
32. Welschmeyer, N.A. Fluorometric analysis of chlorophyll *a* in the presence of chlorophyll *b* and pheopigments. *Limnol. Oceanogr.* **1994**, *39*, 1985–1992. [CrossRef]
33. Liu, H.J.; Xue, B.; Feng, Y.; Zhang, R.; Chen, M.; Sun, J. Size-fractionated Chlorophyll a biomass in the northern South China Sea in summer 2014. *Chin. J. Oceanol. Limnol.* **2016**, *34*, 672–682. [CrossRef]
34. Xu, N.; Lv, S.H.; Chen, J.F.; He, L.S.; Xie, L.C.; Qi, Y.Z. The influence of water temperature and salinity on the growth of *Scrippsiella trochoidea*. *Mar. Environ. Sci.* **2004**, *23*, 36–38. [CrossRef]
35. Ono, K.; Khan, S.; Onoue, Y. Effects of temperature and light intensity on the growth and toxicity of *Heterosigma akashiwo* (Raphidophyceae). *Aquacult. Res.* **2000**, *31*, 427–433. [CrossRef]
36. Guo, Y.J. Studies on *heterosigma akashiwo* (HADA) in the Dalian bight, Liaoning, China. *Oceanol. Limnol. Sin.* **1994**, *25*, 211–215.
37. Wang, Z.H.; Qi, Y.Z.; Chen, J.F.; Xu, N.; Yang, Y.F. Phytoplankton abundance, community structure and nutrients in cultural areas of Daya Bay, South China Sea. *J. Mar. Syst.* **2006**, *62*, 85–94. [CrossRef]
38. Wang, H.Q. The *Heterosigma Akashiwo* red tide and ecological characteristics in Dalian Bight. *Res. Environ. Sci.* **1991**, *4*, 53–59.
39. Finkel, Z.V.; Beardall, J.; Flynn, K.J.; Quigg, A.; Rees, T.A.V.; Raven, J.A. Phytoplankton in a changing world: Cell size and elemental stoichiometry. *J. Plankton Res.* **2009**, *1*, 119–137. [CrossRef]
40. Sun, J.; Liu, D.Y.; Chen, Z.T.; Wei, T.D. Growth of *Platymonas helgolandica var. tsingtaoensis*, *Cylindrotheca closterium* and *Karenia mikimotoi* and their survival strategies under different N/P ratios. *J. Appl. Ecol.* **2004**, *15*, 2122–2126. [CrossRef]
41. Goldman, J.C.; Hansell, D.A.; Dennett, M.R. Chemical characterization of three large oceanic diatoms: Potential impact on water column chemistry. *Mar. Ecol. Prog. Ser.* **1992**, *88*, 257–270. [CrossRef]
42. Beardall, J.; Allen, D.; Bragg, J.; Finkel, Z.V.; Flynn, K.J.; Quigg, A.; Raven, J.A. Allometry and stoichiometry of unicellular, colonial and multicellular phytoplankton. *New Phytol.* **2009**, *181*, 295–309. [CrossRef]
43. Riegman, R.; Mur, L.R. Theoretical considerations on growth kinetics and physiological adaptation of nutrient-limited phytoplankton. *Arch. Microbiol.* **1984**, *140*, 96–100. [CrossRef]
44. Huete-Ortega, M.; Cermeno, P.; Calvo-Díaz, A.; Marañón, E. Isometric size-scaling of metabolic rate and the size abundance distribution of phytoplankton. *Proc. R. Soc. Lond. Ser. B* **2012**, *279*, 1815–1823. [CrossRef]
45. Sunda, W.G.; Hardison, D.R. Evolutionary tradeoffs among nutrient acquisition, cell size, and grazing defense in marine phytoplankton promote ecosystem stability. *Mar. Ecol. Prog. Ser.* **2010**, *401*, 63–76. [CrossRef]
46. Bian, M.M.; Song, J.L.; Feng, J.Y.; Liu, Z.L. Occurrence and ananlysis of red tides in Qinhuangdao sea area. *Heibei Fish.* **2019**, *8*, 26–28.
47. Viličić, D. An examination of cell volume in dominant phytoplankton species of the central and southern Adriatic Sea. *Int. Rev. Gesamten Hydrobiol.* **1985**, *70*, 829–843. [CrossRef]
48. Parsons, T.R.; Stephens, K.; Strickland, J.D.H. On the chemical composition of eleven species of marine phytoplankters. *J. Fish. Res. Board Can.* **1961**, *18*, 1001–1016. [CrossRef]
49. Marañón, E.; Cermeño, P.; López-Sandoval, D.C.; Rodríguez-Ramos, T.; Sobrino, C.; Huete-Ortega, M.; Maria Blanco, J.; Rodríguez, J. Unimodal size scaling of phytoplankton growth and the size dependence of nutrient uptake and use. *Ecol. Lett.* **2013**, *16*, 371–379. [CrossRef]
50. Geider, R.; La Roche, J. Redfield revisited: Variability of C: N: P in marine microalgae and its biochemical basis. *Eur. J. Phycol.* **2002**, *37*, 1–17. [CrossRef]
51. Klausmeier, C.A.; Litchman, E.; Levin, S.A. Phytoplankton growth and stoichiometry under multiple nutrient limitation. *Limnol. Oceanogr.* **2004**, *49*, 1463–1470. [CrossRef]
52. Arrigo, K.R. Marine microorganisms and global nutrient cycles. *Nature* **2005**, *437*, 349–355. [CrossRef] [PubMed]

Article

Characteristics of the Biochemical Composition and Bioavailability of Phytoplankton-Derived Particulate Organic Matter in the Chukchi Sea, Arctic

Bo Kyung Kim, Jinyoung Jung, Youngju Lee, Kyoung-Ho Cho, Jong-Ku Gal, Sung-Ho Kang and Sun-Yong Ha *

Division of Polar Ocean Sciences, Korea Polar Research Institute, 26 Songdomirae-ro, Yeonsu-gu, Incheon 21990, Korea; bkkim@kopri.re.kr (B.K.K.); jinyoungjung@kopri.re.kr (J.J.); yjlee@kopri.re.kr (Y.L.); kcho@kopri.re.kr (K.-H.C.); jkgal@kopri.re.kr (J.-K.G.); shkang@kopri.re.kr (S.-H.K.)
* Correspondence: syha@kopri.re.kr

Received: 21 July 2020; Accepted: 20 August 2020; Published: 21 August 2020

Abstract: Analysis of the biochemical composition (carbohydrates, CHO; proteins, PRT; lipids, LIP) of particulate organic matter (POM, mainly phytoplankton) is used to assess trophic states, and the quantity of food material is generally assessed to determine bioavailability; however, bioavailability is reduced or changed by enzymatic hydrolysis. Here, we investigated the current trophic state and bioavailability of phytoplankton in the Chukchi Sea (including the Chukchi Borderland) during the summer of 2017. Based on a cluster analysis, our 12 stations were divided into three groups: the southern, middle, and northern parts of the Chukchi Sea. A principal component analysis (PCA) revealed that relatively nutrient-rich and high-temperature waters in the southern part of the Chukchi Sea enhanced the microphytoplankton biomass, while picophytoplankton were linked to a high contribution of meltwater derived from sea ice melting in the northern part of the sea. The total PRT accounted for 41.8% (±7.5%) of the POM in the southern part of the sea, and this contribution was higher than those in the middle (26.5 ± 7.5%) and northern (26.5 ± 10.6%) parts, whereas the CHO accounted for more than half of the total POM in the northern parts. As determined by enzymatic hydrolysis, LIP were more rapidly mineralized in the southern part of the Chukchi Sea, whereas CHO were largely used as source of energy for higher trophic levels in the northern part of the Chukchi Sea. Specifically, the bioavailable fraction of POM in the northern part of the Chukchi Sea was higher than it was in the other parts. The findings indicate that increasing meltwater and a low nutrient supply lead to smaller cell sizes of phytoplankton and their taxa (flagellate and green algae) with more CHO and a negative effect on the total concentration of POM. However, in terms of bioavailability (food utilization), which determines the rate at which digested food is used by consumers, potentially available food could have positive effects on ecosystem functioning.

Keywords: particulate organic matter; biochemical composition; phytoplankton; Chukchi Sea; Arctic Ocean

---

1. Introduction

In terms of bottom-up controls, phytoplankton is key organism that serves as a primary producer and primary food source for organisms at higher trophic levels in the foodwebs of aquatic ecosystems. Climate change enhances the sea ice melting in the Arctic Ocean with increasing concerns about primary production and nutrient cycling. Sea ice loss reduces surface albedo and enhances light penetration, creating irregularities on the timing and the duration of phytoplankton blooms [1,2]. These conditions can create discontinuity between the available food resources and the nutritional demands of higher producers [1], including higher trophic level organisms [3], and thus affect the energy flow of the entire arctic food web.

In natural systems, however, the food value of phytoplankton cannot be adequately described by measuring their biomass (chlorophyll a) and primary production. Hence, measures of the main biochemical classes (proteins, PRT; lipids, LIP; carbohydrates, CHO) of organic compounds have been used by various authors to estimate the quality and quantity of food in organic pools (reviewed by Bhavya et al. [4]). It is assumed that other biochemical components comprise negligible weights [5] and that the three major biochemical constituents (PRT, LIP, and CHO) are easier to digest and assimilate [6–8]. In reality, LIP, PRT, and CHO play roles in the structural components and energy storage of marine organisms [9,10], accounting for up to 90% of the weight in algae [11].

Generally, particulate organic matter (POM) is composed of living and dead organisms and refractory organic matter. For POM that consists of mainly phytoplankton-derived materials, the biochemical composition of POM reflects the physiological state of the phytoplankton in response to environmental conditions and phytoplankton energy value [4,12–14]. For example, PRT synthesis is generally promoted in productive areas or the exponential growth phase of phytoplankton [13,15,16], while the biosynthesis of non-nitrogenous storage compounds, such as CHO and LIP, is enhanced under high light intensity [17], low temperatures [18] and low nitrogen conditions [19,20]. LIP contain more calories than PRT and CHO [21]. In addition, the labile fractions of POM are characterized by the activities of enzymes, and their degradation provides insight into how POM is bioavailable to consumer organisms [22–24]. Therefore, changes in the biochemical composition and hydrolysable fractions of phytoplankton-derived POM can be useful for determining the physiological and nutritional conditions of phytoplankton.

Our study area is the Chukchi Sea (including the Chukchi Borderland), which contains pathways of water from the Pacific Ocean that flow poleward through the narrow Bering Strait to the Arctic Ocean [25] and transfer freshwater, heat, and nutrients from the northern Bering Sea (Yang and Bai [26] and reference therein). The southern Chukchi Sea is one of the most productive areas globally (up to $4.7\,\mathrm{g\,C\,m^{-2}\,day^{-1}}$; Korsak [27]), and has an especially high benthic productivity and biodiversity [28,29] because of the nutrients supplied by the inflow of water from the Pacific Ocean. A few studies have estimated the quantity and biochemical composition of POM in the Arctic Ocean [30–33]. However, bioavailable POM food resources created through enzymatic hydrolysis have not been investigated. Hence, the purpose of this study was (i) to investigate the spatial distribution and influence of physical (e.g., salinity, temperature, and meltwater) and chemical (major inorganic nutrients) properties on the biochemical composition of POM and (ii) to estimate potentially bioavailable food for higher trophic levels in the Arctic marine ecosystem using the labile fraction of POM obtained by enzymatic hydrolysis.

2. Materials and Methods

2.1. Field Sampling and Measurements of the Environmental Variables

This study was carried out at 12 stations in the Chukchi Sea onboard the R/V *Araon* icebreaker from 7 to 24 August 2017 (Figure 1A). The potential temperature, salinity, and photosynthetically active radiation (PAR) from the surface to a 100 m depth were measured by a rosette-mounted Sea-Bird conductivity-temperature-depth (CTD) system—1% PAR at the surface light level was defined as the euphotic layer [34] by a Secchi disc using the vertical attenuation coefficient (Kd = 1.7/secchi depth). The meltwater percentage (MW; %) was calculated from the salinity at each sampled depth (S_{meas}) and the greatest depth (either the bottom depth or 100 m in this study; S_{deep}), assuming an average sea ice salinity of 6 [35,36] since the mean salinity at a melt pond in the western Arctic Ocean was 5.9 [37]:

$$\mathrm{MW}(\%) = \left\{1 - \left[\frac{(S_{meas} - 6)}{(S_{deep} - 6)}\right]\right\} \times 100 \qquad (1)$$

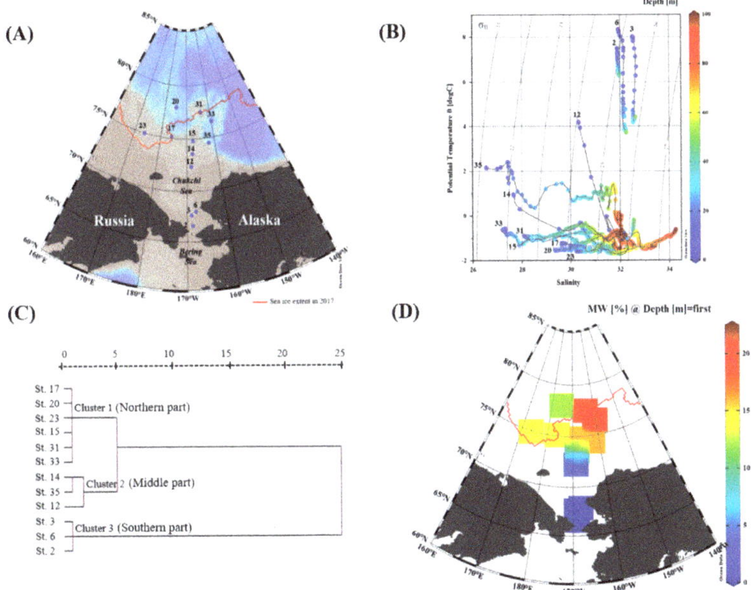

Figure 1. (**A**) Location of sampling area in the Chukchi Sea, August 2017. Sea ice extent for the month of August in 2017 (red line) was obtained from the National Snow and Ice Data Center (NSIDC, Fetterer et al. [38]). (**B**) Temperature-salinity diagram from surface to 100 m depth (labeled with station numbers in bold black at the surface); (**C**) cluster analysis of the surface potential temperature and salinity data allowed identification of 3 types of the regions in the Chukchi Sea, (**D**) MW (%) distributions at surface water in the Chukchi Sea during a summer cruise in August 2017. (Ocean Data View (ODV) version 5.1.0) (AWI, Bremerhaven, Germany, Schlitzer, R.).

The water samples used to determine the dissolved inorganic nutrients, chlorophyll a (chl-a), photosynthetic pigments, and POM (from carbon isotope samples at the surface), were obtained from the surface to the euphotic layer (2–5 depths) using a CTD/rosette sampler with 10-L Niskin bottles (Ocean Test Equipment Inc., Fort Lauderdale, FL, USA). The dissolved inorganic nutrients (nitrate + nitrite, ammonium, silicate, and phosphates) were analyzed onboard using a 4-channel QuAAtro Auto Analyzer (Seal Analytical, Norderstedt, Germany). The concentrations of the nutrients were measured using standard colorimetric methods, and the reference material for nutrients in seawater (Lot. No. "BV", Kanso Technos Co., Ltd., Osaka, Japan) were used in addition to standards for every batch of runs to ensure accurate and comparable measurements during the cruise.

After prefiltration through a 200 μm mesh net to remove large zooplankton, the water samples used to determine total chlorophyll a (chl-a) and accessory pigments were filtered onto GF/F filters (precombusted at 450 °C for 4 h; Whatman, Port Washington, NY, USA) immediately after collection. The filters were stored at −80 °C until the analyses were performed. Size-fractionated chl-a was determined from samples passed sequentially through 20 μm (>20 μm; microphytoplankton), 2 μm (2–20 μm; nanophytoplankton) and Whatman GF/F filters (0.7–2 μm; picophytoplankton). All the chl-a concentrations were calculated by the methods described by Parsons et al. [39] using a Trilogy fluorometer (Turner Designs, San Jose, CA, USA). The phytoplankton community composition was determined with photosynthetic pigments measured by high performance liquid chromatography (HPLC; Agilent 1260 Infinity LC, Agilent Technologies Inc, Santa Clara, CA, USA)-CHEMTAX analyses. For the stable carbon isotope composition of POM, seawater was filtered onto precombusted (450 °C for 4 h) 25 mm GF/F (Whatman, 0.7 μm pore) filters. The filters were immediately stored at −80 °C until further analysis. Stable carbon isotope composition was determined using isotope ratio mass spectrometry (IRMS; visION, Elementar UK, Manchester, UK) in the stable isotope laboratory at the University of Hanyang, Ansan, Korea, after HCl fuming overnight to remove the carbonate. The carbon isotope fractionation, $\delta^{13}C$ (‰), was calculated using the following equation:

$$\delta^{13}C\ (‰)\ =\ \left[\frac{\left(\frac{^{13}C}{^{12}C}\right)_{sample}}{\left(\frac{^{13}C}{^{12}C}\right)_{standard}} - 1\right] \times 1000 \tag{2}$$

where the standard for $\delta^{13}C$ is IAEA-CH-3 [40].

2.2. Biochemical Composition and Enzyme-Hydrolysable Experiments Related to the POM

The water samples ($n = 51$) used to assess the biochemical composition of the POM were obtained from two to six different depths at each site within the euphotic layer, and for each macromolecule (PRT, CHO, and LIP), 0.5–1 L of the seawater sample went through a precombusted 25 mm GF/F filter (at 450 °C for 4 h). The filter was immediately stored at −80 °C until analysis. Analysis of the PRT and CHO was performed using the methods described by Lowry et al. [41] and Dubois et al. [42], respectively. For the total PRT extraction, we added deionized water to a filter and, alkaline copper solution and Folin-Ciocalteu phenol regent to the sample tube. The CHO content was measured by a phenol–sulfuric acid reaction. The LIP were extracted from the filter with chloroform and methanol (1:2; *v:v*) [43], followed by sulfuric acid at 200 °C [44]. The absorbance of the samples, blanks and standards was determined at wavelengths of 750, 490, and 360 nm for the PRT, CHO, and LIP, respectively, using a spectrophotometer (Hitachi, Tokyo, Japan). The concentrations of the macromolecules were determined by comparison to the standard curve created with blank filters (procedural control filters, Whatman GF/F filter). The standard solutions for the PRT, CHO, and LIP were used a protein standard (2 mg mL^{-1}, Albumin from bovine serum, CAS No. 9048-46-8, Sigma-Aldrich, St. Louis, MO, USA), glucose standard (1 mg mL^{-1}, CAS No. 50-99-7, Sigma-Aldrich), and tripalmitin (50 mg in 100 mL chloroform, CAS No. 555-44-2, Sigma-Aldrich), respectively.

For enzyme-hydrolysable experiments, sampling was conducted by randomly selected samples of 35. Three enzymes were used in the enzyme-hydrolysable experiments: proteinase K derived from *Tritirachium album* (CAS No. 39450-01-6), β-glucosidase from almonds (CAS No. 9001-22-3), and lipase from *Rhizopus oryzae* (CAS No. 9001-62-1) (Sigma-Aldrich). Since these enzymes have hydrolytic activities similar to those of natural marine organisms, including autotrophs and heterotrophs [45], proteinase K, β-glucosidase, and lipase were chosen for the hydrolysis of PRT, CHO, and LIP, respectively [22,24,46–49]. The sample filters and blank filters were placed in enzyme solutions (100 mg L^{-1} in 0.1 M sodium phosphate buffer) to react for 2 h (proteinase K), 2 h (β-glucosidase), and 30 min (lipase). After hydrolysis, each filter was rinsed with buffer and deionized water and the concentrations of PRT, CHO, and LIP were determined as previously described. The concentration of the hydrolyzed biochemical fractions was calculated by the difference between before and after treatment of enzyme for each fraction.

2.3. Statistical Analysis

The statistical analyses (*t*-test, Pearson's correlation, and principal component analysis (PCA)) were performed with SPSS statistical software (version 12.0; SPSS Inc., Chicago, IL, USA) and R software (version 3.4). Cluster analysis was performed by using a hierarchical clustering algorithm with Ward's method to identify the groups of sampling stations. A t-test evaluates whether the means of two independent groups are significantly different from each other. The relationships between the depth, nutrients, chl-a, and biochemical components were tested using Pearson's correlation. PCA was used to evaluate the differences in the biochemical components and identify the significance of the environmental factors (e.g., salinity, temperature, density, phytoplankton size, MW (%), and major inorganic nutrient concentrations) among the groups and at each station. The average value of each variable within the euphotic layer was used for PCA (Table S1). We adopted the principle that an eigenvalue >1.0 can be used to determine the number of principal components.

3. Results

3.1. Physicochemical and Biological Characteristics During the Sampling Period

The potential temperature and salinity diagram reveal different hydrodynamic conditions during the sampling periods (Figure 1B). Based on Gong and Pickart's work [50], the summertime water mass properties of Stations 2, 3, and 6 were mainly composed of Alaskan coastal water (potential temperature (T) $\geq$ 3 °C and salinity (S) $\geq$ 0). Chukchi summer water (-1 °C < T < 3 °C and 30 < S < 33.6), and Pacific winter water (T < -1 °C and S > 31.5) were found in the other stations (Figure 1B). Since sea-surface temperature and salinity are strongly affected by sea ice and related meltwaters, brine rejection, continental runoff, and the heat flux in the Arctic Ocean [51], we assumed that the temperature and salinity at the surface were representative of the ambient water conditions. As a result, the cluster analysis of the surface potential temperature and salinity data allowed the identification of the three types of regions in the Chukchi Sea: cluster 1 (hereafter, the southern part; Stations 2, 3, and 6) was located at a latitude of approximately 66–70 °N; cluster 2 (hereafter, the northern part; Stations 15, 17, 20, 23, 31, and 33) was located at a latitude of 74.7–78 °N and included the Chukchi Borderland; cluster 3 (hereafter, the middle part; Stations 12, 14, and 35) was located between two areas of the Chukchi Sea (the southern and northern parts) (Figure 1C and Table 1).

Table 1. Description of sampling stations in the Chukchi Sea, 2017. Euphotic depth is the depth of the 1% light level. All samples were collected from two to six different depths at each site within euphotic depth.

Station	Date (dd-mm-yyyy)	Latitude (°N)	Longitude (°W)	Bottom Depth (m)	Euphotic Depth (m)	Sampling Depth (m)	Group
2	07-08-2017	66.6298	168.6874	43	20	0, 10, 20	Southern part
3	07-08-2017	67.6699	168.9601	48	31	0, 10, 18, 30	Southern part
6	07-08-2017	68.0130	167.8668	50	17	0, 10	Southern part
12	09-08-2017	72.3601	168.6668	48	35	0, 10, 15, 20, 30	Middle part
14	09-08-2017	73.5803	168.2824	119	44	0, 10, 20, 30, 44	Middle part
35	24-08-2017	74.5003	162.2487	1536	43	0, 10, 20, 30, 40	Middle part
15	10-08-2017	74.7987	167.8904	192	54	0, 10, 15, 20, 30, 44	Northern part
17	11-08-2017	75.1509	176.0166	319	33	0, 10, 13, 18, 30	Northern part
20	12-08-2017	77.9999	174.9342	1672	54	0, 10, 14, 22, 24, 30	Northern part
23	18-08-2017	75.0008	173.6090	135	39	0, 10, 20, 25, 34	Northern part
31	22-08-2017	77.4722	164.1178	267	40	0, 10, 15, 20, 30, 40	Northern part
33	23-08-2017	76.5254	159.9693	2102	73	0, 10, 20, 30	Northern part

The potential temperature at the surface was approximately 8 °C in the southern part, while in the northern part, it fell further, to below 0 °C, ranging from -1.6 to -0.6 °C. The salinity at the surface in the southern Chukchi Sea (shallow continental shelf) was above 31.9, with the maximum value (32.5) recorded at Station 3, while the salinity in the northern part was below 30.3, with the minimum value (27.2) recorded at Station 33 (Figure 1B). Overall, the northern part of the Chukchi Sea is characterized by a relatively cold temperature and low salinity, while we found higher temperatures and salinities in the surface water in the southern part (Figure 1B). Hydrodynamic characteristics are subject to the considerable influence of sea ice. The meltwater percentage (MW; %) in the euphotic layer of the study area ranged from 0 to 21.1%, with large spatial variations. Such a situation is specific to the northern part, with an average MW (%) ranging from 4.6 to 18.4 and a mean of 12.8% (SD = $\pm$ 3.6). Based on the sea ice extent, the MW (%) accounted for <15% of the surface water at the inner stations (Stations 17, 20, and 23) while at the outer stations (Stations 15, 31, 33, and 35), the MW accounted for more than 15% of the surface water (t-test, $p < 0.05$; Figure 1D). This result suggests that the salinity was greatly influenced by the regional melting of sea ice.

The concentrations of the dissolved inorganic nitrate + nitrite + ammonium (DIN), silicate (DSi) and phosphate (DIP) are shown in Figure 2. In the sampling period, the DSi and DIP concentrations from the surface to the euphotic layer ranged from 1.9 to 29.0 µM and 0.2 to 1.7 µM with means of 7.8 (SD = ±6.4 µM) and 0.8 (SD = ±0.3 µM), respectively (Figure 2A). The concentration of DIN, which was generally depleted (<1 µM) at the surface layer throughout our study area, was in the range of 0–13.2 µM, with an average of 1.5 µM (SD = ±3.0 µM) (Figure 2B). All the mean nutrient concentrations decreased from the southern to the northern parts of this region.

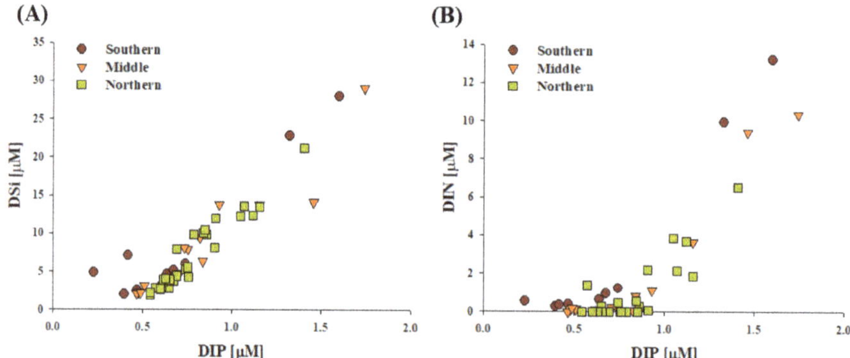

Figure 2. The stoichiometric (**A**) dissolved inorganic silicate (DSi) and dissolved inorganic (DIP) and (**B**) dissolved inorganic nitrate + nitrite + ammonium (DIN) and dissolved inorganic phosphate (DIP) from the surface to euphotic layer at sampling stations.

The average total chl-a concentration of phytoplankton from the surface to the euphotic depth ranged from 0.04 to 5.3 µg L^{-1} with a mean of 0.8 µg L^{-1} (SD = ±1.3 µg L^{-1}) at all stations, decreasing northward (Figure 3A). The phytoplankton community was dominated by picophytoplankton, which accounted for 46.2% (SD = ±15.0%) of the total chl-a concentration, followed by nanophytoplankton (mean ± SD = 27.8 ± 10.0%) and microphytoplankton (mean ± SD = 26.0 ± 17.3%) in the northern part of the Chukchi Sea (Figure 3B). In the southern and middle parts, microphytoplankton were dominant (mean ± SD = 80.1 ± 5.9% for the southern part and mean ± SD = 35.0 ± 34.5% for the middle part) within the euphotic layer (Figure 3B).

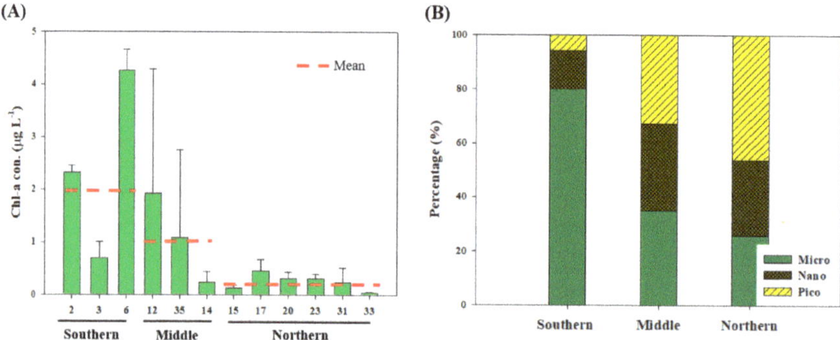

Figure 3. Average chlorophyll a (chl-a; µg L^{-1}) concentration of phytoplankton within euphotic layer (**A**) in the study stations of the Chukchi Sea. (**B**) Relative chl-a (%) for size fraction of phytoplankton (0.7–2 µm, 2–20 µm and >20 µm; i.e., pico-, nano- and micro-phytoplankton, respectively). Data were sorted by station depths and divided into southern, middle, and northern. Error bar indicated standard deviation (*n* = 2–5).

3.2. Biochemical Composition (PRT, LIP, and CHO) of POM

The LIP and PRT concentrations in the POM ranged from 5.4 to 169.1 µg L^{-1} (mean ± SD = 32.4 ± 32.8 µg L^{-1}) and 9.7 to 573.8 µg L^{-1} (mean ± SD = 61.6 ± 101.0 µg L^{-1}) within the euphotic layer, respectively (Figure 4). The CHO concentration ranged from 29.9 to 406.4 µg L^{-1} with a mean of 86.6 µg L^{-1} (SD = ± 67.9 µg L^{-1}) (Figure 4A). The vertical distribution of the LIP, PRT, and CHO concentrations did not show a specific trend ($p > 0.05$) but was characterized by significant spatial changes (Figure 4A). In the southern part of the Chukchi Sea, the average PRT concentration (198.8 µg L^{-1}) was approximately 5.1 and 6.9 times higher than those of the stations in the middle (39.2 µg L^{-1}) and northern parts (28.7 µg L^{-1}) (t-test, $p < 0.05$), respectively. Similarly, at the southern stations, the average LIP (80.3 µg L^{-1}) concentration was approximately 3.4 and 3.8 times higher than the average LIP concentrations in the middle and northern parts, respectively, while the average CHO (162.7 µg L^{-1}) concentration was approximately 1.8 and 2.7 times higher than the average CHO concentrations in the middle and northern parts (Figure 4A).

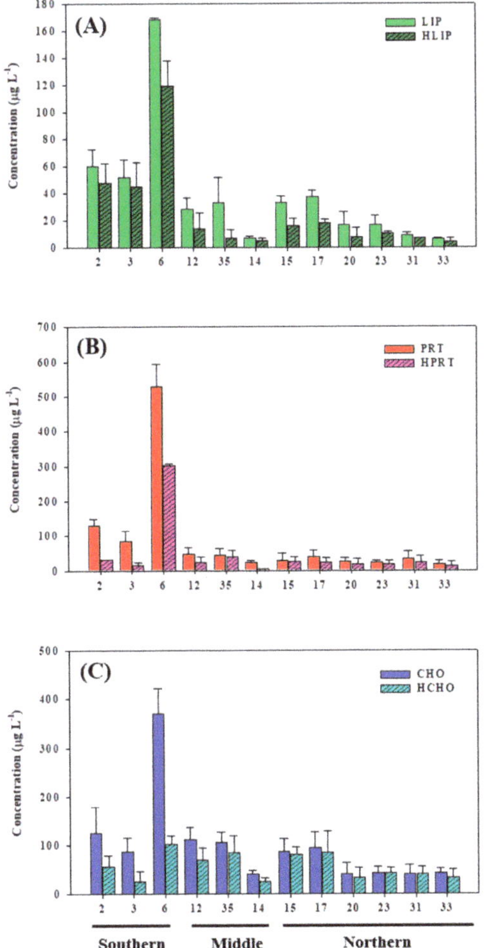

Figure 4. Average total ($n = 51$) and hydrolysable ($n = 35$) (**A**) lipids, (**B**) proteins, and (**C**) carbohydrates concentration of particulate organic matter (POM) within euphotic layer. Error bar indicated standard deviation ($n = 2$–6).

The food material (FM) is represented by the sum of PRT, CHO, and LIP concentrations in POM ([13] and reference therein) and concentration of each biochemical constituent (PRT, CHO, and LIP) covaried with the FM, as expected. The FM ranged from 53.7 to 1074.2 μg L^{-1}, with an average of 180.5 μg L^{-1} (SD = $\pm$ 195.3 μg L^{-1}), and the FM decreased northward in this study.

3.3. Hydrolysable Compounds of POM

The concentrations of hydrolysable compounds (hydrolysable PRT, HPRT; hydrolysable LIP, HLIP; hydrolysable PRT, HPRT) in the POM were different among the groups (Figure 4A). In the southern part, the concentrations of HPRT ranged from 10.6 to 306.0 μg L^{-1} (mean $\pm$ SD = 93.4 $\pm$ 129.4 μg L^{-1}), and the concentrations of HLIP ranged from 33.4 to 132.2 μg L^{-1} (mean $\pm$ SD = 64.5 $\pm$ 36.7 μg L^{-1}) (Figure 4A). The HCHO ranged from 8.3 to 113.8 μg L^{-1}, with a mean of 56.5 $\pm$ 36.3 μg L^{-1} (Figure 4A). HLIP represented 79.0% of the total LIP value, followed by HCHO, which represented 34.2% of the total CHO value and HPRT, which represented 31.0% of the total PRT value. In comparison, the HPRT concentrations in the middle and northern parts ranged from 4.3 to 59.3 μg L^{-1} (mean $\pm$ SD = 25.5 $\pm$ 20.4 μg L^{-1}) and from 0.1 to 44.2 μg L^{-1} (mean $\pm$ SD = 20.6 $\pm$ 13.1 μg L^{-1}), respectively (Figure 4A). In the middle and northern parts, the HLIP concentrations ranged from 3.2 to 22.1 μg L^{-1} (mean $\pm$ SD = 8.6 $\pm$ 7.0 μg L^{-1}) and 0.9 to 23.9 μg L^{-1} (mean $\pm$ SD = 10.6 $\pm$ 6.2 μg L^{-1}), respectively, and the HCHO concentrations ranged from 19.7 to 124.6 μg L^{-1} (mean $\pm$ SD = 63.7 $\pm$ 35.6 μg L^{-1}) and 28.0 to 114.3 μg L^{-1} (mean $\pm$ SD = 52.8 $\pm$ 24.7 μg L^{-1}), respectively. Consistent with this observation, HCHO accounted for 72.1% (middle part) and 89.3% (northern part) of the overall value, which was more than the contributions of HLIP or HPRT (Figure 4). The concentrations of the hydrolysable compounds except for HCHO were higher in the southern part than in the middle or northern parts. Overall, the average concentrations of HLIP, HPRT, and HCHO at all the stations were 22.5 μg L^{-1} (SD = $\pm$ 29.1 μg L^{-1}), 38.2 μg L^{-1} (SD = $\pm$ 67.5 μg L^{-1}), and 55.9 μg L^{-1} (SD = $\pm$ 29.2 μg L^{-1}), respectively. The contributions of the hydrolysable components of POM to the total value were 56.1 $\pm$ 25.5% for LIP, 54.0 $\pm$ 31.3% for PRT, and 73.2 $\pm$ 26.6% for CHO.

In this study, the bioavailable fraction of POM (BFM, as the sum of HPRT, HLIP, and HCHO concentrations) can be considered the actual nutritional constituents and/or potentially available food for consumers that are able to be digested. In FM, the remaining values (excluding BFM) are expressed as a non-bioavailable form (N-BFM). In our study, similar to FM, the BFM concentration was much higher (mean $\pm$ SD = 214.4 $\pm$ 194.5 μg L^{-1}) in the southern than in the middle (mean $\pm$ SD = 97.8 $\pm$ 52.6 μg L^{-1}) and northern (mean $\pm$ SD = 84.1 $\pm$ 36.7 μg L^{-1}) parts. Similarly, the average N-BFM (256.0 μg L^{-1}) at the southern stations was approximately 4.3 and 8.4 times greater than that at the middle and northern stations, respectively. These results show that the positive effect of a large amount of FM is influenced by the quantity of BFM and that the majority of POM is not actually composed of bioavailable PRT, CHO, and LIP.

3.4. Multivariate Statistical Analysis

PCA was performed to determine the similarity among the environmental variables between stations. The PCA ordination of the sampled stations according to the measured environmental parameters is plotted in Figure 5 with eigenvalues presented in Table 2. The first two principal components (PC1 and PC2) accounted for 60.2% and 25.0% of the total variability, respectively. The temperature, salinity, density, and microphytoplankton (%) (eigenvectors of 0.931, 0.882, 0.785, and 0.891, respectively) were differentiated from the MW (%) and picophytoplankton (%) (eigenvectors of -0.907 and -0.880, respectively) by PC1, while PC2 was positively correlated with the major inorganic nutrient variables (eigenvectors $\geq$ 0.8). The analysis indicated general latitudinal groupings of stations in terms of their physical, chemical, and biological characteristics. The southern part was distinguished from the northern part by relatively high nutrient concentrations, temperatures, salinity, densities and relative contribution (%) of microphytoplankton. The northern part was characterized by a high MW (%) and relative contribution (%) of picophytoplankton values. The diagonal trajectory of the stations in the middle part within the ordination indicated that they represented a combination of PC1 and PC2.

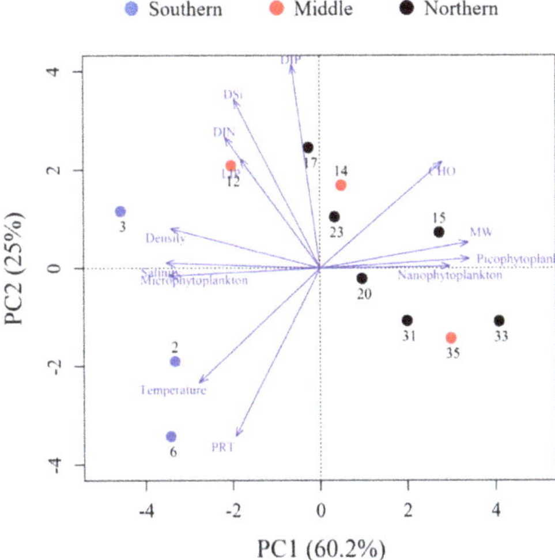

Figure 5. The principal component analysis performed from sampling stations. The environmental variables taken into consideration are temperature, salinity, density, nutrients (DIP; dissolved inorganic phosphate, DSi; dissolved inorganic silicate, DIN; dissolved inorganic nitrogen, nitrite + nitrate + ammonium), MW (meltwater, %), relative contribution of phytoplankton size classes (microphytoplankton, picophytoplankton, and nanophytoplankton), and relative contribution of biochemical pools (carbohydrates, CHO; proteins, PRT; lipids, LIP). Rotated eigenvectors for each parameter are indicated by arrows.

Table 2. A summary of eigenvectors of each environmental variable, eigenvalues, percentage (%) of variance explained by the first two axes resulted from the PCA. DIP; dissolved inorganic phosphate, DSi; dissolved inorganic silicate, DIN; dissolved inorganic nitrogen (nitrogen, nitrite + nitrate + ammonium).

Component	Principal Component	
	PC1	**PC2**
Eigenvalue	7.8	3.3
Percentage	60.2	25.0
Cumulative percent	60.2	85.2
Eigenvector		
Variables	**Eigenvectors**	
Salinity	0.8882	0.450
Temperature	0.931	−0.138
Density	0.785	0.579
DIP	−0.242	0.920
DSi	0.159	0.936
DIN	0.287	0.805
MW	−0.907	−0.303
Microphytoplankton (%)	0.891	0.389
Nanophytoplankton (%)	−0.753	−0.351
Picophytoplankton (%)	−0.880	−0.373
LIP (%)	0.240	0.673
PRT (%)	0.824	−0.463
CHO (%)	−0.919	0.103

4. Discussion

4.1. Origin and Quantity of POM

Our POM samples were collected by filtration and consisted of a variety of complex mixtures of compounds. Many studies have reported that chemical markers, such as chl-a, natural abundance of the stable isotopes of carbon ($\delta^{13}C$), and the C:N ratio, can be used to distinguish phytoplankton, as live components, from POM [9,12,30,52]. In our samples, the respective concentrations of PRT, LIP, and CHO in the POM had a linear relationship to the chl-a concentration (r = 0.689, 0.714, 0.724, $n = 47$, $p < 0.01$ for PRT, LIP, and CHO, respectively), which was used as a proxy for phytoplankton biomass. The $\delta^{13}C$ value of the POM ranged from −28.5 to −22.1‰ (mean ± SD = −26.2 ± 2.8‰); our values were within the range previously reported in phytoplankton samples (Kim et al. [30]; Ahn et al. [33]; reference therein). Kim et al. [30] and Ahn et al. [33] reported that POM is mainly derived from phytoplankton during summer in the Arctic Ocean based on $\delta^{13}C$ and the C:N ratio. In addition, during the sampling period, the low DIN concentration (<1 μM) and salinity distribution in the surface water suggest that the POM was greatly influenced by regional sea ice rather than a riverine source (terrigenous). Therefore, in our study, the POM was considered to have mainly come from a marine phytoplankton origin.

A field study has shown a large spatial variability in the concentration of FM in the Chukchi Sea. Kim et al. [30] reported that FM concentrations ranged from 80.5 to 698.8 μg L^{-1}, with an average of 294.4 μg L^{-1} (SD = ± 228.1 μg L^{-1}) in the euphotic layer of the Chukchi Sea, a value that was approximately 1.6 times higher than that in this study (mean ± SD = 180.5 ± 195.3 μg L^{-1}). Yun et al. [31] also found concentrations of FM similar to results from a previous study in this area in this area that similar to results from a previous study in this area, ranged from 89.7 to 362.4 μg L^{-1} with an average of 156.4 μg L^{-1} in the euphotic layer during summer. These variations are thought to result from spatial and temporal variations in the biomass, composition, and productivity of phytoplankton which are common in the Arctic Ocean [53].

Early studies indicated that primary production is higher in the southern Chukchi Sea than in the northern Chukchi Sea, which is consistent with chl-a abundance [27,54–57]. Based on ^{13}C uptake in the southern Chukchi Sea, Lee et al. [54] estimated a daily production of 0.6 g C m^{-2} day^{-1} (0.1 to 1.5 g C m^{-2} day^{-1}). In comparison, the estimated averages of the daily primary production rates for the southern Chukchi Sea are 1.6 g C m^{-2} day^{-1} and 1.7 g C m^{-2} day^{-1} from Zeeman [58] and Korsak [27], respectively. The mean daily production in the northern Chukchi Sea measured by Yun et al. [56] was somewhat lower (mean ± SD = 0.14 ± 0.10 g C m^{-2} day^{-1}) than the rate (0.66 ± 0.62 g C m^{-2} day^{-1}) in the southern region, which is consistent with the findings from Lee et al. [54] (0.16 ± 0.16 g C m^{-2} day^{-1}) and Lee et al. [55] (mean ± SD = 0.18 ± 0.07 g C m^{-2} day^{-1}). Similarly, the mean chl-a concentration (2.0 μg L^{-1}) in the southern part of the Chukchi Sea during the summer of 2017 was approximately one order of magnitude higher than the average value (0.2 μg L^{-1}) in the northern part (Figure 3). These results suggest that the regional differences in quantitative POM may have resulted from the different levels of phytoplankton biomass in the Chukchi Sea.

4.2. Biochemical Composition in Relation to Environmental Parameters

Overall, CHO accounted for 53.3% of the POM for all the survey stations, followed by PRT (29.2%) and LIP (17.5%) (Figure 6A), which led to a low PRT:CHO ratio (0.6). Consistent with this observation, the DIN:DIP (mean ± SD = 1.3 ± 2.0) molar ratio within the euphotic layer was also low compared with the N:P Redfield ratio of 16 [59], indicating substantial nitrogen limitation in this region (Figure 2B). However, interestingly, the PCA revealed that there were significant differences in the compounds among the groups (Figure 5). More specifically, the biochemical composition of the POM was dominated by PRT (41.8%); in the southern part, there was PRT:CHO ratio of 1.2 despite a low DIN:DIP ratio (mean ± SD = 2.7 ± 3.0), while a CHO-dominant (>50%) system was found in the northern Chukchi Sea with a PRT:CHO ratio of 0.5. In general, the PRT fraction was greater than the CHO and LIP fractions under sufficient nitrogen conditions and growth stages of phytoplankton, which could lead to a high (>1) PRT:CHO ratio [13,15,16]. Fogg and Thake [60] and Hu [61] reported that as prolonged stressful conditions (such as nitrogen limitation) occur, metabolic changes in synthesizing enzyme systems can convert CHO into LIP synthesis. Thus, our results suggest that at least in the southern region, nitrogen limitation was not severe enough to limit phytoplankton growth. The northern part has not been exposed to nitrogen stress for a long time.

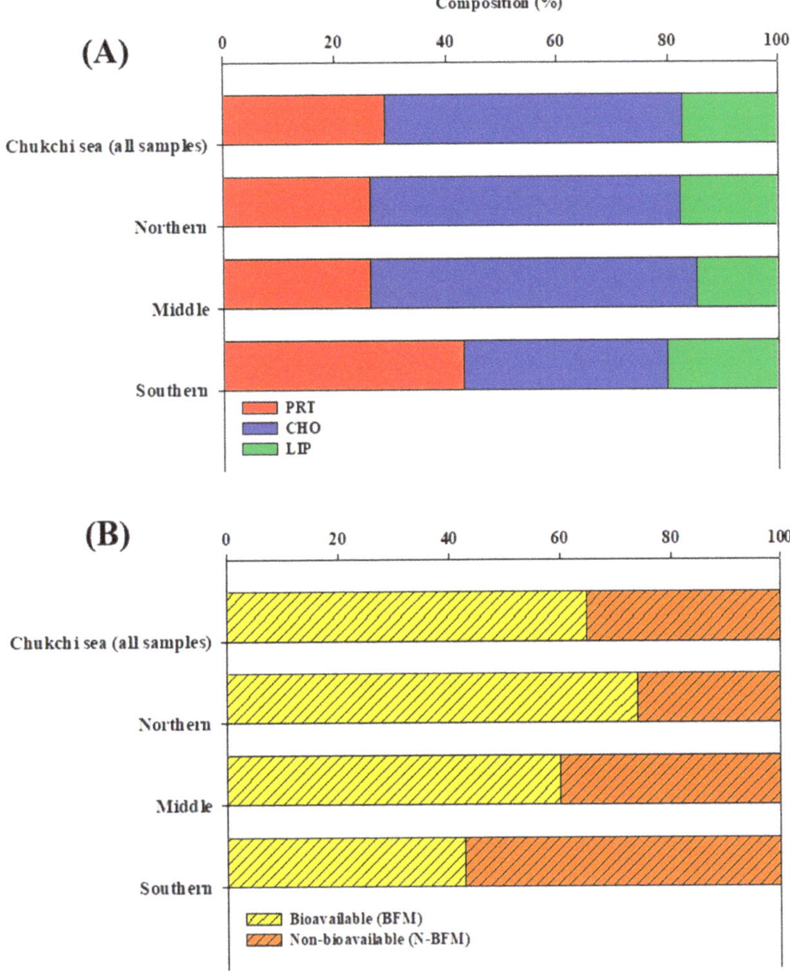

Figure 6. Spatial distribution of (**A**) specific biochemical compositions (PRT, CHO, and LIP) and (**B**) percentage non-bioavailable (N-BFM; non-hydrolysable) and bioavailable (BFM; hydrolysable, sum of hydrolysable PRT (HPRT), hydrolysable CHO (HCHO), and hydrolysable LIP (HLIP) concentration) fractions in each group (southern, middle, and northern part) and all samples (Chukchi Sea). Composition of the hydrolysable pool was deduced by subtraction of the non-hydrolysable pools from those of the total POM.

In addition, the results of the biplots (Figure 5) based on PCA revealed that microphytoplankton were influenced by relatively nutrient-replete conditions and had maximum chl-a and PRT values, while the picophytoplankton were more sensitive to nutrient deficiency and the MW (%), and were characterized by a high CHO contribution. These conditions were situated between two distinct areas (the southern and the northern parts) (Figure 5). Similar phenomena have been described by Jin et al. [62] and Li et al. [63] in relation to dominant species, nutrient depletion, and ice cover conditions in the western Arctic Ocean. Li et al. [63] suggested that small cells (<2 μm) thrive as a result of low nitrate availability and a strong stratification since pico-sized cells have a large surface area to volume ratio compared to that of larger cells, which allows effective nutrient and photon acquisition. CHEMTAX pigment analysis revealed that changes in temperature (caused by the timing of sea ice retreat) influence phytoplankton community structure [64]. Thus, it seems that the variation in biochemical compounds discovered among the two different groups (i.e., the southern and northern groups) could be the result of environmentally (such as the level of nutrients and freshwater) induced differences in the size classes and communities of phytoplankton in the Chukchi Sea.

Generally, the analysis of photosynthetic marker pigments (e.g., fucoxanthin, diadinoxanthin and diatoxanthin for diatoms, zeaxanthin for cyanobacteria, chlorophyll b and prasinoxanthin for green algae, 19′ butanoyloxy fucoxanthin and 19′ hexanoyloxy fucoxanthin for flagellates) can be useful biomarkers for phytoplankton biomass and species [65]. In our study, thirteen pigments except chl-a were identified through the (HPLC)-CHEMTAX analyses (Figure S1). As shown by the abundance of specific phytoplankton groups based on their corresponding biomarker pigments, the southern part was dominated by diatoms (88%), whereas pigments associated with flagellates and green algae made up approximately 44% of the total accessory pigment concentration and diatoms (53%) were observed in the northern Chukchi Sea (Figure S1). Several studies of phytoplankton have documented that species-specific characteristics, such as the cell wall structure and functional characteristics, play a significant role in the variation in biochemical components of phytoplankton [10,33,66–69]. Haug et al. [66] found that in diatoms, the concentration of PRT was generally much higher than that of CHO and LIP, whereas dinoflagellates have abundant CHO within their cell walls. Yun et al. [69] also reported that there were significantly close relationships between flagellates and the LIP fraction and diatoms and the PRT proportion in the Chukchi Sea. According to Shifrin and Chisholm [67], green algae and diatoms contained an average of 17.1% and 24.5% LIP (% of total dry weight), respectively, during log-phase growth in 30 cultured phytoplankton species. Therefore, the distributions and the relative contribution of these different dominant species and/or taxa of phytoplankton might also largely affect changes in the biochemical composition in the region.

4.3. Bioavailability of POM

Even now, the FM concentration, is used to represent the quantity of food in POM in relation to indicators of energy and material transfer to higher trophic levels [8,13,14,30]. However, FM is ideal when POM is made only of bioavailable constituents. In reality, POM contains bioavailable and non-bioavailable (refractory or less labile) fractions.

Bioavailability is a pivotal term for nutritional effectiveness, and the contribution of BFM to FM (%; nutritional efficiency) was used to assess POM bioavailability in this study. The nutritional efficiency ranged between 33.1 and 89.7%, with an average of 64.1% in the Chukchi Sea. More interestingly, the nutritional efficiency in the northern Chukchi Sea (74.0%) was approximately 1.2 times higher (60.0%) than it was in the middle part, while a lower mean value (42.7%) was observed in the southern part (Figure 6B). These results may have contributed to the different hydrolysis rates among the components, for which a greater presence is also an important factor. For example, the POM in the southern Chukchi Sea had a high contribution from PRT (41.8%) but a low level of HPRT (approximately 31.0% of their total pool), whereas a high level of HLIP (approximately 79% of their total pool) were observed despite a low contribution of LIP to the POM (20.5%). In the northern part, a large contribution from the HCHO (>80% of their total pool) was observed, with CHO accounting, for more than 50%, on average, of the POM in the northern part. In the middle part, HCHO accounted for 72.1% of the total CHO pool, followed by HPRT (55.4% of their total pool) and HLIP (40% of their total pool).

However, our findings, except for the middle part, are contradictory to the conclusions of Handa and Tominaga [70] and to the results obtained by Dawson and Liebezeit [71], Christian and Karl [72], and Fabiano et al. [8]. These reports suggested that cellular and proteinous amino acids were lost more rapidly than extractable sugars and particulate CHO. Such contrasting results suggest that different sizes [62] and species [73] of phytoplankton likely influence bioavailability. In our study, we found that the bioavailable contribution was negatively correlated with the relative amount of microphytoplankton (r = −0.652, $p < 0.05$, $n = 20$) and positively correlated with the relative amount of picophytoplankton (r = 0.668, $p < 0.05$, $n = 20$) (Figure S2); this result is consistent with the results from Jin et al. [62], who reported that picophytoplankton is more likely to mineralize and degrade in the upper ocean layers. In addition, CHO and amino acids are more enriched in intracellular materials than in cell wall materials [74]. Liebezeit [73] showed somewhat lower CHO degradation (38%) at stations dominated by diatoms than at stations dominated by Haptophyceae (86%) in the upper 100 m of the water column in the Bransfield Strait. Diatoms are characterized by silica shells (frustules) that are resistant to acid conditions (reviewed in DeNicolar [75]) and crushing forces [76]. In this sense, inherent structural differences in phytoplankton might also affect enzymatic hydrolysis, because phytoplankton was the major source of organic matter in our study. Taken together, although these results cannot be explained simply, different enzymatically hydrolysable efficiencies among the three different regions in the Chukchi Sea resulted from a selective loss of labile compounds and different communities of phytoplankton. Therefore, the higher POM bioavailability in the northern part of the Chukchi Sea could be caused by the different biochemical structures of the dominant picophytoplankton community compared to those of the microphytoplankton and diatom dominated community in the southern part of the Chukchi Sea. Clearly, a higher POM bioavailability provides more effective food materials for potential consumers in the northern part of the Chukchi Sea despite their lower biomass and lower primary productivity.

5. Conclusions

The biochemical composition of POM in the regions of the Chukchi Sea studied was due to differences in both environmental variables and the structure of the phytoplankton community. We also expect the observed results of the biochemical composition of POM to influence the nutritional quality of the available food. For instance, changes in the size, quantity and bioavailability of prey (phytoplankton) could affect the feeding, growth, reproduction and survival of predators [1,77,78]. In particular, in the Arctic Ocean, recent studies have indicated warming and decreased salinity of the water, with concomitant small phytoplankton sizes and decreased primary production [56,63]. If the sea ice continues to melt, then the quantity, quality, and labile level of POM will change, and consequently, the ecosystem structure, such as the trophic chain and microbial loop efficiency, will change in Arctic ecosystems. Therefore, further studies are needed to better understand the recent potential food materials under rapidly changing environmental conditions in the Arctic Ocean and picophytoplankton trophic roles in the microbial foodweb process.

Supplementary Materials: The following are available online at http://www.mdpi.com/2073-4441/12/9/2355/s1, Figure S1: Relative contribution of accessory pigments to total accessory pigment (wt:wt) in euphotic layer of southern, middle, and northern part of the Chukchi Sea, Figure S2: The relationship between relative contribution of micro (red dot) and picophytoplankton (green dot) fraction to total phytoplankton biomass (chl-a) and POM bioavailability. Solid lines indicate the fitted regression lines of the raw data points, Table S1: Average environmental parameters (± SD) within euphotic layer at each station in the Chukchi Sea.

Author Contributions: S.-Y.H. conceived of the study, participated in its design and helped to draft the manuscript; B.K.K. drafted the manuscript and performed the field and laboratory experiments; J.J. and Y.L. carried out the analysis of the nutrients, chl-a, $\delta^{13}C$, and pigments; K.-H.C. processed the CTD data; J.-K.G. critically reviewed the manuscript; S.-H.K. was the leader of the Korean Arctic Research Program and provided scientific advice. All authors have read and agreed to the published version of the manuscript.

Funding: This study was supported by Ministry of Oceans and Fisheries (MOF) and undertaken part of "Korea-Arctic Ocean Observing System (K-AOOS; 20160245)".

Acknowledgments: We thank the captain, officers, and crew of the R/V *Araon* for their valuable assistance at field work.

Conflicts of Interest: The authors declare that the research was conducted in the absence of any commercial or financial relationships that could be construed as a potential conflict of interest.

References

1. Søreide, J.E.; Leu, E.; Berge, J.; Graeve, M.; Falk-Petersen, S. Timing of blooms, algal food quality and *Calanus glacialis* reproduction and growth in a changing Arctic. *Glob. Chang. Boil.* **2010**, *16*. [CrossRef]

2. Engel, A.; Bracher, A.; Dinter, T.; Endres, S.; Grosse, J.; Metfies, K.; Peeken, I.; Piontek, J.; Salter, I.; Nöthig, E.-M. Inter-annual variability of organic carbon concentration in the eastern Fram Strait during summer (2009–2017). *Front. Mar. Sci.* **2019**, *6*, 1–17. [CrossRef]

3. Wassmann, P.; Duarte, C.M.; Agustí, S.; Sejr, M.K. Footprints of climate change in the Arctic marine ecosystem. *Glob. Chang. Boil.* **2011**, *17*, 1235–1249. [CrossRef]

4. Bhavya, P.S.; Kim, B.K.; Jo, N.; Kim, K.; Kang, J.J.; Lee, J.H.; Lee, D.; Lee, J.H.; Joo, H.; Ahn, S.H.; et al. A review on the macromolecular compositions of phytoplankton and the implications for aquatic biogeochemistry. *Ocean Sci. J.* **2018**, *54*, 1–14. [CrossRef]

5. Kreeger, D.; Goulden, C.; Kilham, S.; Lynn, S.; Datta, S.; Interlandi, S. Seasonal changes in the biochemistry of lake seston. *Freshw. Boil.* **1997**, *38*, 539–554. [CrossRef]

6. Fichez, R. Composition and fate of organic matter in submarine cave sediments; implications for the biogeochemical cycle of organic carbon. *Oceanol. Acta* **1991**, *14*, 369–377.

7. Danovaro, R.; Fabiano, M.; Croce, N.D. Labile organic matter and microbial biomasses in deep-sea sediments (Eastern Mediterranean Sea). *Deep. Sea Res. Part I Oceanogr. Res. Pap.* **1993**, *40*, 953–965. [CrossRef]

8. Fabiano, M.; Danovaro, R.; Crisafi, E.; La Ferla, R.; Povero, P.; Acosta-Pomar, L.; Ferla, R. Particulate matter composition and bacterial distribution in Terra Nova Bay (Antarctica) during summer 1989–1990. *Polar Boil.* **1995**, *15*, 393–400. [CrossRef]

9. Parsons, T.R.; Stephens, K.; Strickland, J.D.H. On the chemical composition of eleven species of marine phytoplankters. *J. Fish. Res. Board Can.* **1961**, *18*, 1001–1016. [CrossRef]

10. Brown, M.R. The amino-acid and sugar composition of 16 species of microalgae used in mariculture. *J. Exp. Mar. Boil. Ecol.* **1991**, *145*, 79–99. [CrossRef]

11. Richmond, A.; Hu, Q. *Handbook of Microalgal Culture: Applied Phycology and Biotechnology*, 2nd ed.; John Wiley and Sons: Hoboken, NJ, USA, 2011. [CrossRef]

12. Fabiano, M.; Povero, P.; Danovaro, R. Distribution and composition of particulate organic matter in the Ross Sea (Antarctica). *Polar Boil.* **1993**, *13*, 525–533. [CrossRef]

13. Danovaro, R.; Dell'Anno, A.; Pusceddu, A.; Marrale, D.; Croce, N.D.; Fabiano, M.; Tselepides, A. Biochemical composition of pico-, nano- and micro-particulate organic matter and bacterioplankton biomass in the oligotrophic Cretan Sea (NE Mediterranean). *Prog. Oceanogr.* **2000**, *46*, 279–310. [CrossRef]

14. Kim, B.K.; Lee, S.; Ha, S.-Y.; Jung, J.; Kim, T.-W.; Yang, E.J.; Jo, N.; Lim, Y.J.; Park, J.; Lee, S.H. Vertical distributions of macromolecular composition of particulate organic matter in the water column of the amundsen sea polynya during the summer in 2014. *J. Geophys. Res. Oceans* **2018**, *123*, 1393–1405. [CrossRef]

15. Hecky, R.E.; Kilham, P. Nutrient limitation of phytoplankton in freshwater and marine environments: A review of recent evidence on the effects of enrichment. *Limnol. Oceanogr.* **1988**, *33*, 796–822. [CrossRef]

16. Kilham, S.; Kreeger, D.; Goulden, C.; Lynn, S. Effects of nutrient limitation on biochemical constituents of *Ankistrodesmus falcatus*. *Freshw. Boil.* **1997**, *38*, 591–596. [CrossRef]

17. Lee, S.H.; Whitledge, T.E.; Kang, S.-H. Carbon uptake rates of sea ice algae and phytoplankton under different light intensities in a landfast sea ice zone, Barrow, Alaska. *Arctic* **2008**, *61*, 281–291. [CrossRef]

18. Smith, A.E.; Morris, I. Pathways of carbon assimilation in phytoplankton from the Antarctic Ocean1. *Limnol. Oceanogr.* **1980**, *25*, 865–872. [CrossRef]

19. Takagi, M.; Watanabe, K.; Yamaberi, K.; Yoshida, T. Limited feeding of potassium nitrate for intracellular lipid and triglyceride accumulation of *Nannochloris* sp. UTEX LB1999. *Appl. Microbiol. Biotechnol.* **2000**, *54*, 112–117. [CrossRef]

20. Li, Y.; Horsman, M.E.; Wang, B.; Wu, N.; Lan, C.Q. Effects of nitrogen sources on cell growth and lipid accumulation of green alga *Neochloris oleoabundans*. *Appl. Microbiol. Biotechnol.* **2008**, *81*, 629–636. [CrossRef]

21. Winberg, G.G. *Symbols, Units and Conversion Factors in Study of Fresh Waters Productivity*; International Biological Programme: London, UK, 1971; p. 23.

22. Dell'Anno, A.; Fabiano, M.; Mei, M.; Danovaro, R. Enzymatically hydrolysed protein and carbohydrate pools in deep-sea sediments: Estimates of the potentially bioavailable fraction and methodological considerations. *Mar. Ecol. Prog. Ser.* **2000**, *196*, 15–23. [CrossRef]

23. Pusceddu, A.; Dell'Anno, A.; Danovaro, R.; Manini, E.; Sarà, G.; Fabiano, M. Enzymatically hydrolyzable protein and carbohydrate sedimentary pools as indicators of the trophic state of detritus sink systems: A case study in a Mediterranean coastal lagoon. *Estuaries* **2003**, *26*, 641–650. [CrossRef]

24. Misic, C.; Harriague, A.C.; Mangoni, O.; Aulicino, G.; Castagno, P.; Cotroneo, Y. Effects of physical constraints on the lability of POM during summer in the Ross Sea. *J. Mar. Syst.* **2017**, *166*, 132–143. [CrossRef]

25. Woodgate, R.A.; Aagaard, K.; Weingartner, T.J. Monthly temperature, salinity, and transport variability of the Bering Strait through inflow. *Geophys. Res. Lett.* **2005**, *32*, L04601. [CrossRef]

26. Yang, Y.; Bai, X. Summer changes in water mass characteristics and vertical thermohaline structure in the Eastern Chukchi Sea, 1974–2017. *Water* **2020**, *12*, 1434. [CrossRef]

27. Korsak, M.N. Primary production of organic matter. In *Results of the Third Joint US-USSR Bering and Chukchi Seas Expedition (BERPAC): Summer 1988*; Nagel, P.A., Ed.; US Fish and Wildlife Service: Washington, DC, USA, 1992; pp. 215–218.

28. Grebmeier, J.; Cooper, L.; Feder, H.M.; Sirenko, B.I. Ecosystem dynamics of the Pacific-influenced Northern Bering and Chukchi Seas in the Amerasian Arctic. *Prog. Oceanogr.* **2006**, *71*, 331–361. [CrossRef]

29. Grebmeier, J.M. Shifting patterns of life in the Pacific arctic and Sub-Arctic seas. *Annu. Rev. Mar. Sci.* **2012**, *4*, 63–78. [CrossRef]

30. Kim, B.K.; Lee, J.H.; Yun, M.S.; Joo, H.; Song, H.J.; Yang, E.J.; Chung, K.H.; Kang, S.-H.; Lee, S.H. High lipid composition of particulate organic matter in the northern Chukchi Sea, 2011. *Deep. Sea Res. Part II Top. Stud. Oceanogr.* **2015**, *120*, 72–81. [CrossRef]

31. Yun, M.S.; Lee, D.B.; Kim, B.K.; Kang, J.J.; Lee, J.H.; Yang, E.J.; Park, W.G.; Chung, K.H.; Lee, S.H. Comparison of phytoplankton macromolecular compositions and zooplankton proximate compositions in the northern Chukchi Sea. *Deep. Sea Res. Part II Top. Stud. Oceanogr.* **2015**, *120*, 82–90. [CrossRef]

32. Yun, M.S.; Joo, H.T.; Park, J.W.; Kang, J.J.; Kang, S.-H.; Lee, S.H. Lipid-rich and protein-poor carbon allocation patterns of phytoplankton in the northern Chukchi Sea, 2011. *Cont. Shelf Res.* **2018**, *158*, 26–32. [CrossRef]

33. Ahn, S.H.; Whitledge, T.E.; Stockwell, D.A.; Lee, J.H.; Lee, H.W.; Lee, S.H. The biochemical composition of phytoplankton in the Laptev and East Siberian seas during the summer of 2013. *Polar Boil.* **2018**, *42*, 133–148. [CrossRef]

34. Marra, J.F.; Lance, V.P.; Vaillancourt, R.D.; Hargreaves, B.R. Resolving the ocean's euphotic zone. *Deep. Sea Res. Part I Oceanogr. Res. Pap.* **2014**, *83*, 45–50. [CrossRef]

35. Catalano, G.; Povero, P.; Fabiano, M.; Benedetti, F.; Goffart, A. Nutrient utilisation and particulate organic matter changes during summer in the upper mixed layer (Ross Sea, Antarctica). *Deep. Sea Res. Part I Oceanogr. Res. Pap.* **1997**, *44*, 97–112. [CrossRef]

36. Rivaro, P.F.; Abelmoschi, M.L.; Grotti, M.; Ianni, C.; Magi, E.; Margiotta, F.; Massolo, S.; Saggiomo, V. Combined effects of hydrographic structure and iron and copper availability on the phytoplankton growth in Terra Nova Bay Polynya (Ross Sea, Antarctica). *Deep. Sea Res. Part I Oceanogr. Res. Pap.* **2012**, *62*, 97–110. [CrossRef]

37. Bates, N.R.; Garley, R.; Frey, K.E.; Shake, K.L.; Mathis, J.T. Sea-ice melt CO_2–carbonate chemistry in the western Arctic Ocean: Meltwater contributions to air–sea CO_2 gas exchange, mixed-layer properties and rates of net community production under sea ice. *Biogeosciences* **2014**, *11*, 6769–6789. [CrossRef]

38. Fetterer, F.; Knowles, K.; Meier, W.N.; Savoie, M.; Windnagel, A.K. *Updated Daily. Sea Ice Index, Version 3. [Indicate Subset Used]*; National Snow and Ice Data Center (NSIDC): Boulder, CO, USA, 2017. [CrossRef]

39. Parsons, T.R.; Maita, Y.; Lalli, C.M. *A Manual of Chemical and Biological Methods for Seawater Analysis*; Pergamon Press: Oxford, UK, 1984; p. 173.

40. Coplen, T.B.; Brand, W.A.; Gehre, M.; Gröning, M.; Meijer, H.A.J.; TOMAN, B.; Verkouteren, R.M. New guidelines for δ13C measurements. *Anal. Chem.* **2006**, *78*, 2439–2441. [CrossRef]

41. Lowry, O.H.; Rosebrough, N.J.; Farr, A.L.; Randall, R.J. Protein measurement with the Folin phenol reagent. *J. Boil. Chem.* **1951**, *193*, 265–275.

42. Dubois, M.; Gilles, K.A.; Hamilton, J.K.; Rebers, P.A.; Smith, F. Colorimetric method for determination of sugars and related substances. *Anal. Chem.* **1956**, *28*, 350–356. [CrossRef]

43. Bligh, E.G.; Dyer, W.J. A rapid method of total lipid extraction and purification. *Can. J. Biochem. Physiol.* **1959**, *37*, 911–917. [CrossRef]

44. Marsh, J.B.; Weinstein, D.B. Simple charring method for determination of lipids. *J. Lipid Res.* **1966**, *7*, 574–576.

45. Dall, W.; Moriaty, J.W. The biology of crustacea. In *Functional Aspects of Nutrition and Digestion*; Mantel, L.H., Ed.; Academic Press: New York, NY, USA, 1983; pp. 215–261.

46. Gordon, D.C., Jr. A microscopic study of organic particles in the North Atlantic Ocean. *Deep. Sea Res. Oceanogr. Abstr.* **1970**, *17*, 175–185. [CrossRef]

47. Mayer, L.M.; Linda, S.L.; Sawyer, T.; Plante, C.J.; Jumars, P.A.; Self, R.L.; Schick, L.L. Bioavailable amino acids in sediments: A biomimetic, kinetics based approach. *Limnol. Oceanogr.* **1995**, *40*, 511–520. [CrossRef]

48. Fabiano, M.; Pusceddu, A. Total and hydrolizable particulate organic matter (carbohydrates, proteins and lipids) at a coastal station in Terra Nova Bay (Ross Sea, Antarctica). *Polar Boil.* **1998**, *19*, 125–132. [CrossRef]

49. Pusceddu, A.; Bianchelli, S.; Danovaro, R. Quantity and biochemical composition of particulate organic matter in a highly trawled area (Thermaikos Gulf, Eastern Mediterranean Sea). *Adv. Oceanogr. Limnol.* **2015**, *6*, 21–32. [CrossRef]

50. Gong, D.; Pickart, R.S. Summertime circulation in the eastern Chukchi Sea. *Deep. Sea Res. Part II Top. Stud. Oceanogr.* **2015**, *118*, 18–31. [CrossRef]

51. Stroh, J.N.; Panteleev, G.; Kirillov, S.; Makhotin, M.; Shakhova, N. Sea-surface temperature and salinity product comparison against external in situ data in the Arctic Ocean. *J. Geophys. Res. Ocean.* **2015**, *120*, 7223–7236. [CrossRef]

52. Volkman, J.K.; Tanoue, E. Chemical and biological studies of particulate organic matter in the ocean. *J. Oceanogr.* **2003**, *25*, 265–279. [CrossRef]

53. Yun, M.S.; Chung, K.H.; Zimmermann, S.; Zhao, J.; Joo, H.M.; Lee, S.H. Phytoplankton productivity and its response to higher light levels in the Canada Basin. *Polar Boil.* **2012**, *35*, 257–268. [CrossRef]

54. Lee, S.H.; Whitledge, T.E.; Kang, S.-H. Recent carbon and nitrogen uptake rates of phytoplankton in Bering Strait and the Chukchi Sea. *Cont. Shelf Res.* **2007**, *27*, 2231–2249. [CrossRef]

55. Lee, S.H.; Yun, M.S.; Kim, B.K.; Saitoh, S.-I.; Kang, C.-K.; Kang, S.-H.; Whitledge, T. Latitudinal carbon productivity in the Bering and Chukchi Seas during the summer in 2007. *Cont. Shelf Res.* **2013**, *59*, 28–36. [CrossRef]

56. Yun, M.S.; Whitledge, T.E.; Stockwell, D.; Son, S.H.; Lee, J.H.; Lee, D.B.; Lee, S.H.; Park, J.W.; Park, J. Primary production in the Chukchi Sea with potential effects of freshwater content. *Biogeosciences* **2016**, *13*, 737–749. [CrossRef]

57. Nishino, S.; Kikuchi, T.; Fujiwara, A.; Hirawake, T.; Aoyama, M. Water mass characteristics and their temporal changes in a biological hotspot in the southern Chukchi Sea. *Biogeosciences* **2016**, *13*, 2563–2578. [CrossRef]

58. Zeeman, S.I. The importance of primary production and CO_2. In *Results of the Third Joint US-USSR Bering and Chukchi Seas Expedition (BERPAC), Summer 1988*; U.S. Fish and Wildlife Service: Washington, DC, USA, 1992; pp. 39–49.

59. Redfield, A.C.; Ketchum, B.H.; Richards, F.A. *The Influence of Organisms on the Composition of Sea-Water in the Sea*; Hill, M.N., Ed.; Springer: New York, NY, USA, 1963; pp. 26–77.

60. Fogg, G.E.; Thake, B. Cultures of limited volume. In *Algal Cultures and Phytoplankton Ecology*; University of Wisconsin Press: Maddison, WI, USA, 1987; pp. 12–42.

61. Hu, Q. Environmental effects on cell composition. In *Handbook of Microalgal Culture: Applied Phycology and Biotechnology*; Richmond, A., Ed.; Oxford University Press: Oxford, UK, 2004; pp. 114–122.

62. Jin, H.; Zhuang, Y.; Li, H.; Chen, J.; Gao, S.; Ji, Z.; Zhang, Y. Response of phytoplankton community to different water types in the western Arctic Ocean surface water based on pigment analysis in summer 2008. *Acta Oceanol. Sin.* **2017**, *36*, 109–121. [CrossRef]

63. Li, W.K.W.; McLaughlin, F.A.; Lovejoy, C.; Carmack, E.C. Smallest algae thrive as the arctic ocean freshens. *Science* **2009**, *326*, 539. [CrossRef] [PubMed]

64. Fujiwara, A.; Hirawake, T.; Suzuki, K.; Imai, I.; Saitoh, S.-I. Timing of sea ice retreat can alter phytoplankton community structure in the western Arctic Ocean. *Biogeosciences* **2014**, *11*, 1705–1716. [CrossRef]
65. Jeffrey, S.W.; Mantoura, R.F.C.; Bjørnland, T. Data for the identification of 47 key phytoplankton pigments. In *Phytoplankton Pigments in Oceanography: Guidelines to Modern Methods. Monographs on Oceanographic Methodology*; Jeffrey, S.W., Mantoura, R.F.C., Wright, S.W., Eds.; UNESCO: Paris, France, 1997; pp. 449–559.
66. Haug, A.; Myklestad, S.; Sakshaug, E. Studies on the phytoplankton ecology of the Trondheimsfjord. I. The chemical composition of phytoplankton populations. *J. Exp. Mar. Boil. Ecol.* **1973**, *11*, 15–26. [CrossRef]
67. Shifrin, N.S.; Chisholm, S.W. Phytoplankton lipids: Interspecific differences and effects of nitrate, silicate and light-dark cycles1. *J. Phycol.* **2008**, *17*, 374–384. [CrossRef]
68. Jo, N.; Kang, J.J.; Park, W.G.; Lee, B.R.; Yun, M.S.; Lee, J.H.; Kim, S.M.; Lee, D.; Joo, H.; Lee, J.H.; et al. Seasonal variation in the biochemical compositions of phytoplankton and zooplankton communities in the southwestern East/Japan Sea. *Deep. Sea Res. Part II Top. Stud. Oceanogr.* **2017**, *143*, 82–90. [CrossRef]
69. Yun, M.S.; Joo, H.M.; Kang, J.J.; Park, J.W.; Lee, J.H.; Kang, S.-H.; Sun, J.; Lee, S.H. Potential implications of changing Photosynthetic end-products of Phytoplankton caused by sea ice conditions in the Northern Chukchi Sea. *Front. Microbiol.* **2019**, *10*, 2274. [CrossRef]
70. Handa, N.; Tominaga, H. A detailed analysis of carbohydrates in marine particulate matter. *Mar. Boil.* **1969**, *2*, 228–235. [CrossRef]
71. Dawson, R.; Liebezeit, G. Biochemical compounds in the pelagic and sedimentary environment of Antarctic waters. *Publs. Cent. Natn. Exploit. Océans (Sér. Act. Colloques)* **1982**, *14*, 67–86.
72. Christian, J.R.; Karl, D.M. Bacterial ectoenzymes in marine waters: Activity ratios and temperature responses in three oceanographic provinces. *Limnol. Oceanogr.* **1995**, *40*, 1042–1049. [CrossRef]
73. Liebezeit, G. Particulate carbohydrates in relation to phytoplankton in the euphotic zone of the Bransfield Strait. *Polar Boil.* **1984**, *2*, 225–228. [CrossRef]
74. Cowie, G.L.; Hedges, J.I. Digestion and alteration of the biochemical constituents of a diatom (*Thalassiosira weisflogii*) ingested by an herbivorous zooplankton (*Calanus pacificus*). *Limnol. Oceanogr.* **1996**, *41*, 581–594. [CrossRef]
75. DeNicola, D.M. A review of diatoms found in highly acidic environments. *Hydrobiology* **2000**, *433*, 111–122. [CrossRef]
76. Hamm, C.; Merkel, R.; Springer, O.; Jurkojc, P.; Maier, C.; Prechtel, K.; Smetacek, V. Architecture and material properties of diatom shells provide effective mechanical protection. *Nature* **2003**, *421*, 841–843. [CrossRef]
77. Vargas, C.A.; Escribano, R.; Poulet, S. Phytoplankton food quality determines time windows for successful zooplankton reproductive pulses. *Ecology* **2006**, *87*, 2992–2999. [CrossRef]
78. Friedrichs, L.; Hornig, M.; Schulze, L.; Bertram, A.; Jansen, S.; Hamm, C. Size and biomechanic properties of diatom frustules influence food uptake by copepods. *Mar. Ecol. Prog. Ser.* **2013**, *481*, 41–51. [CrossRef]

Article

Contribution of Small Phytoplankton to Primary Production in the Northern Bering and Chukchi Seas

Jung-Woo Park [1,*], Yejin Kim [2], Kwan-Woo Kim [2], Amane Fujiwara [3], Hisatomo Waga [1,4], Jae Joong Kang [2], Sang-Heon Lee [2], Eun-Jin Yang [5] and Toru Hirawake [1,6]

[1] Faculty/Graduate School of Fisheries Sciences, Hokkaido University, Hakodate 041-8611, Japan; hwaga@alaska.edu (H.W.); hirawake.toru@nipr.ac.jp (T.H.)
[2] Department of Oceanography, Pusan National University, Busan 46241, Korea; yejini@pusan.ac.kr (Y.K.); goanwoo7@pusan.ac.kr (K.-W.K.); jaejung@pusan.ac.kr (J.J.K.); sanglee@pusan.ac.kr (S.-H.L.)
[3] Research Institute for Global Change, Japan Agency for Marine-Earth Science and Technology, Yokosuka 237-0061, Japan; amane@jamstec.go.jp
[4] International Arctic Research Center, University of Alaska Fairbanks, Fairbanks, AK 99775, USA
[5] Division of Polar Ocean Sciences, Korea Polar Research Institute, Incheon 21990, Korea; ejyang@kopri.re.kr
[6] National Institute of Polar Research, Tachikawa 190-8518, Japan
*   Correspondence: park@salmon.fish.hokudai.ac.jp

Citation: Park, J.-W.; Kim, Y.; Kim, K.-W.; Fujiwara, A.; Waga, H.; Kang, J.J.; Lee, S.-H.; Yang, E.-J.; Hirawake, T. Contribution of Small Phytoplankton to Primary Production in the Northern Bering and Chukchi Seas. *Water* **2022**, *14*, 235. https://doi.org/10.3390/w14020235

Academic Editor: Michele Mistri

Received: 30 November 2021
Accepted: 8 January 2022
Published: 14 January 2022

Abstract: The northern Bering and Chukchi seas are biologically productive regions but, recently, unprecedented environmental changes have been reported. For investigating the dominant phytoplankton communities and relative contribution of small phytoplankton (<2 μm) to the total primary production in the regions, field measurements mainly for high-performance liquid chromatography (HPLC) and size-specific primary productivity were conducted in the northern Bering and Chukchi seas during summer 2016 (ARA07B) and 2017 (OS040). Diatoms and *phaeocystis* were dominant phytoplankton communities in 2016 whereas diatoms and Prasinophytes (Type 2) were dominant in 2017 and diatoms were found as major contributors for the small phytoplankton groups. For size-specific primary production, small phytoplankton contributed 38.0% (SD = ±19.9%) in 2016 whereas 25.0% (SD = ±12.8%) in 2017 to the total primary productivity. The small phytoplankton contribution observed in 2016 is comparable to those reported previously in the Chukchi Sea whereas the contribution in 2017 mainly in the northern Bering Sea is considerably lower than those in other arctic regions. Different biochemical compositions were distinct between small and large phytoplankton in this study, which is consistent with previous results. Significantly higher carbon (C) and nitrogen (N) contents per unit of chlorophyll-*a*, whereas lower C:N ratios were characteristics in small phytoplankton in comparison to large phytoplankton. Given these results, we could conclude that small phytoplankton synthesize nitrogen-rich particulate organic carbon which could be easily regenerated.

Keywords: Bering Sea; Chukchi Sea; HPLC; small phytoplankton; primary productivity

1. Introduction

The biologically productive northern Bering Sea and the Chukchi Sea are important conduit of water masses and organic matters from the North Pacific Ocean transported into the Arctic Ocean and biologically productive regions [1–5]. Over the past few decades, many environmental changes have been reported in the regions [4,6–8]. Unprecedented high sea surface temperature was reported in the Bering Sea in 2014 and persisted in 2018 and 2019 [8; refs therein]. The Pacific origin freshwater flux with increasing northward volume transport into the Arctic Ocean had been increased over the 1991–2015 period [9]. Moreover, seasonal sea ice cover has been retreating earlier and forming later in the Pacific Arctic region over the last decade [10]. These current and ongoing changes in environmental conditions could subsequently cause changes in biogeochemical processes and consequently alter marine ecosystem structure in the northern Bering and Chukchi seas [11,12]. Indeed, the prior studies indicate that the variation in primary productivity of phytoplankton is mainly governed by freshwater content variability in the Pacific Arctic region [13,14]. Moreover, the seasonal sea ice cover could largely influence phytoplankton community composition [15], phytoplankton bloom period [16] and primary productivity [17].

Refs. [13,18–20] reported that pico-phytoplankton increased whereas larger cells declined in the Arctic Ocean because of stronger stratification and consequently lower nutrient supply into the upper water column caused by freshening surface waters. Based on the phytoplankton size classes derived from satellite ocean color data in the northern Bering and southern Chukchi seas [21], observed increasing trends in pico-phytoplankton in the Chirikov and St. Lawrence Island Polynya regions whereas an increasing trend in micro-phytoplankton in the southeastern Chukchi Sea from 1998 to 2016. The physiological conditions and subsequently photosynthetic end-products of phytoplankton affected by the recent environmental conditions were also previously reported in the northern Bering and southern Chukchi seas [21–23]. Phytoplankton as important primary producers in marine ecosystems can be a good indicator of environmental changes. These long-term changes in the functional phytoplankton group are strongly related to increasing annual sea surface temperature [13]. Therefore, monitoring the phytoplankton community responses such as shifts in dominant phytoplankton species and biomass to the current environmental changes is crucial to observe marine ecosystem alterations in the northern Bering and Chukchi seas [12,18–20].

Especially, the contribution of small phytoplankton could be necessary to understand potential impacts on the total primary production and, thus, whole marine ecosystems [12,20,21]. Moreover, the biochemical characteristics of phytoplankton such as C:N ratio are critical for understanding marine biogeochemical processes responding to environmental conditions. Ref. [24] reported higher C:N ratio related with low chlorophyll-*a* concentration and lower C:N ratio to high chlorophyll-*a* concentration in the Arctic Ocean. The C:N ratio could differ in various environmental conditions related to nutrients. However, little information on the small phytoplankton contribution to the total primary production and their biochemical traits such as C:N ratio is currently available in the northern Bering and Chukchi seas.

In this study, our objectives are to investigate the dominant phytoplankton communities and to assess the relative contribution of small phytoplankton (0.7–2.0 μm; pico-phytoplankton) to the total primary production and their biochemical characteristics (e.g., C:N ratio) in the northern Bering and Chukchi seas.

2. Materials and Methods

2.1. Study Area and Water Sampling

The ARA07B cruise was conducted in the northern Bering Sea and the Chukchi Sea during 5–19 August, 2016 onboard the Icebreaker R/V *Araon* (Figure 1; Table 1). As a total of 16 stations during the ARA07B cruises, only one station (st. 1) was located in the northern Bering Sea and 15 stations were in the Chukchi Sea. Water was sampled by Niskin bottles on conductivity-temperature-depth (CTD)/rosette sampler for the total chlorophyll-*a* and size-fractionated chlorophyll-*a* concentration. Euphotic depths were measured by a Secchi disk [25]. The OS040 cruise was executed mostly in the northern Bering Sea (8 stations) and partly in the southern Chukchi Sea (2 stations) during 9–21 July, 2017 onboard T/S *Oshoro-Maru* (Figure 1; Table 1). Physical properties and water samples were collected by CTD/rosette with Niskin bottles. The euphotic depths were calculated by comparing downward irradiance and surface irradiance measured by compact optical profiling system (C-OPS; Biospherical instrument Inc., San Diego, CA, USA).

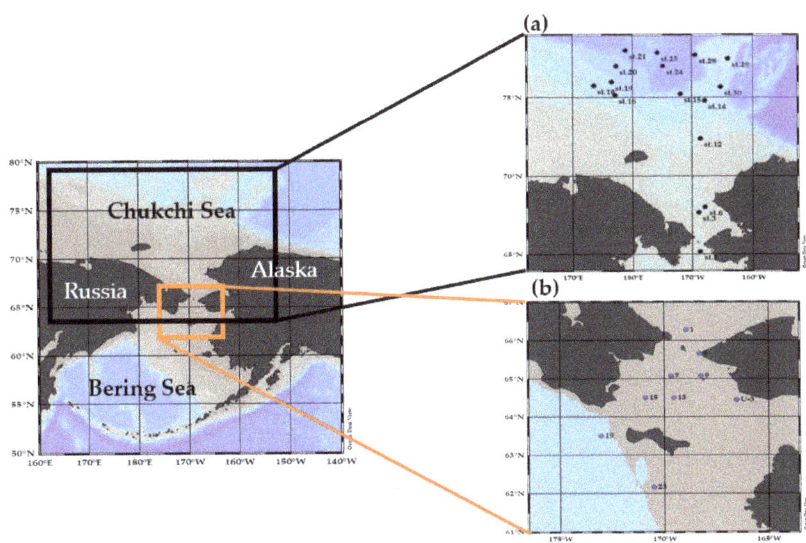

Figure 1. Sampling locations of (**a**) ARA07B and (**b**) OS040 cruises.

Table 1. Sampling locations in the Northern Bering and Chukchi Seas.

	Station	Date	Latitude (°N)	Longitude (°E)	Bottom depth (m)
	st. 1	5 Aug 2016	65.17	−168.69	49
	st. 3	6 Aug 2016	67.67	−168.96	50
	st. 6	6 Aug 2016	68.01	−167.87	52.3
	st. 12	8 Aug 2016	72.36	−168.67	55
	st. 14	9 Aug 2016	74.80	−167.90	223
	st. 15	17 Aug 2016	75.24	−171.98	512
	st. 16	10 Aug 2016	75.15	−176.00	331.57
ARA07B	st. 18	11 Aug 2016	75.77	177.07	486
	st. 19	12 Aug 2016	76.00	173.60	282
	st. 20	12 Aug 2016	77.00	176.57	1232
	st. 21	13 Aug 2016	78.00	177.28	1693
	st. 23	15 Aug 2016	77.87	−175.91	1564
	st. 24	16 Aug 2016	77.00	−175.00	2008
	st. 28	18 Aug 2016	77.70	−169.50	1756
	st. 29	18 Aug 2016	77.47	−164.12	280
	st. 30	19 Aug 2016	76.58	−165.38	987
	1	9 Jul 2017	66.28	−168.90	57
	5	11 Jul 2017	65.66	−168.26	45
	7	12 Jul 2017	65.06	−169.64	51
	9	14 Jul 2017	65.07	−168.19	42
OS040	U−3	16 Jul 2017	64.44	−166.52	27
	13	17 Jul 2017	64.50	−169.52	42
	15	18 Jul 2017	64.50	−170.89	46
	19	19 Jul 2017	63.50	−173.02	66
	23	21 Jul 2017	62.17	−170.50	46

2.2. Chlorophyll-a Analysis

The water samples were obtained from 6 different light depths (100%, 50%, 30%, 12%, 5% and 1% of the surface photosynthetically active radiation (PAR) for measuring the chlorophyll-*a* concentration. For the total chlorophyll-*a* concentration, 300 mL of seawater was filtered through 25 mm glass fiber filter (GF/F; Whatman). To obtain size-fractionated chlorophyll-*a* concentration, 500 mL seawater was filtered through 20 μm and 2 μm pore size membrane filters and then 47 mm GF/F sequentially. After the filtration was done, the filters were wrapped with aluminum foil and stored at −80°C freezer until analysis at the home laboratory. Chlorophyll-*a* extractions were followed by [26] and the concentrations were measured with a fluorometer (Turner Designs 10AU).

2.3. High-Performance Liquid Chromatography Analysis for Accessory Pigment Concentration

For high-performance liquid chromatography (HPLC) analysis, the water from 3 light depths (100%, 30% and 1%) were sampled during the ARA07B and OS040 cruises. Seawater (0.8–2.5 L) was passed through 2 μm membrane filter and 47 mm diameter GF/F filters to measure pigments concentration of small size phytoplankton (<2 μm) under gentle vacuum pressure (<100 mmHg). Seawater (0.5–1.5 L) was filtered onto GF/F for pigments of total phytoplankton during the ARA07B. For the OS040, samples were obtained only for total phytoplankton. For avoiding degradation, the filters for HPLC analysis were immediately frozen and stored in liquid nitrogen at −80°C freezer until analysis at home laboratory. In the laboratory, the filter samples were broken into small pieces and then soaked in 3 mL of N′N-dimethylformamide (DMF) with canthaxanthin served as an internal standard. After 20 min of sonication, the filters were extracted at 4°C in dark for 24 h and then extracts were filtered through a 0.45 μm pore membrane filter to remove GF/F particles. For minimizing photo-degradation of pigments, all the procedures were conducted under a low light condition. Pigments were analyzed using HPLC (Agilent Infinite 1260 in operation by JAMSTEC, Mutsu, Japan) with a ternary linear gradient system to separate each pigment. The pigment concentrations were calculated by the function of peak area, standard response factors and peak area of the internal standard following [27]. All the standards for each pigment were purchased from DHI in Denmark.

The CHEMTAX software based on a factorization program was used for estimating the relative contributions of different phytoplankton communities to the total chlorophyll-*a* concentration [28]. The ratios of accessory pigments to chlorophyll-*a* for each phytoplankton taxon for the CHEMTAX program were based on marker pigment concentrations of algal groups present in the Arctic Ocean [13,29] (Table 2). Since our two research cruises were in different periods and years, the final ratio matrix was separated for phytoplankton communities (Table 2). The contributions of Diatoms. Dinoflagellates, Cryptophytes, Pelagophytes, Prasinophytes (Type 2 and 3), Chlorophytes, Haptophytes and Phaeosystis were estimated by the CHEMTAX program. Small phytoplankton community was estimated from HPLC results by the equations described in the literature [28,29]. The relative proportions of the three size classes are derived from the concentrations of phytoplankton diagnostic pigments for the Chukchi and Bering seas using the equations described in [30,31].

Table 2. Pigment:chlorophyll-*a* ratios for nine algal groups referred to [32]. CHEMTAX initial ratio matrix and final pigment ratios obtained by CHEMTAX on the pigment data.

Class	chl-b	chl-c3	fucox	perid	allox	19butfu	19hexfu	chl-c	neox	prasinox	lut
Initial ratio matrix											
Diatoms	0	0	0.425	0	0	0	0	0.171	0	0	0
Dinoflagellates	0	0	0	0.6	0	0	0	0	0	0	0
Cryptophytes	0	0	0	0	0.673	0	0	0	0	0	0
Chryso-Pelago	0	0.114	0.285	0	0	0.831	0	0.285	0	0	0
Prasino-2	0.812	0	0	0	0	0	0	0	0.033	0	0.096
Prasino-3	0.764	0	0	0	0	0	0	0	0.078	0.248	0.009
Chlorophytes	0.339	0	0	0	0	0	0	0	0.036	0	0.187
Phaeocystis	0	0.208	0.35	0	0	0	0	0	0	0	0
Hapto-7	0	0.171	0.259	0	0	0.013	0.491	0.276	0	0	0

Table 2. *Cont.*

Class	chl-b	chl-c3	fucox	perid	allox	19butfu	19hexfu	chl-c	neox	prasinox	lut
ARA07B Final ratio matrix											
Diatoms	0	0	0.785	0	0	0	0	0.395	0	0	0
Dinoflagellates	0	0	0	0.6	0	0	0	0	0	0	0
Cryptophytes	0	0	0	0	0.673	0	0	0	0	0	0
Chryso-Pelago	0	0.114	0.285	0	0	0.831	0	0.285	0	0	0
Prasino-2	0.593	0	0	0	0	0	0	0	0.007	0	0.007
Prasino-3	4.006	0	0	0	0	0	0	0	0.166	0.803	0.027
Chlorophytes	0.339	0	0	0	0	0	0	0	0.036	0	0.187
Phaeocystis	0	0.1791	0.301	0	0	0	0	0	0	0	0
Hapto-7	0	0.171	0.259	0	0	0.013	0.508	0.276	0	0	0
OS040Final ratio matrix											
Diatoms	0	0	0.722	0	0	0	0	0.328	0	0	0
Dinoflagellates	0	0	0	1.409	0	0	0	0	0	0	0
Cryptophytes	0	0	0	0	0.673	0	0	0	0	0	0
Chryso-Pelago	0	0.114	0.285	0	0	0.831	0	0.285	0	0	0
Prasino-2	0.812	0	0	0	0	0	0	0	0.033	0	0.096
Prasino-3	0.280	0	0	0	0	0	0	0	0.107	0.471	0.011
Chlorophytes	0.643	0	0	0	0	0	0	0	0.029	0	0.969
Phaeocystis	0	0.558	1.457	0	0	0	0	0	0	0	0
Hapto-7	0	0.171	0.259	0	0	0.013	0.617	0.276	0	0	0

Abbreviations: chlorophyll-*b* (chl-b), chlorophyll-*c*3 (chl-c3), fucoxanthin (fucox), peridinin (period), alloxanthin (allox), 19′-butanoyloxyfucoxanthin (19butfu), 19′-hexanoyloxyfucoxanthin(19hexfu), chlorophyll-*c*1+*c*2 (chl-c), neoxanthin (neox), prasinoxanthin (prasinox), lutein (lut). Chrysophytes and Pelagophytes (Cryso-pelago). Prasinophytes type 2 (Prasino-2), Prasinophytes type 3 (Prasino-3), Haptophytes (Hapto-7).

2.4. Particulate Organic Carbon and Primary Productivity

The water samples for particulate organic carbon (POC) and primary productivity were obtained from 6 light depths (100, 50, 30, 12, 5 and 1% of PAR). 300 mL of seawater was filtered through 0.7 μm GF/F (pre-combusted at 450 °C for 4 h) for total POC and 500 mL was passed through 2 μm pore size membrane filter and then filtered onto GF/F filter for small POC (0.7–2 μm). Carbon and nitrogen uptake experiments were conducted using a ^{13}C-^{15}N dual isotope tracer technique previously reported from the Chukchi Sea [3,33]. After a 4 h incubation on deck, 300 mL water was filtered onto pre-combusted GF/F for total primary productivity and 500 mL water was filtered through 2 μm pore size membrane filter and sequentially onto GF/F filter for small phytoplankton productivity (0.7–2 μm). The filters were immediately preserved and stored in a freezer (−20 °C) until further mass spectrometric analysis using a Delta V+ Isotope Ratio Mass Spectrometers of Alaska Stable Isotope Facility at the University of Alaska Fairbanks, USA for ARA07B samples and using a 20–22 Isotope Ratio Mass Spectrometer (SERCON) at Japan Agency for Marine-Earth Science and Technology (JAMSTEC, Mutsu, Japan) for OS040 samples after HCl fuming overnight to remove carbonate. The carbon and nitrogen uptake rates were calculated based on [34].

2.5. Statistical Analysis

Student's t-test was applied to verify correlations among factors and differences between the mean values of POC:chlorophyll-*a* ratio, PON:chlorophyll-*a* ratio, C:N ratio of each cruise and size group. The agglomerative hierarchical clustering (AHC) with Ward's method (XLSTAT software, Addinsoft, Boston, MA, USA) was performed to calculate the dissimilarity in observed 20 variables; temperature and salinity), size-fractionated primary productivity, particulate organic carbon of each size, size-fractionated chlorophyll-*a* and accessory pigments, among stations.

3. Results and Discussion

3.1. Spatial Distribution of Temperature and Salinity

The temperature and salinity ranged from −1.5 °C to 9.2 °C (mean ± standard deviation (SD) = 0 ± 2.7 °C) and from 26.5 to 32.3 (mean ± SD = 29.9 ± 1.6) during the ARA07B cruise (Figure 2). The temperature during the OS040 were from −1.1 to 13.3 °C (mean ± SD = 6.2 ± 3.6 °C) and the salinity ranged from 28.9 to 32.9 (mean ± SD = 31.7 ± 0.9). Water mass at the most stations in the northern Chukchi corresponded to melting glacier water, which called Ice melt water (IMW; temperature < 2.0 °C and salinity < 30.0) and Bering Chukchi winter water (BCWW; −2–0 °C and <30–33.5 for temperature and salinity; [35]) during the ARA07B cruise. Other stations during the ARA07B were influenced by Bering shelf water (BSW; 0.0–10.0 °C and 31.8–33.0 for temperature and salinity). During the OS040 cruise, the relatively warm and low salinity Alaskan coastal water (ACW; 2.0–13.0 °C and <31.9 for temperature and salinity) and the warm and saline Bering shelf water (BSW) were predominant (Figure 2). The Bering shelf Anadyr water (BSAW; −1.0–2.0 °C and 31.8–33.0 for temperature and salinity), which is a mixed BSW with cold/saline Anadyr water (AW; [36,37]), was observed at some stations for the OSO40 cruise.

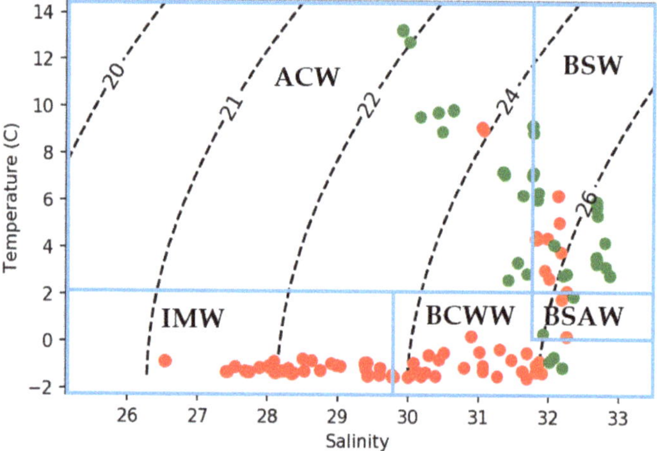

Figure 2. T–S diagram in the ARA07B (Red) and OS040 (Green). Alaskan coastal water (ACW), Bering Shelf water (BSW), ice melt water (IMW), Bering–Chukchi winter water (BCWW), Bering Sea Anadyr water (BSAW).

3.2. Chlorophyll-a Concentration and Different Size Chlorophyll-a Compositions in the Northern Bering and Chukchi Seas

The average euphotic depths were 45.6 m (SD = ±22.2 m) for the ARA07B cruise and 23.8 m (SD = ±9.1 m) for the OS040 cruise, respectively. In ARA07B, Chlorophyll-*a* concentrations were 0.02–1.3 mg chl-*a* m^{-3} (mean ± SD = 0.2 ± 0.3 mg chl-*a* m^{-3}) at surface, 0.02–15.0 mg chl-*a* m^{-3} (mean ± SD = 1.0 ± 2.5 mg chl-*a* m^{-3}) for euphotic layer. In OS040, Chlorophyll-*e* concentrations were 0.002–5.5 mg chl-*a* m^{-3} (mean ± SD = 0.7 ± 1.4 mg chl-*a* m^{-3}) at surface, 0.002–5.5 mg chl-*a* m^{-3} (mean ± SD = 1.6 ± 2.2 mg chl-*a* m^{-3}) for euphotic layer. Within the euphotic zone, integral chlorophyll-*a* concentrations were 3.2–172.1 mg chl-*a* m^{-2} (mean ± SD = 34.2 ± 48.0 mg chl-*a* m^{-2}) during the ARA07B and 12.3–107.8 mg chl-*a* m^{-2} (mean ± SD = 45.4 ± 34.1 mg chl-a m^{-2}) for the OS040, respectively (Figure 3). The average euphotic-depth integral chlorophyll-*a* concentrations in this study are within the range reported previously in the northern Bering Sea and the Chukchi Sea [3,14,21].

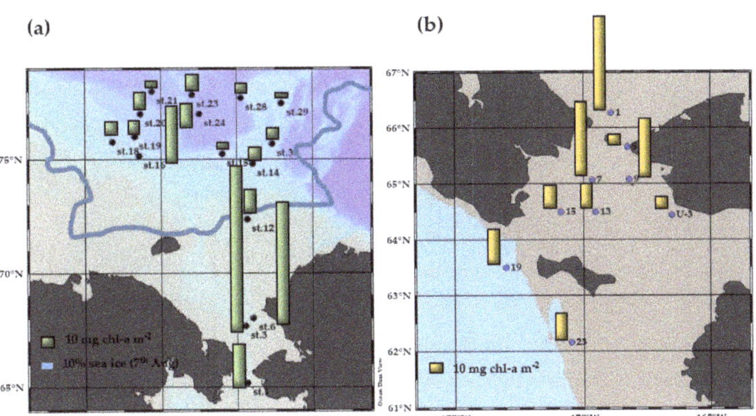

Figure 3. Spatial distributions of column-integrated chlorophyll-*a* concentration of (**a**) ARA07B and (**b**) OS040.

The chlorophyll-*a* contributions of each size phytoplankton (pico-, nano- and micro-phytoplankton) to the total phytoplankton were plotted in Figure 4 for the three different depths (100, 30 and 1% of light depths) at every station of the ARA07B and only surface for the OS040. The contributions of small phytoplankton to the total chlorophyll-*a* concentrations were found largely variable among the stations during both cruises.

The contributions of small phytoplankton to the total chlorophyll-*a* concentrations ranged from 2.9% to 71.1% with a depth-integrated average of 32.2% (SD = ±23.1%) during the ARA07B. In the ARA07B, the dominant size group of phytoplankton was micro-phytoplankton (mean ± SD = 43.5 ± 29.7% of chlorophyll-*a* concentration) followed by pico-phytoplankton (32.1 ± 23.1%) and nano-phytoplankton (24.3 ± 9.1%) during the observation period. In the Chukchi Sea, large phytoplankton are generally dominant although the areal distribution of their contribution mostly depends on local water masses in different nutrient conditions [3,21]. Normally, large phytoplankton growing under nutrient-enriched conditions are predominant in AW or BSW, whereas small phytoplankton are dominant in nutrient-depleted ACW [3,21]. Our average contribution of small phytoplankton is relatively higher than that (24.8 ± 23.0%) previously reported by [21] in the Chukchi Sea during the middle of August to early September, 2004. By contrast, our average contribution of small phytoplankton is relatively lower than that (55.1 ± 26.8%) from the study by [38] that was conducted in the northern Chukchi Sea during mid-July–mid-August, 2012. This difference among the studies could be caused by different regions with non-homogeneous nutrient conditions and different observation periods with a seasonal phytoplankton succession. The relative contribution of small phytoplankton could be caused by freshwater content in the Chukchi Sea since the nutrient concentrations and primary production rates of phytoplankton are largely governed by the nutrient-depleted freshwater content in the Chukchi Sea [14,39].

In comparison to the Chukchi Sea, the contributions of small phytoplankton were 0.7–80% (mean ± SD = 37.2 ± 31.0%) to the total chlorophyll-*a* concentration in the northern Bering Sea for the OS040 in this study. The proportions of different size chlorophyll-*a* were 40.2% (±35.4%), 22.5% (±10.5%) and 37.2% (±31.3%) for micro-, nano- and pico-phytoplankton, respectively, during our observation period in 2017. In the northern Bering Sea, the dominant size groups of phytoplankton are generally nano- and micro-phytoplankton based on phytoplankton size class results derived from satellite ocean color data from 1998 to 2016 [12]. The overall dominant size of phytoplankton is composed of nano-phytoplankton (49.0 ± 9.6%), followed by micro-phytoplankton (34.9 ± 8.0%) and pico-phytoplankton (16.1 ± 7.3%) in the Chirikov Basin of the northern Chukchi Sea [12]. However, the chlorophyll-*a* contributions of small phytoplankton are largely variable among different seasons [40]. The average contribution of small phytoplankton was 14.8% in late May to early June during the phytoplankton bloom period and largely increased up to 50.0% in middle June after the bloom [41]. Consistently, [13] found a seasonal increasing contribution of small phytoplankton in the northern Bering Sea (around Chirikov Basin) from May (5.2%) to July (31.8%). In addition to the seasonal variation, spatially the biochemical environmental conditions in the northern Bering Sea are also generally influenced by northward advection of AW, BSW and ACW [3,5]. Over recent decades, several environmental changes have been reported in the northern Bering Sea [4,5]. A steady increasing trend in the annual contribution of small phytoplankton is distinct in the Chirikov Basin from 1998 to 2016, although no significantly strong relationship was observed between the annual contribution of small phytoplankton and sea surface temperature [12]. Long-term changes in dominant phytoplankton communities should be monitored for Arctic marine ecosystems under ongoing environmental changes. Especially, the contribution of small phytoplankton could be used as one of indicators for changing marine ecosystems.

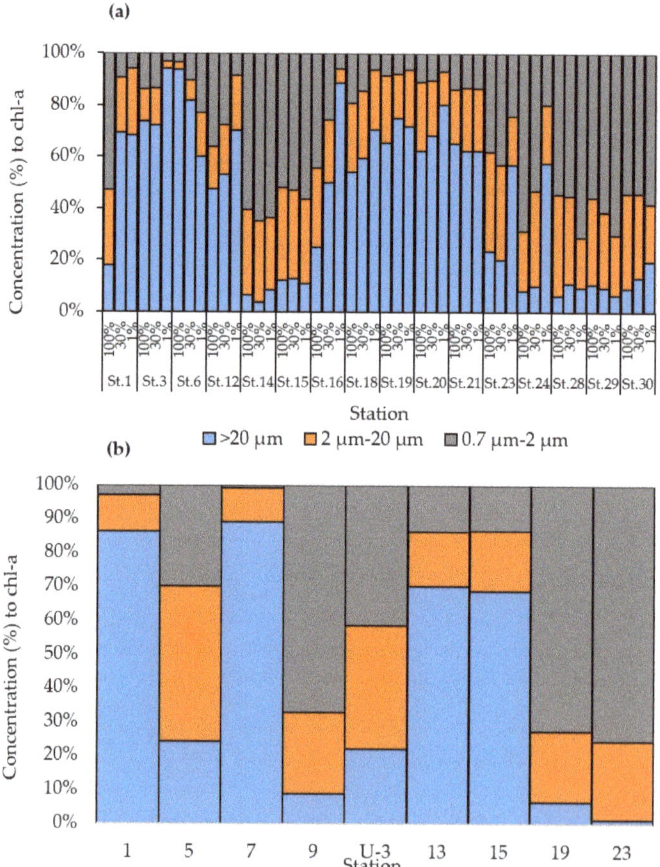

Figure 4. Total chlorophyll-*a* concentration of (**a**) ARA07B and (**b**) OS040.

3.3. Pigment Composition and Major Dominant Phytoplankton Groups

The euphotic depth-integral concentrations of marker pigments from the two cruises are shown in Figure 5. Fucoxanthin (a marker pigment of diatoms), chlorophyll-*c1+c2* and chlorophyll-*b* (a marker pigment of chlorophytes) were major accessory pigments during the ARA07B, although the pigment compositions spatially varied significantly across the stations. Among the pigments, fucoxanthin was the most dominant pigment with an average value of 12.58 ± 21.8 mg m^{-2} and the second and third predominant pigments were chlorophyll-c1+c2 (4.04 ± 4.83 mg m^{-2}) and chlorophyll-b (2.64 ± 2.53 mg m^{-2}). Previous studies reported that fucoxanthin dominating the Chukchi shelf is a typical characteristic during fall [13,31]. For the small phytoplankton group for the ARA07B (data not shown), major predominant pigments were chlorophyll-*b* (1.59 ± 1.83 mg m^{-2}), fucoxanthin (1.46 ± 1.47 mg m^{-2}) and chlorophyll-c1+c2 (0.65 ± 0.62 mg m^{-2}). In comparison, fucoxanthin, chlorophyll-c1+c2 and peridinin (a marker pigment of dinoflagellates) were major accessory pigments for the OS040. Fucoxanthin was the most dominant pigment with an average value of 23.03 ± 19.89 mg m^{-2}, followed by chlorophyll-c1+c2 (9.35 ± 7.23 mg m^{-2}) and peridinin (7.54 ± 9.89 mg m^{-2}) for the OS040. High proportions of diatom-related pigments (fucoxanthin, chlorophyll-*c1+c2*) were observed in both cruise periods. Small diatoms appeared to be major phytoplankton communities for the small phytoplankton group during the ARA07B, based on the high proportions of chlorophyll-b and fucoxanthin. No pigment data were available for the small phytoplankton during the OS040 cruises.

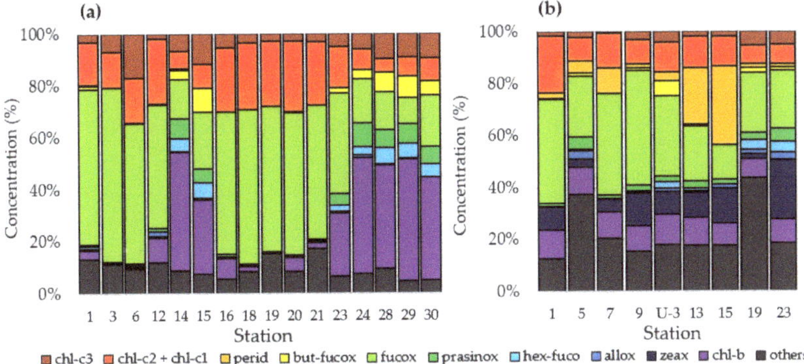

Figure 5. Pigment compositions of total phytoplankton in the (**a**) ARA07B (**b**) OSO040.

Based on the CHEMTAX results, eight major phytoplankton communities were identified in the study area (Figure 6). Diatoms ($43.1\% \pm 31.5\%$) and Phaeocystis ($33.2\% \pm 14.9\%$) were co-dominated during the ARA07B. In comparison, Diatoms were the most dominant community ($46.1 \pm 17.3\%$) and the second dominant community was Prasinophyte (Type 2) ($11.8\% \pm 5.3\%$) for the OS040. Micro-phytoplankton communities were most dominant ($59.7 \pm 30.5\%$), followed by nano-phytoplankton ($11.5 \pm 9.7\%$) and pico-phytoplankton ($28.9 \pm 23.5\%$) during the ARA07B. For the OS040, micro-phytoplankton contributed 51.5% ($\pm 18.2\%$) of the total chlorophyll a concentration. In comparison, nano-phytoplankton and pico-phytoplankton contributed 11.0% ($\pm 10.5\%$) and 37.5% ($\pm 15.7\%$), respectively. The relative proportions of the three size classes based on the diagnostic pigments from HPLC are different from those of the size-fractionated chlorophyll-*a* concentrations (Figure 4). This is probably due to a simple assumption that diatom-related pigments belong to the micro-phytoplankton although small diatoms (<2 μm) could contribute to the phytoplankton group.

3.4. Primary Production Contribution of Small Phytoplankton and Their Ecological Roles

The daily primary productivities of total phytoplankton which were integrated over the euphotic zone at each station were 33.9–811.8 mg C m^{-2} d^{-1} (mean $\pm$ SD = 142.6 ± 205.7 mg C m^{-2} d^{-1}) for the ARA07B and 202.1–3100.1 mg C m^{-2} d^{-1} (mean $\pm$ SD = 942.1 ± 969.9 mg C m^{-2} d^{-1}) for the OS040 (Figure 7). In comparison, the daily primary productivities of small phytoplankton ranged from 4.9 to 227.7 (mean $\pm$ SD = 42.3 ± 53.1 mg C m^{-2} d^{-1}) and 56.1 to 322.2 mg C m^{-2} d^{-1} (mean $\pm$ SD = 152.8 ± 85.2 mg C

$m^{-2} d^{-1}$) for the ARA07B and the OS040, respectively (Figure 8). The contribution of small phytoplankton to the total primary productivity ranged from 8.1 to 71.7% (mean $\pm$ SD = 38.0 $\pm$ 19.9%) for the ARA07B and from 6.0 to 40.3% (mean $\pm$ SD = 25.0 $\pm$ 12.8%) for the OS040 (Figure 9).

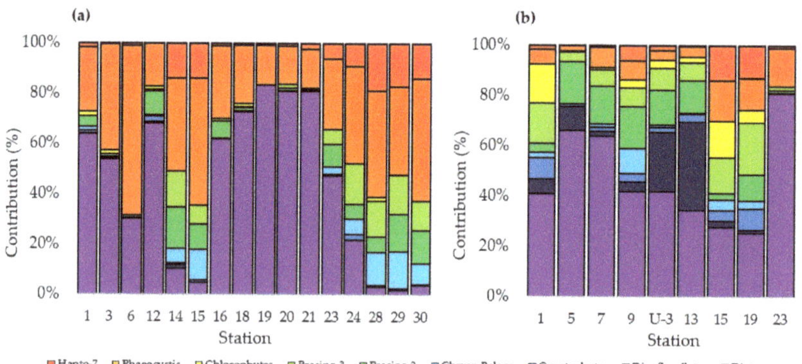

Figure 6. Phytoplankton community compositions of (**a**) ARA07B and (**b**) OS040.

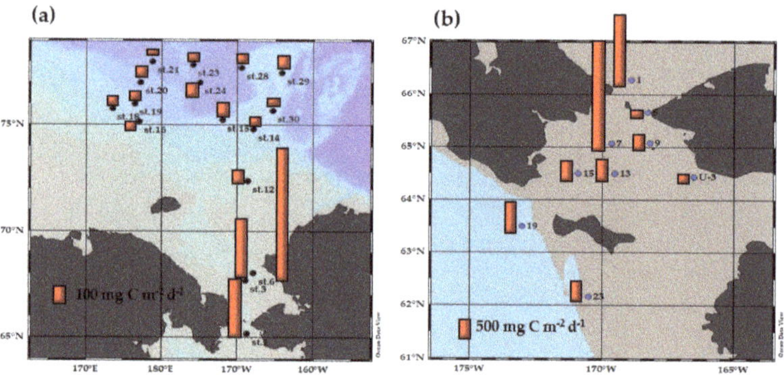

Figure 7. Primary production of total phytoplankton during the (**a**) ARA07B and (**b**) OS040.

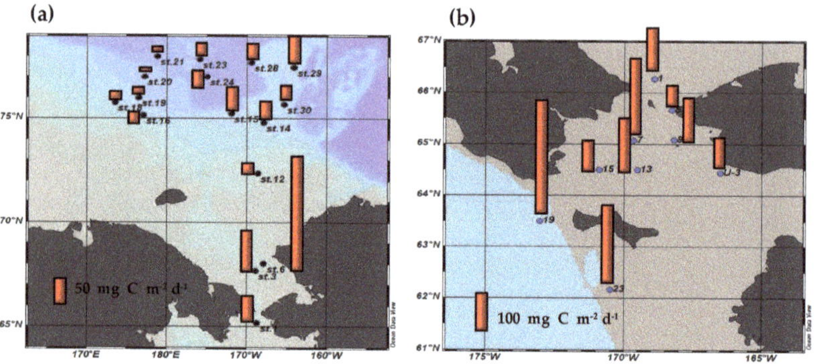

Figure 8. Primary production of small phytoplankton during the (**a**) ARA07B and (**b**) OS040.

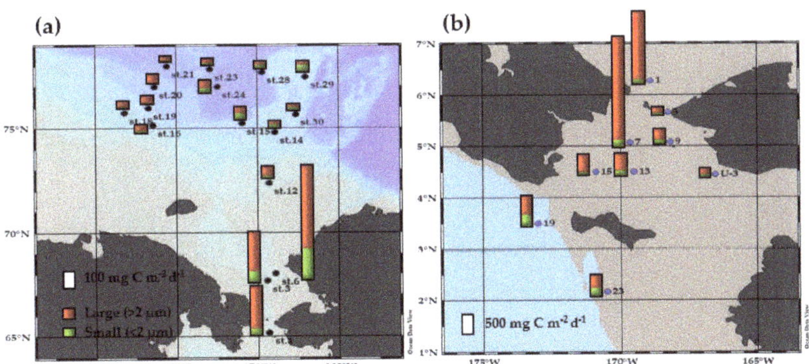

Figure 9. Primary production of small phytoplankton of (**a**) ARA07B and (**b**) OS040.

Overall, the primary productions of total and small phytoplankton communities during the study period were different depending on the sea area. Indeed, agglomerative hierarchical clustering (AHC) analysis based on 25 stations and phytoplankton size-related variables sorted stations into four distinct groups (Figure 10; Table 3). Cluster 1 include station 1 of OS040 that was high primary productive region (1992.9 mg C m^{-2} d^{-1}) near Bering strait. Cluster 1 had a relative low contribution of small phytoplankton in primary productivity (5.9%) and surface chlorophyll-*a* (2.9%). Cluster 2 was station 7 of OS040 that was an extremely high productive station (3100.0 mg C m^{-2} d^{-1}) and represented the lowest contribution of small phytoplankton among clusters. Small phytoplankton contribution to primary production was 6.9% and the contribution to surface chlorophyll-*a* concentration was only 0.7% for clusters 2. The physical properties of Cluster 1 (3.5 °C and 32.7 psu) and 2 (5.5 °C and 32.7 psu) were similar. These two clusters are influence by BSW [3,21]. Cluster 3 contains all the stations of the northern Chukchi Sea and two stations of the Bering Sea. The stations form Cluster 3 had a lower productivity and lower concentration of surface chlorophyll-*a*. In Cluster 3, small phytoplankton contribution was the highest among the clusters. 40.5% of primary production, 39.1% of surface chlorophyll-*a* and 40.9% of POC were contributed by small phytoplankton. Dominant water mass, IMW can explain the high contribution of small phytoplankton in Cluster 3 because IMW has nutrient-depleted water from sea ice melting [34]. Cluster 4 includes most of the stations in the Bering Sea and 3 stations of the southern Chukchi sea in ARA07B. Cluster 4 had a lower productivity (559.2 mg C m^{-2} d^{-1}) than cluster 1 and 2 but higher than cluster 3. Cluster 4 seems to be affected by nutrient-depleted ACW but not too low productivity for Cluster 4. This suggests that water masses that had an effect on Cluster 4 were not only ACW but also other source such as mixed water of AW, ACW and BSW.

Table 3. Mean values of properties for the 4 clusters classified by the AHC.

Cluster	T (°C)	S (psu)	Small Contribution to PP	Small Contribution to Surface chl-a	Small Contribution to POC	PP (mg C m^{-2} d^{-1})	Chl-a (mg m^{-3})	POC (mg m^{-3})
1	3.5	32.7	6.0%	2.9%	25.0%	1992.9	66.3	349.0
2	5.5	32.7	6.9%	0.7%	18.6%	3100.1	128.8	479.5
3	0.1	29.7	40.5%	39.1%	40.9%	79.6	13.9	177.7
4	5.6	31.7	26.7%	39.1%	43.9%	559.2	60.7	236.1

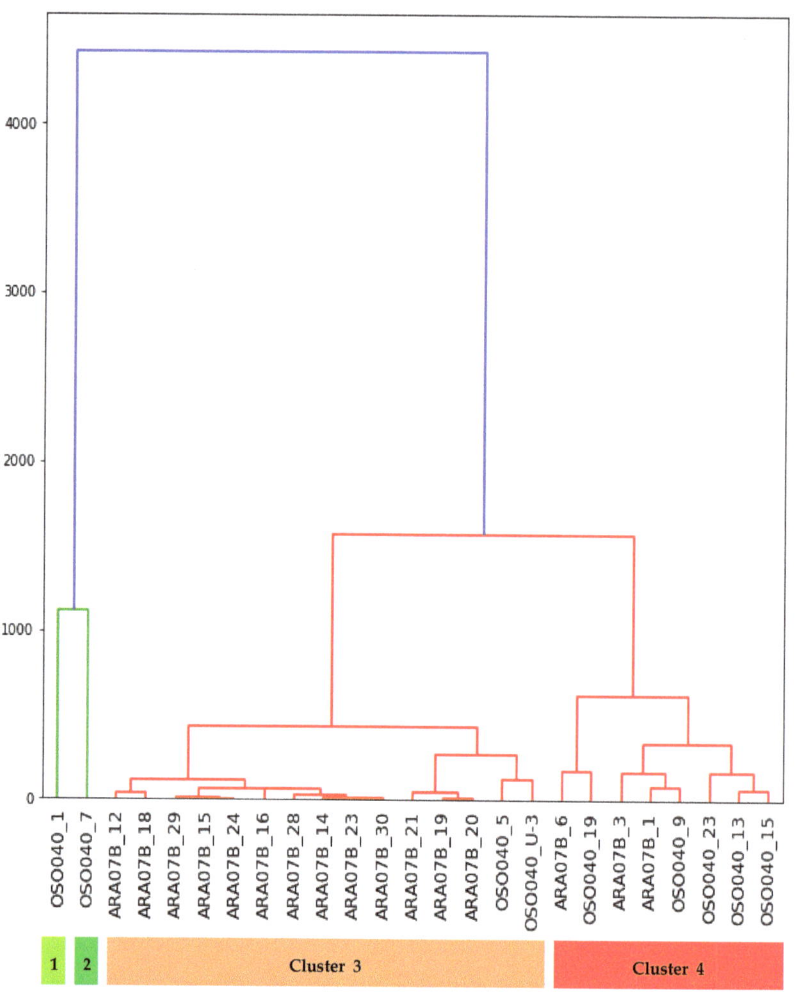

Figure 10. Dendrogram stands for sampling stations were divided into four clusters by agglomerative hierarchical clustering (AHC).

The primary production contributions of small phytoplankton are rather different from their chlorophyll-*a* contributions in this study. Normally, the contributions of small phytoplankton are higher to primary production in comparison to those in chlorophyll-a concentrations in the polar oceans [21,41] and temperate oceans [42]. This is probably due to the considerably higher POC contribution of small phytoplankton (and consequently higher production contributions of small phytoplankton) than the chlorophyll-*a* contribution [21,40,42]. We also observed the higher POC:chlorophyll-*a* ratio in small phytoplankton than large phytoplankton during both cruises (Figure 11) as discussed later. However, the case in the northern Bering Sea in this study is against the general pattern previously reported. The lower contribution of small phytoplankton was observed in the primary production rather than chlorophyll-*a* concentration in the northern Bering Sea. This indicates higher standing stock (represented by chlorophyll-*a* concentrations) of small phytoplankton but their significantly lower contribution to the primary productions in the northern Bering Sea during this study than in other studies. Ref. [20] argued that seasonal increasing contribution of small phytoplankton is not caused by their increasing biomass and photosynthetic rate but caused by relative declining in biomass and photosynthetic rate of large phytoplankton in the Amundsen Sea, Antarctic Ocean. Based on these results, the biomass of large phytoplankton could have had decreased faster than their photosynthetic rate in the northern Bering Sea during our observation period.

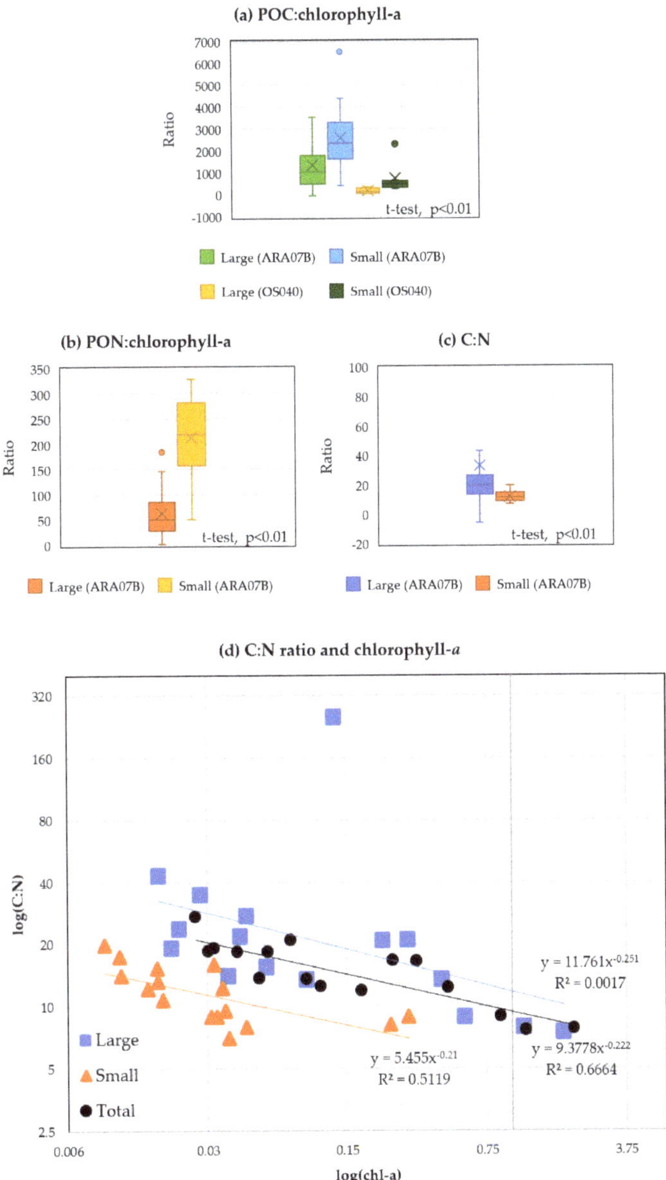

Figure 11. Comparison of (**a**) POC:chlorophyll-*a* ratios, (**b**) PON:chlorophyll-*a* ratios and (**c**) C:N ratios between small and large phytoplankton in the northern Bering and Chukchi seas. Only POC:chlorophyll-*a* data available for the OS040. (**d**) C:N ratio and chlorophyll-a of each size group.

The regional contributions of small phytoplankton to the primary production are summarized at various regions in the Arctic Ocean (Table 4). The average contribution of small phytoplankton in this study is comparable to the previous results in the Chukchi Sea. However, it is considerably lower than those (average ± SD = 56.7 ± 20.0%) in the Kara, Laptev and East Siberian Seas [43]. Similarly, reference [41] found a similar contribution (average ± SD = 60 ± 7.9%) of small phytoplankton in the high northern Chukchi Sea and Canada Basin. Because of no data in the northern Bering Sea, the small phytoplankton contribution to the primary production in this study could not be compared. Regionally, the primary production contribution of small phytoplankton in the northern Bering Sea (average ± SD = 25.0 ± 12.8%) is considerably lower than those in others (Table 4). At this point, we do not know whether this is a latitudinal pattern (i.e., increasing contribution of small phytoplankton in higher latitude) or simply seasonal difference among the different regions in the Pacific Arctic Ocean. Indeed, [12] found a seasonal patterns of different phytoplankton size compositions with increasing contribution of small phytoplankton in the northern Bering Sea. Since the seasonal contribution of small phytoplankton to the primary production would be different, further seasonal observations on the small phytoplankton contribution to the primary production will be warranted for better understanding their ecological roles in the Bering and Chukchi Seas.

Table 4. Small phytoplankton contributions to the total primary production in the Arctic Ocean.

Study Area	Year	Season	Smallcontribution	Methods	Size	References
Northern Chukchi Sea and Canada Basin	2008	August–September	19.8–60.3%	In situ	<5 μm	[41]
Bering Strait and Chukchi Sea	2004	August–September	31.72 ± 23.59%	In situ	<5 μm	[22]
Kara, Laptev and East Siberian Sea	2013	August–September	52.7–71.2%	In situ	<5 μm	[43]
Barents Sea	2003–2005	Early to late bloom period	31–87%	In situ	<10 μm	[44]
North water polynya	1998	April–July	19%	In situ	<5 μm	[45]
Chukchi Sea and Bering Strait	2016	August	38.0 ± 19.9%	In situ	<2 μm	This study
Northern Bering Sea and Bering Strait	2017	July	25.0 ± 12.8%	In situ	<2 μm	

Biochemical compositions (POC:chlorophyll-*a*, PON:chlorophyll-*a* and C:N ratios) were compared between small and large phytoplankton in Figure 11. Large phytoplankton group has relatively lower POC:chlorophyll-*a* ratios (*t*-test, $p < 0.01$) which were 78.0–3549.0 (mean ± SD = 1358.6 ± 1170.8) for the ARA07B and 41.4–340.2 (mean ± SD = 173.8 ± 110.4) for the OS040 (Figure 9a). In comparison, POC:chlorophyll-*a* ratios of small phytoplankton were 408.5–6547.4 (mean ± SD = 2590.2 ± 1523.0) for ARA07B and 274.9–2303.6 (mean ± SD = 623.4 ± 639.2) for the OS040. The PON:chlorophyll-*a* ratio of large phytoplankton was 1.9–184.2 (mean ± SD = 62.4 ± 48.7) whereas the ratio of small phytoplankton ranged from 50.0 to 328.7 (mean ± SD = 211.9 ± 88.3) for the ARA07B (no data for OS040). The C:N ratios were 7.5–251.9 (mean ± SD = 34.1 ± 58.9) for large phytoplankton and 7.0–19.9 (mean ± SD = 11.9 ± 3.8) for small phytoplankton during the ARA07B cruise. Small phytoplankton showed a comparatively higher POC:chlorophyll-*a* ratio than large phytoplankton during both cruises (Figure 11). This result is consistent with the previous result in the Chukchi Sea, which suggests that higher carbon contents per unit of chlorophyll-*a* concentration in small phytoplankton in comparison to large phytoplankton [21]. In the Antarctic Ocean, [41] observed the consistent results in non-polynya and polynya regions in the Amundsen Sea. A similar pattern was observed for the PON:chlorophyll-*a* ratio in this study. However, the C:N ratios of small phytoplankton were lower than those of large phytoplankton in this study. Similarly, the overall C:N assimilation ratio of small phytoplankton was previously reported as significantly lower than that of large phytoplankton [21]. These results are consistent with the result in the Gulf of St. Lawrence, Canada [44]. In the Antarctic Ocean, the similar result was obtained in the Amundsen Sea [41]. The C:N ratios were negatively correlated with chlorophyll-*a* concentrations for small and large phytoplankton in this study ($R^2 > 0.6$). However, there was no statistically significant difference in the relationship between small and large phytoplankton ($p > 0.05$; Figure 11).

4. Summary and Conclusions

For determining the dominant phytoplankton communities and the relative contribution of small phytoplankton (<2 μm) to the total primary production, two arctic research cruises were conducted in the Chukchi Sea onboard the icebreaker R/N *Araon* in 2016 (ARA07B) and mainly in the northern Bering Sea onboard T/S *Oshoro-Maru* in 2017 (OS040) for this study. The dominant phytoplankton communities were diatoms and phaeocystis during the ARA07B, whereas diatoms and Prasinophyte (Type 2) during the OS040. Based on the AHC analysis, the primary productions of total and small phytoplankton communities were different depending on the sea area. Overall, high primary productions and low contributions of small phytoplankton during both study periods were distributed in the Bering Strait region which was affected by nutrient-enriched BSW. Different biochemical compositions between small and large phytoplankton were observed in this study. The small phytoplankton group had a higher POC:chlorophyll-*a* (*t*-test, $p < 0.01$) and PON:chlorophyll-*a* ratio than large phytoplankton in this study, which suggests that small phytoplankton have higher carbon and nitrogen contents per unit of chlorophyll-a concentration [21]. In addition, small phytoplankton had lower C:N ratios than large phytoplankton in this study. Together with these results, we could conclude that small phytoplankton incorporate more nitrogen in relation to carbon into their bodies and thus produce nitrogen-rich organic matters [43] which could be relatively faster regenerated than carbon-rich organic matters such as carbohydrates [46]. Therefore, the study for small phytoplankton which could be an important basic food source in the Arctic ecosystem should be further conducted under the current warming ocean scenario.

Author Contributions: Conceptualization, J.-W.P., T.H. and S.-H.L.; methodology, J.-W.P., A.F., Y.K., K.-W.K. and H.W.; software, J.-W.P.; validation, J.-W.P. and T.H.; formal analysis, J.-W.P., A.F., J.J.K. and T.H.; investigation, K.-W.K., J.-W.P. and E.-J.Y.; writing—original draft, J.-W.P.; writing—review and editing, T.H. and J.J.K.; visualization, J.-W.P., Y.K. and K.-W.K.; supervision, T.H. All authors have read and agreed to the published version of the manuscript.

Funding: This study was supported by the Arctic Challenge for Sustainability (ArCS; JPMXD1300000000) and Arctic Challenge for Sustainability II (ArCS-II; JPMXD1420318865) Project of the Ministry of Education, Culture, Sports, Science and Technology of Japan (MEXT) and by the Global Change Observation Mission-Climate (GCOM-C; 17RSTK001714) project of the Japan Aerospace Exploration Agency (JAXA). This research also was part of the projected titled "Korea-Arctic Ocean Warming and Responses of Ecosystem (K-AWARE, KOPRI, 1525011760)", funded by the Ministry of Oceans and Fisheries, Korea.

Institutional Review Board Statement: Not applicable.

Informed Consent Statement: Not applicable.

Data Availability Statement: Data Availability Statements in section "MDPI Research Data Policies" at https://www.mdpi.com/ethics, accessed on 7 January 2022.

Acknowledgments: The austhors thank the captains and crew members for our field observations. In addition, We thanks to ArCS-II project for funding this publication.

Conflicts of Interest: The authors declare no conflict of interest.

References

1. Sambrotto, R.N.; Goering, J.J.; McRoy, C.P. Large yearly production of phytoplankton in the western Bering Sea. *Science* **1984**, *225*, 1147–1150. [CrossRef] [PubMed]
2. Springer, A.M.; McRoy, C.P. The paradox of pelagic food webs in the northern Bering Sea. III. Patterns of primary production. *Cont. Shelf Res.* **1993**, *13*, 575–599. [CrossRef]
3. Lee, S.H.; Whitledge, T.E.; Kang, S.H. Recent carbon and nitrogen uptake rates of phytoplankton in Bering Strait and the Chukchi Sea. *Cont. Shelf Res.* **2007**, *27*, 2231–2249. [CrossRef]
4. Bluhm, B.; Gradinger, R. Regional Variability in Food Availability for Arctic Marine Mammals. *Ecol. Appl.* **2008**, *18*, S77–S96. [CrossRef] [PubMed]
5. Grebmeier, J.M.; Bluhm, B.A.; Cooper, L.W.; Denisenko, S.G.; Iken, K.; Kędra, M.; Serratos, C. Time-series benthic community composition and biomass and associated environmental characteristics in the Chukchi Sea during the RUSALCA 2004–2012 Program. *Oceanography* **2015**, *28*, 116–133. [CrossRef]
6. Overland, J.; Stabeno, P. Is the Climate of the Bering Sea Warming and Affecting the Ecosystem? *EOS Trans. AGU* **2004**, *85*, 309–316. [CrossRef]
7. Grebmeier, J.; Cooper, L.; Feder, H.; Sirenko, B. Ecosystem Dynamics of the Pacific-Influenced Northern Bering and Chukchi Seas in the Amerasian Arctic. *Prog. Oceanogr.* **2006**, *71*, 331–361. [CrossRef]
8. Danielson, S.L.; Ahkinga, O.; Ashjian, C.; Basyuk, E.; Cooper, L.W.; Eisner, L.; Farley, E.; Iken, K.B.; Grebmeier, J.M.; Juranek, L.; et al. Manifestation and consequences of warming and altered heat fluxes over the Bering and Chukchi Sea continental shelves. *Deep-Sea Res. II Top. Stud. Oceanogr.* **2020**, *177*, 104871. [CrossRef]
9. Woodgate, R. Increases in the Pacific inflow to the Arctic from 1990 to 2015, and insights into seasonal trends and driving mechanisms from year-round Bering Strait mooring data. *Prog. Oceanogr.* **2018**, *160*, 124–154. [CrossRef]
10. Frey, K.E.; Moore, G.W.K.; Cooper, L.W.; Grebmeier, J.M. Divergent patterns of recent sea ice cover across the Bering, Chukchi, and Beaufort seas of the Pacific Arctic Region. *Prog. Oceanogr.* **2015**, *136*, 32–49. [CrossRef]
11. Grebmeier, J.M. Shifting Patterns of Life in the Pacific Arctic and Sub-Arctic Seas. *Annu. Rev. Mar. Sci.* **2012**, *4*, 63–78. [CrossRef]
12. Lee, S.H.; Ryu, J.; Lee, D.; Park, J.W.; Kwon, J.I.; Zhao, J.; Son, S. Spatial variations of small phytoplankton contributions in the Northern Bering Sea and the Southern Chukchi Sea. *GISci. Remote Sens.* **2019**, *56*, 794–810. [CrossRef]
13. Coupel, P.; Matsuoka, A.; Ruiz-Pino, D.; Gosselin, M.; Marie, D.; Tremblay, J.-É.; Babin, M. Pigment signatures of phyto- plankton communities in the Beaufort Sea. *Biogeosciences* **2015**, *12*, 991–1006. [CrossRef]
14. Yun, M.S.; Whitledge, T.E.; Stockwell, D.; Son, S.H.; Lee, J.H.; Park, J.W.; Lee, D.B.; Park, J.; Lee, S.H. Primary production in the Chukchi Sea with potential effects of freshwater content. *Biogeosciences* **2016**, *13*, 737–749. [CrossRef]
15. Fujiwara, A.; Hirawake, T.; Suzuki, K.; Eisner, L.; Imai, I.; Nishino, S.; Kikuchi, T.; Saitoh, S.-I. Influence of timing of sea ice retreat on phytoplankton size during marginal ice zone bloom period on the Chukchi and Bering shelves. *Biogeosciences* **2016**, *13*, 115–131. [CrossRef]
16. Kahru, M.; Brotas, V.; Manzano-Sarabia, M.; Mitchell, B.G. Are phytoplankton blooms occurring earlier in the Arctic? *Glob. Chang. Biol.* **2011**, *17*, 1733–1739. [CrossRef]
17. Arrigo, K.R.; van Dijken, G.L. Continued increases in Arctic Ocean primary production. *Prog. Oceanogr.* **2015**, *136*, 60–70. [CrossRef]
18. Li, W.; McLaughlin, F.; Lovejoy, C.; Carmack, E. Smallest Algae Thrive as the Arctic Ocean Freshens. *Science* **2009**, *326*, 539. [CrossRef]
19. Neeley, A.; Harris, L.; Frey, K. Unraveling phytoplankton community dynamics in the Northern Chukchi Sea under sea-ice covered, and sea-ice-free conditions. *Geophys. Res. Lett.* **2018**, *45*, 7663–7671. [CrossRef]
20. Lim, Y.J.; Kim, T.W.; Lee, S.; Lee, D.; Park, J.; Kim, B.K.; Lee, S.H. Seasonal Variations in the Small Phytoplankton Contribution to the Total Primary Production in the Amundsen Sea, Antarctica. *J. Geophys. Res. Ocean.* **2019**, *124*, 8324–8341. [CrossRef]
21. Lee, S.H.; Yun, M.S.; Kim, B.K.; Joo, H.T.; Kang, S.H.; Kang, C.K.; Whitledge, T.E. Contribution of small phytoplankton to total primary production in the Chukchi Sea. *Cont. Shelf Res.* **2013**, *68*, 43–50. [CrossRef]
22. Joo, H.T.; Park, J.W.; Son, S.H.; Noh, J.H.; Jeong, J.Y.; Kwak, J.H.; Saux-Picart, S.; Choi, J.H.; Kang, C.K.; Lee, S.H. Long-term annual primary production in the Ulleung Basin as a biological hot spot in the East/Japan Sea. *J. Geophys. Res. Ocean.* **2014**, *119*, 5. [CrossRef]
23. Yun, M.S.; Joo, H.T.; Park, J.W.; Kang, J.J.; Kang, S.H.; Lee, S.H. Lipid-rich and protein-poor carbon allocation patterns of phytoplankton in the northern Chukchi Sea, 2011. *Cont. Shelf Res.* **2018**, *158*, 26–32. [CrossRef]
24. Frigstad, H.; Andersen, T.; Bellerby, R.G.J.; Silyakova, A.; Hessen, D. Variation in the seston C:N ratio of the Arctic Ocean and pan-Arctic shelves. *J. Mar. Syst.* **2014**, *129*, 214–223. [CrossRef]
25. Kirk, J.T.O. Effects of suspensoids (turbidity) on penetration of solar radiation in aquatic ecosystems, in Perspectives in Southern Hemisphere. *Hydrobiologia* **1985**, *28*, 195–208. [CrossRef]
26. Parsons, T.R.; Maita, Y.; Lalli, C.M. A manual of chemical & biological methods for seawater analysis. *Mar. Pollut. Bull.* **1984**, *1*, 101–112.
27. Suzuki, K.; Hinuma, A.; Saito, H.; Kiyosawa, H.; Liu, H.; Saino, T.; Tsuda, A. Responses of phytoplankton and heterotrophic bacteria in the northwest subarctic Pacific to in situ iron fertilization as estimated by HPLC pigment analysis and flow cytometry. *Prog. Oceanogr.* **2005**, *64*, 167–187. [CrossRef]

28. Wright, S.W.; Jeffrey, S.W.; Mantoura, R.F.C.; Llewellyn, C.A.; Bjørnland, T.; Repeta, D.; Welschmeyer, N. Improved HPLC method for the analysis of chlorophylls and carotenoids from marine phytoplankton. *Mar. Ecol. Prog. Ser.* **1991**, *77*, 183–196. [CrossRef]

29. Vidussi, F.; Roy, S.; Lovejoy, C.; Gammelgaard, M.; Thomsen, H.A.; Booth, B.; Tremblay, J.-É.; Mostajir, B. Spatial and temporal variability of the phytoplankton community structure in the North Water Polynya, investigated using pigment biomarkers. *Can. J. Fish. Aquat. Sci.* **2004**, *61*, 2038–2052. [CrossRef]

30. Fujiwara, A.; Hirawake, T.; Suzuki, K.; Saitoh, S.-I. Remote sensing of size structure of phytoplankton communities using optical properties of the Chukchi and Bering Sea shelf region. *Biogeosciences* **2011**, *8*, 3567–3580. [CrossRef]

31. Fujiwara, A.; Hirawake, T.; Suzuki, K.; Imai, I.; Saitoh, S.-I. Timing of sea ice retreat can alter phytoplankton community structure in the western Arctic Ocean. *Biogeosciences* **2014**, *11*, 1705–1716. [CrossRef]

32. Ardyna, M.; Mundy, C.J.; Mills, M.M.; Oziel, L.; Grondin, P.-L.; Lacour, L.; Verin, G.; Dijken, G.; Ras, J.; Alou-Font, E.; et al. Environmental drivers of under-ice phytoplankton bloom dynamics in the Arctic Ocean. *Elem. Sci. Anthr.* **2020**, *8*, 30. [CrossRef]

33. Lee, S.H.; Kim, H.J.; Whitledge, T.E. High incorporation of carbon into proteins by the phytoplankton of the Bering Strait and Chukchi Sea. *Cont. Shelf Res.* **2009**, *29*, 1689–1696. [CrossRef]

34. Hama, T.; Miyazaki, T.; Ogawa, Y.; Iwakuma, T.; Takahashi, M.; Otsuki, A.; Ichmura, S. Measurement of photosynthetic production of a marine phytoplankton population using a sTable 13C isotope. *Mar. Biol.* **1983**, *73*, 31–36. [CrossRef]

35. Danielson, S.A.; Eisner, L.; Ladd, C.; Mordy, C.; Sousa, L.; Weingartner, T.J. A comparison between late summer 2012 and 2013 water masses, macronutrients, and phytoplankton standing crops in the northern Bering and Chukchi Seas. *Deep-Sea Res. Part II* **2017**, *135*, 7–26. [CrossRef]

36. Coachman, L.K.; Aagaard, K.; Trip, R.B. *Bering Strait: The Regional Physical Oceanography*; University of Washington Press: Seattle, WA, USA, 1975.

37. Springer, A.M.; McRoy, C.P.; Turco, K.R. The paradox of pelagic food webs in the northern Bering Sea—II. Zooplankton communities. *Cont. Shelf Res.* **1989**, *9*, 359–386. [CrossRef]

38. Yun, M.S.; Kim, B.K.; Joo, H.T.; Yang, E.J.; Nishino, S.; Chung, K.H.; Kang, S.H.; Lee, S.H. Regional productivity of phytoplankton in the Western Arctic Ocean during summer in 2010. *Deep Sea Res. II* **2015**, *120*, 61–71. [CrossRef]

39. Yun, M.S.; Whitledge, T.E.; Kong, M.; Lee, S.H. Low primary production in the Chukchi Sea shelf, 2009. *Cont. Shelf Res.* **2014**, *76*, 1–11. [CrossRef]

40. Lee, S.H.; Joo, H.M.; Yun, M.S.; Whiteledge, T.E. Recent phytoplankton productivity of the northern Bering Sea during early summer in 2007. *Polar Biol.* **2012**, *35*, 83–98. [CrossRef]

41. Lee, S.H.; Kim, B.K.; Lim, Y.J.; Joo, H.; Kang, J.J.; Lee, D.; Park, J.; Ha, S.-Y.; Lee, S.H. Small phytoplankton contribution to the standing stocks and the total primary production in the Amundsen Sea. *Biogeosciences* **2017**, *14*, 3705–3713. [CrossRef]

42. Lee, S.H.; Joo, H.T.; Lee, J.H.; Lee, J.H.; Kang, J.J.; Lee, H.W.; Lee, D.B.; Kang, C.K. Seasonal carbon uptake rates of phytoplankton in the northern East/Japan Sea. *Deep Sea Res. Part II Top. Stud. Oceanogr.* **2017**, *143*, 45–53. [CrossRef]

43. Bhavya, P.S.; Lee, J.H.; Lee, H.W.; Kang, J.J.; Lee, J.H.; Lee, D.B.; Ahn, S.H.; Stockwell, D.; Whitledge, T.E.; Lee, S.H. First in situ estimations of small phytoplankton carbon and nitrogen uptake rates in the Kara, Laptev, and East Siberian seas. *Biogeosciences* **2018**, *15*, 5503–5517. [CrossRef]

44. Hodal, H.; Kristiansen, S. The importance of small-celled phytoplankton in spring blooms at the marginal ice zone in the northern Barents Sea. *Deep Sea Res. Part II Top. Stud. Oceanogr.* **2008**, *55*, 2176–2185. [CrossRef]

45. Mei, Z.-P.; Legendre, L.; Gratton, Y.; Tremblay, J.-É.; LeBlanc, B.; Klein, B.; Goseelin, M. Phytoplankton production in the North Water Polynya: Size-fractions and carbon fluxes, April to July 1998. *Mar. Ecol. Prog. Ser.* **2003**, *256*, 13–27. [CrossRef]

46. Kim, J.-M.; Lee, K.; Suh, Y.-S.; Han, I.-S. Phytoplankton Do Not Produce Carbon-Rich Organic Matter in High CO_2 Oceans. *Geophys. Res. Lett.* **2018**, *45*, 4189–4197. [CrossRef]

 *water*

Article

Picocyanobacterial Contribution to the Total Primary Production in the Northwestern Pacific Ocean

Ho-Won Lee [1], Jae-Hoon Noh [1], Dong-Han Choi [1], Misun Yun [2], P. S. Bhavya [3], Jae-Joong Kang [4], Jae-Hyung Lee [5], Kwan-Woo Kim [4], Hyo-Keun Jang [4] and Sang-Heon Lee [4,*]

[1] Marine Ecosystem Research Center, Korea Institute of Ocean Science and Technology, Busan 49111, Korea; howonlee@kiost.ac.kr (H.-W.L.); jhnoh@kiost.ac.kr (J.-H.N.); dhchoi@kiost.ac.kr (D.-H.C.)

[2] College of Marine and Environmental Sciences, Tianjin University of Science and Technology, Tianjin 300457, China; misunyun@tust.edu.cn

[3] Department of Biological Oceanography, Leibniz Institute for Baltic Sea Research Warnemuende, 18119 Rostock, Germany; bhavya.panthalil@io-warnemuende.de

[4] Department of Oceanography, Pusan National University, Busan 46241, Korea; jaejung@pusan.ac.kr (J.-J.K.); goanwoo7@pusan.ac.kr (K.-W.K.); janghk@pusan.ac.kr (H.-K.J.)

[5] South Sea Fisheries Research Institute, National Institute of Fisheries Science, Yeosu-si 59780, Korea; jhlee88@korea.kr

* Correspondence: sanglee@pusan.ac.kr

Citation: Lee, H.-W.; Noh, J.-H.; Choi, D.-H.; Yun, M.; Bhavya, P.S.; Kang, J.-J.; Lee, J.-H.; Kim, K.-W.; Jang, H.-K.; Lee, S.-H. Picocyanobacterial Contribution to the Total Primary Production in the Northwestern Pacific Ocean. *Water* **2021**, *13*, 1610. https://doi.org/10.3390/w13111610

Academic Editor: Maria Moustaka-Gouni

Received: 16 April 2021
Accepted: 4 June 2021
Published: 7 June 2021

Publisher's Note: MDPI stays neutral with regard to jurisdictional claims in published maps and institutional affiliations.

Abstract: Picocyanobacteria (*Prochlorococcus* and *Synechococcus*) play an important role in primary production and biogeochemical cycles in the subtropical and tropical Pacific Ocean, but little biological information on them is currently available in the North Pacific Ocean (NPO). The present study aimed to determine the picocyanobacterial contributions to the total primary production in the regions in the NPO using a combination of a dual stable isotope method and metabolic inhibitor. In terms of cell abundance, *Prochlorococcus* were mostly dominant (95.7 ± 1.4%) in the tropical Pacific region (hereafter, TP), whereas *Synechococcus* accounted for 50.8%–93.5% in the subtropical and temperate Pacific region (hereafter, SP). Regionally, the averages of primary production and picocyanobacterial contributions were 11.66 mg C m^{-2}·h^{-1} and 45.2% (±4.8%) in the TP and 22.83 mg C m^{-2}·h^{-1} and 70.2% in the SP, respectively. In comparison to the carbon, the average total nitrogen uptake rates and picocyanobacterial contributions were 10.11 mg N m^{-2}·h^{-1} and 90.2% (±5.3%) in the TP and 4.12 mg N m^{-2}·h^{-1} and 63.5%, respectively. These results indicate that picocyanobacteria is responsible for a large portion of the total primary production in the region, with higher contribution to nitrogen uptake rate than carbon. A long-term monitoring on the picocyanobacterial variability and contributions to primary production should be implemented under the global warming scenario with increasing ecological roles of picocyanobacteria.

Keywords: cyanobacteria; *Prochlorococcus*; *Synechococcus*; primary production; northwestern Pacific Ocean

1. Introduction

Phytoplankton are major biological components as primary producers in marine ecosystems. Marine phytoplankton not only account for a significant proportion of global primary production, but also are an important food source in marine ecosystems and a potential moderator of global carbon cycle at the ocean–atmosphere interface [1]. Distribution, abundance, and diversity of phytoplankton differ greatly among dominant water masses in the various oceanic regions, which are closely related to physiochemical properties. In addition, long-term research on the limiting factors (e.g., temperature, nutrients, and light regime) of phytoplankton has reported that biological and ecological changes resulted from variations of these factors such as increasing of seawater temperature and reinforcement of stratification [2,3]. Primary production is widely used as one of key biological factors for understanding the regional differences in basic environmental and biological conditions such as thermohaline properties, nutrients, chlorophyll-*a*, etc. [4–8].

The Pacific Ocean, due to its vastness extending from tropical regions to both the boundaries of polar oceans, is subjected to have distinctive climatic conditions at its various regions [4]. In the northwestern Pacific Ocean (NPO), the physico-chemical conditions are mainly influenced by North Equatorial Current, Kuroshio Current, Tsushima Warm Current, and pelagic/coastal water intrusions at the coastal zones in the East China Sea (ECS). In terms of phytoplankton community, autotrophic picoplankton communities were more abundant in the NPO than large-sized phytoplankton and heterotrophic bacteria [9,10]. Lee et al. [10] reported that autotrophic plankton (mainly pico-sized phytoplankton) comprised up to 80% of the total phytoplankton biomass in the euphotic zone, whereas the contribution of heterotrophic bacteria was 6–21% of phytoplankton biomass in the NPO. Furthermore, a few research works reported that picoplankton including pico-sized cyanobacteria (*Prochlorococcus* and *Synechococcus*) have been a significant component of biomass and primary production in the subtropical and tropical Pacific Ocean [11–17].

In general, *Prochlorococcus* exhibits a wide adaptation for the variability in light or nutrient conditions, whereas they are often observed to be limited by high temperature in the water column [16,18–23]. Other cyanobacterial species such as *Synechococcus* have relatively eurythermal characteristics and extend to low salinity waters. Hence, *Synechococcus* are widely distributed around the world ocean from tropical to polar waters with a high biomass in the upper euphotic zone [22,24]. Recently, it was reported that abundance and distribution of the small-sized autotrophic plankton communities including cyanobacteria, *Prochlorococcus*, and *Synechococcus*, increase in various oceans with global warming, which indicates that this issue is not limited to a local scale anymore [25,26].

Normally, the primary production by picophytoplankton (i.e., picocyanobacteria and picoeukaryotes) is estimated through filter fractionation [7,27–31]. However, it is difficult to distinguish carbon and/or nitrogen uptake rates between picocyanobacteria (*Prochlorococcus* and *Synechococcus*) and picoeukaryotes. Moreover, the fractionation in natural samples makes it difficult to physically separate picophytoplankton from heterotrophic bacteria, in case of nitrogen uptake [32]. Previous studies used metabolic inhibitors to partition the relative contributions of eukaryotes and prokaryotes in marine systems [33–35]. For example, cycloheximide inhibits the function of the 80-S ribosome of eukaryotes [36], whereas streptomycin specifically inhibits protein synthesis on the 70-S ribosome in bacteria [37]. Thus, these metabolic inhibitors could be effective in separating target organisms from non-target organisms [32]. Middelburg and Nieuwenhuize [34,35] successfully partitioned autotrophic and heterotrophic activity using metabolic inhibitors. Fouilland et al. [32] also applied metabolic inhibitors to partition the uptakes of nitrate, ammonium, and urea between prokaryotic and eukaryotic phytoplankton. As a result, they quantitatively reported the contribution of heterotrophic bacteria to nitrogen uptake [32].

In present study, a metabolic inhibitor (cycloheximide) based on the method of Fouilland et al. [32] was applied to measure picocyanobacterial contribution to the total primary production, since the inhibitor can remove the eukaryotes and directly determine only carbon and nitrogen uptake rates by picocyanobacteria in the samples. The objectives of this study were as follows: (1) to determine carbon and nitrogen uptake rates by picocyanobacteria and (2) to evaluate picocyanobacterial contribution to the primary production in the regions (subtropical-temperate Pacific region and tropical Pacific region) in the NPO.

2. Materials and Methods

2.1. Study Area and Sample Collection

The present study was conducted at 9 stations in the NPO during the POSEIDON cruise (13 May–4 June 2014) (Figure 1). In order to understand characteristics of primary productivity under the different environmental conditions in the NPO, our productivity stations were divided into two regions; the subtropical and temperate Pacific region (SP; A89 and A50), mainly affected by the coastal water of the ECS and tropical Pacific region

(TP; F10, F06, F03, F01, P03, P07, and P11), which are influenced by the Tsushima Warm Current, North Equatorial Current, and Kuroshio Current. Seawater samples from water column up to 1% light depths were collected using 10 L Niskin sampling bottles on the R/V Onnuri of the Korea Institute of Ocean Science and Technology (KIOST, Busan, Korea). The physical parameters (temperature and salinity) were determined using a Sea-Bird 911plus system (Sea-Bird, Inc., Brooklyn, NY, USA).

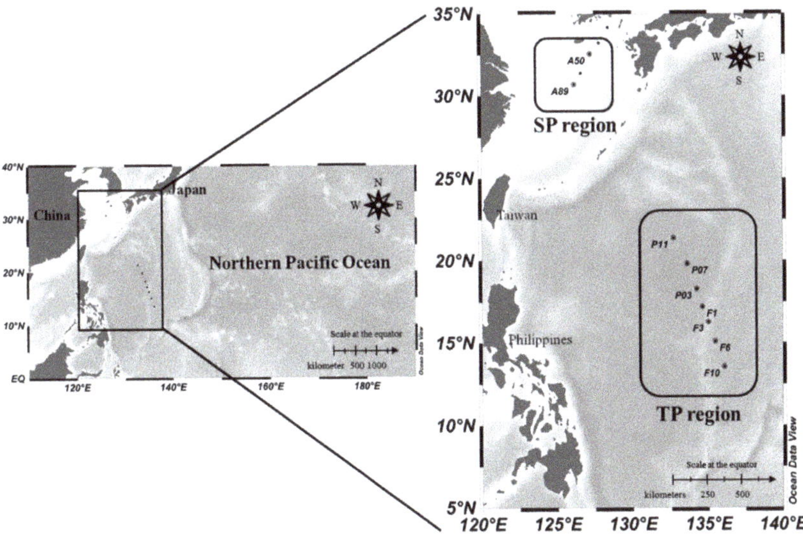

Figure 1. Sampling station in the two sampled regions of the northwestern Pacific Ocean; TP: tropical Pacific, SP: subtropical and temperate Pacific.

2.2. Measurements for Biomass and Abundance of Phytoplankton and Nutrient Concentrations

Chlorophyll-*a* (Chl-*a*) and phytoplankton abundance, as well as nutrient concentrations were measured at the 9 productivity stations. One liter of seawater for Chl-*a* concentrations presenting for phytoplankton biomass was filtered onto 25 mm GF/F filters. The filters were stored in a deep freezer and extracted within a month using 6 mL of 95% acetone by the method of Parsons et al. [38]. The final extracts were analyzed using a 10 AU fluorometer (Turner Design Inc., San Jose, CA, USA). Seawater samples for the enumeration and identification of major pico-sized phytoplankton groups (<2 µm) were counted by flow cytometry (BD Accury C6, BD Biosciences Inc., Mountain View, CA, USA) after staining with mixture of yellow-green and UV beads by the method of Olson et al. [39]. Nutrient data were provided by KIOST based on the standard colorimetric procedure [38].

2.3. Carbon and Nitrogen Uptake Rate Measurements

Total carbon and nitrogen uptake rates were measured at the 9 different stations using a ^{13}C-^{15}N dual isotope tracer technique that has been applied in various oceans [27,40–43]. Seawater samples at 6 light depths (100%, 50%, 30%, 12%, 5%, and 1% of light intensity at surface) were collected from Niskin samplers to 1 L polycarbonate bottles covered with different LEE film screens (LEE Filters, Inc., Hampsire, UK) that corresponded to the different light levels. Further, the water samples were injected with enriched solutions of ^{13}C (NaH^{13}CO$_3$) and ^{15}N (K^{15}NO$_3$ or ^{15}NH$_4$Cl) (less than 10% of the ambient concentrations) followed by deck incubation for 4 h. Hourly picocyanobacterial carbon and nitrogen uptake rates were measured at all the stations except station A50 in the SP using the dual isotope technique. For measuring the picocyanobacterial carbon and nitrogen uptake rates, the autotrophic eukaryotes were inhibited by a metabolic inhibitor (cycloheximide), which blocks the cytoplasmic protein biosynthesis in 80-S ribosome of phytoplankton (eukaryotes) [32]. All the bottles were incubated in deck incubators along with primary productivity sample bottles for 4 h.

After incubation, seawater samples (0.5 L) for the carbon and nitrogen uptake rates were filtered onto the pre-combusted 25 mm GF/F filters. The filters were immediately frozen in the deep freezer until the analysis. Prior to the mass spectrometric analysis, samples were thawed, dried overnight, and packed in tin capsules. Particulate organic carbon (POC)/nitrogen (PON) and the amount of ^{13}C and ^{15}N were determined by Finnigan Delta + XL mass spectrometer at the Stable Isotope Facility, University of Alaska Fairbanks (UAF), USA after HCl fuming during 24 h for removing carbonate. Samples of analyzed total carbon and nitrogen uptake rates were calculated by using the methods of Hama et al. [44] and Dugdale and Goering [45]. Dark carbon uptake rates were subtracted for considering the heterotrophic bacterial process [46]. Because the carbon uptake rates from dark bottles were subtracted from the light bottles for removal of heterotrophic productivity without light, we assumed that only the contributions of autotrophic bacterial (i.e., picocyanobacterial) communities were obtained for the primary productivity.

3. Results

3.1. Physiochemical Structures in Water Column

Vertical profiles of temperature and salinity from all the stations in the NPO are shown in Figure 2. Surface temperature and salinity at the stations in the TP were higher than those in the SP. The average temperature and salinity in the upper water column were 17.3 °C and 33.2 psu in the SP, respectively, whereas they were 29.1 °C (S.D. = ±0.92 °C) and 34.8 psu (S.D. = ±0.52 psu) for TP, respectively.

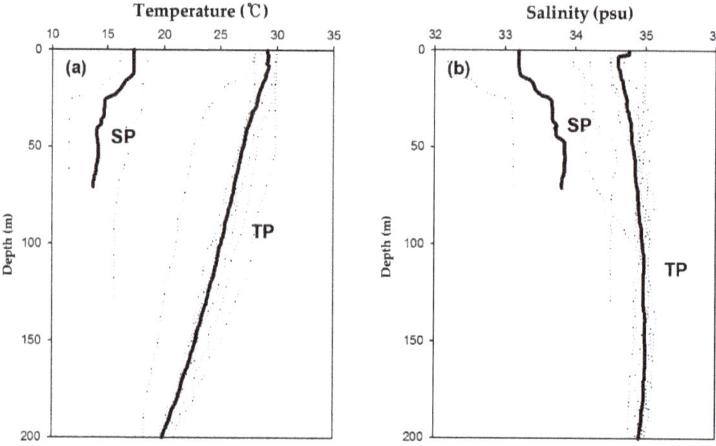

Figure 2. Vertical profiles of temperature (**a**), salinity (**b**) in the northwestern Pacific Ocean. Solid line represents average temperature and salinity in the TP and SP regions, respectively.

Generally, nutrient concentrations were depleted in both TP and SP regions except for 1% light depths (Figure 3). The mean nitrate concentrations within the euphotic zone were 0.13 (S.D. = ±0.35 µM) and 0.84 µM (S.D. = ±1.80 µM) in the TP and the SP, respectively. Ammonium concentrations were consistently low at euphotic zones of all the stations. The mean ammonium concentrations in the TP and the SP were 0.14 (S.D. = ±0.07 µM) and 0.18 µM (S.D. = ±0.03 µM), respectively. The euphotic depths at the stations in the TP were deeper than those in the SP (*t*-test, $p < 0.05$). The mean euphotic depths were 127.4 m (S.D. = ±16.5 m) and 35.0 m in the TP and the SP, respectively (Table 1).

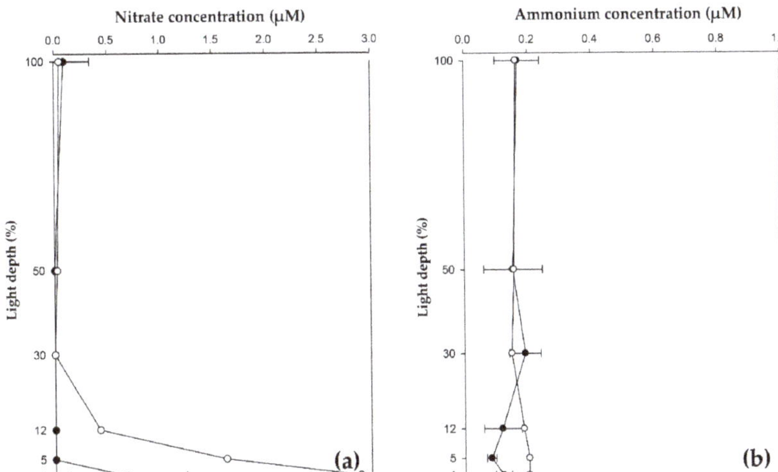

Figure 3. Vertical profiles of nitrate (**a**) and ammonium (**b**) concentrations averaged from each region (TP, closed circles; SP, open circles). SDs are shown by bars. Data courtesy of KIOST.

Table 1. The environmental conditions in the TP and SP regions of the northwestern Pacific Ocean.

	TP		SP	
	Mean	S.D.	Mean	S.D.
Temperature in the surface (°C)	29.1	0.9	17.3	-
Temperature in the euphotic depth (°C)	26.7	2.3	15.3	2.3
Salinity in the surface (psu)	34.8	0.5	33.2	-
Salinity in the euphotic zone (psu)	34.8	0.6	33.9	0.6
Nitrate in the euphotic zone (µM)	0.13	0.35	0.84	1.8
Ammonium in the euphotic depth (µM)	0.14	0.07	0.18	0.03
Euphotic depth (m)	127.4	16.5	35	-

3.2. Distribution of Phytoplankton in Water Column

The average euphotic depth-integral total Chl-*a* concentrations were 15.0 (S.D. = ±6.6 mg Chl-*a* m^{-2}) and 18.1 mg Chl-*a* m^{-2} in the TP and SP, respectively (Table 2). Although the integral total Chl-*a* concentrations were not significantly different between the TP and SP locations (Table 2), the vertical distributions of Chl-*a* were obviously different between the two locations (Figure 4). Deep chlorophyll maximum (DCM) layers, in which the Chl-*a* concentrations were significantly (*t*-test, $p < 0.01$) higher compared to those at the surface, were observed at the bottom (1% light depth) of the euphotic zone in the TP. However, no substantial DCM layers were found in the SP (Figure 4).

The cell abundance of autotrophic plankton, including picocyanobacteria (*Synechococcus* and *Prochlorococcus*) and picoeukaryotes, were different between the TP and the SP (Figure 5). The average depth-integral abundance *of Synechococcus, Prochlorococcus,* and picoeukaryotes in the TP were 1.85 × 10^{11} (S.D. = ±0.64 × 10^{11} cells m^{-2}), 0.64 × 10^{13} (S.D. = ±0.10 × 10^{13} cells m^{-2}), and 0.96 × 10^{11} cells m^{-2} (S.D. = ±0.31 × 10^{11} cells m^{-2}), respectively (Figure 5). In the SP, the cell abundance of *Synechococcus* and picoeukaryotes were 14.4 × 10^{11} and 4.28 × 10^{11} cells m^{-2}, respectively. No *Prochlorococcus* were generally found in the SP except some at 46 m of A89 (Figure 5). Consequently, *Prochlorococcus* (mean ± S.D. = 95.7 ± 1.4%), *Synechococcus* (mean ± S.D. = 2.8 ± 1.0%), and picoeukaryotes (mean ± S.D. = 1.4 ± 0.4%) contributed the plankton community in the TP. In contrast, *Synechococcus* accounted for 93.5% and 50.8%, whereas picoeukaryotes were 5.2% and 49.2% at A89 and A50 in the SP, respectively.

Table 2. Chlorophyll-*a*, C/N ratio, *f*-ratio, carbon, and nitrogen (nitrate and ammonium) uptake rates by total phytoplankton communities in the TP and SP regions of the northwestern Pacific Ocean.

	TP			SP		
	Mean	**S.D.**	*n*	**Mean**	**S.D.**	*n*
Integrated total Chlorophyll-a (mg Chl-*a* m^{-2})	15	6.6	7	18.1	-	2
C/N ratio (atom/atom)	11	1.8	7	9.8	-	2
Carbon specific uptake (h^{-1})	0.001508	0.001034	42	0.004951	0.004069	12
Carbon absolute uptake (mg C m^{-3}·h^{-1})	0.099	0.068	42	0.688	0.653	12
Integrated carbon uptake (mg C m^{-2}·h^{-1})	11.66	4.8	7	20.85	-	2
Nitrate specific uptake (h^{-1})	0.000632	0.000435	42	0.001097	0.001096	12
Nitrate absolute uptake (mg NO$_3^-$ m^{-3}·h^{-1})	0.007987	0.006853	42	0.022084	0.024058	12
Integrated nitrate uptake (mg NO$_3^-$ m^{-2}·h^{-1})	1.06	0.68	7	0.69	-	2
Ammonium specific uptake (h^{-1})	0.006756	0.003664	42	0.006355	0.003179	12
Ammonium absolute uptake (mg NH$_4^+$ m^{-3}·h^{-1})	0.072252	0.044304	42	0.120235	0.077651	12
Integrated ammonium uptake (mg NH$_4^+$ m^{-2}·h^{-1})	9.05	3.1	7	4.05	-	2
Nitrogen specific uptake (h^{-1})	0.007388	0.004099	42	0.007452	0.004255	12
Nitrogen absolute uptake (mg N m^{-2}·h^{-1})	0.08	0.047	42	0.142	0.1	12
Integrated nitrogen uptake (mg N m^{-2}·h^{-1})	10.11	2.49	7	4.74	-	2
f-ratio	0.1	0.03	7	0.13	-	2

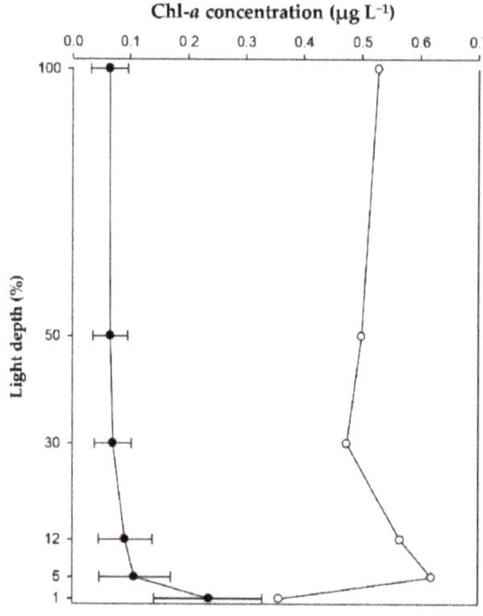

Figure 4. Vertical profiles of chlorophyll-*a* concentrations in the TP (closed circles) and SP (open circles) regions. SDs are shown by bars.

3.3. Total Carbon and Nitrogen Uptake Rates in the NPO

The largest carbon uptake rate was at 100% light depth at each station in the SP, whereas in the TP, the largest rate was observed at 30–50% light depths (Figure 6a). The lowest carbon uptake rate was found at the chlorophyll-maximum layer corresponding to 1% light depth in the SP. The average rates of carbon uptake at each light depth were significantly higher in the SP (*t*-test, $p < 0.05$) than in the TP. The ranges of depth-integrated carbon uptake rates in the TP and SP were 3.29–16.89 mg C m^{-2}·h^{-1} with an average of 11.66 mg C m^{-2}·h^{-1} and 9.17–32.54 mg C m^{-2}·h^{-1} with an average of 20.85 mg C m^{-2}·h^{-1}, respectively (Figure 7a and Table 2). Based on our dark carbon uptake rates in this study, the heterotrophic contributions to the total primary productions were 1.5% (S.D. = ±0.7%) and 8.7% (S.D. = ±12.8%) for the SP and the TP, respectively.

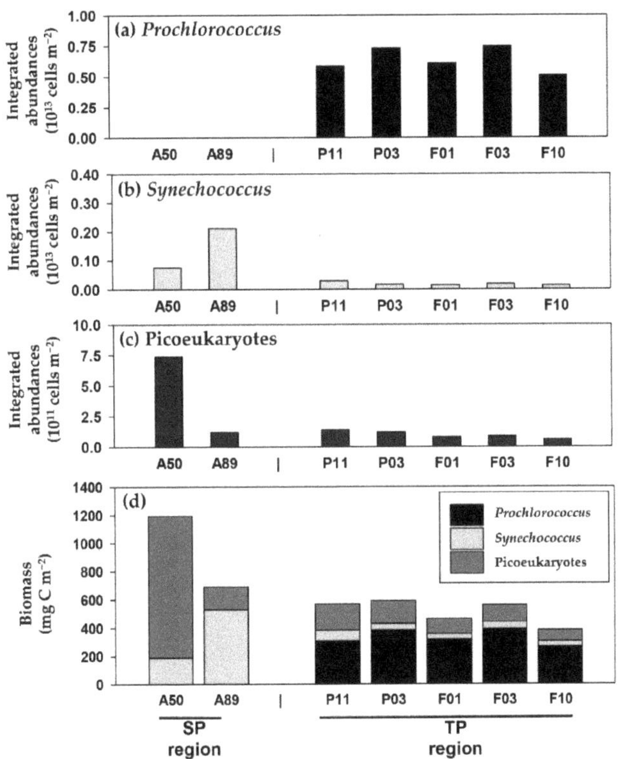

Figure 5. Integrated abundance of *Prochlororoccus* (**a**), *Synechococcus* (**b**), and picoeukaryotes (**c**) in the euphotic zone. Biomass of *Prochlororoccus*, *Synechococcus*, and picoeukaryotes in the euphotic zone (**d**) in the euphotic zone.

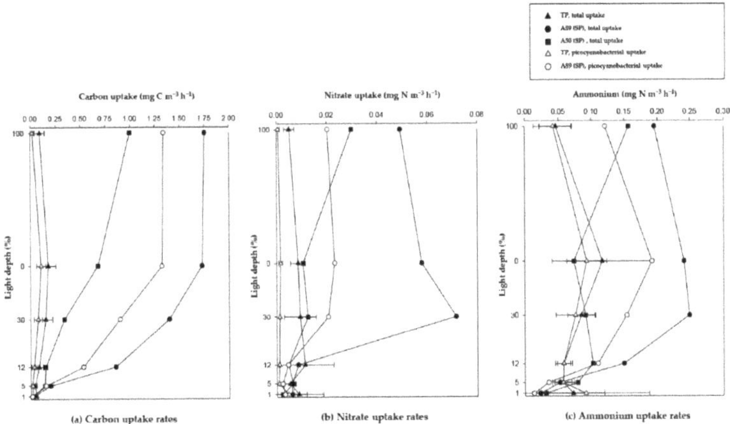

Figure 6. Vertical profiles of the total carbon and nitrogen absolute uptake rates (TP, closed triangles; A89, closed circles; A50, closed squares) and picocyanobacterial carbon and nitrogen absolute uptake rates (TP, open triangles; A89, open circles) in the TP and SP regions in the North Pacific Ocean. SDs are shown by bars. Carbon uptake rates (**a**), Nitrate uptake rates (**b**), and Ammonium uptake rates (**c**).

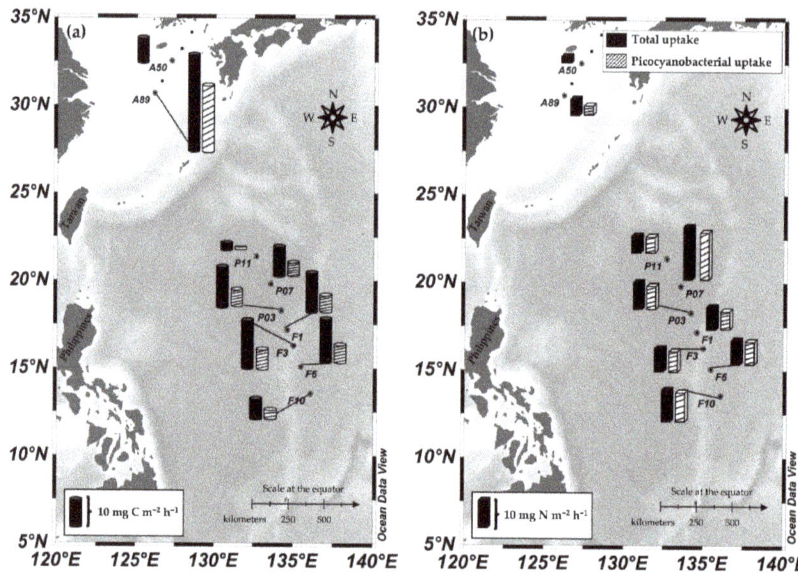

Figure 7. Regional distribution of total carbon and nitrogen uptake rates ((**a**), left) and picocyanobacterial carbon and nitrogen uptake rates ((**b**), right) in the northwestern Pacific Ocean. Bars with diagonal stripes indicate carbon and nitrogen uptake rates of picocyanobacterial communities.

Nitrogen uptake rates did not show any significant pattern with light depths as carbon uptake rates (Figure 6b,c). The depth-integrated nitrogen (nitrate+ammonium) uptake rates in the TP and SP ranged from 6.52 to 17.96 mg N m^{-2}·h^{-1} with an average of 10.11 mg N m^{-2}·h^{-1} and from 2.98 mg N m^{-2}·h^{-1} to 6.50 mg N m^{-2}·h^{-1} with an average of 4.74 mg N m^{-2}·h^{-1}, respectively (Figure 7b and Table 2). In detail, the mean of nitrate and ammonium uptake rates in the TP were 1.06 mg N m^{-2}·h^{-1} and 9.05 mg N m^{-2}·h^{-1}, respectively, whereas those in the SP were 0.69 mg N m^{-2}·h^{-1} and 4.05 mg N m^{-2}·h^{-1}, respectively. Ammonium uptake rates were substantially higher than nitrate uptake rates in both regions.

3.4. Picocyanobacterial Carbon and Nitrogen Uptakes in the NPO

The average rates of picocyanobacterial carbon uptakes showed similar trends like vertical abundance profiles of these predominant species (Figure 8). Vertical profiles of picocyanobacterial carbon, nitrate, and ammonium uptake rates showed similar trends as those of the uptake rates by total phytoplankton community at each light depth (Figure 6). Picocyanobacterial carbon uptake rates integrated from the euphotic depths were 5.31 mg C m^{-2}·h^{-1} (S.D. = ±2.16 mg C m^{-2}·h^{-1}) in the TP, whereas the integrated carbon uptake rates by picocyanobacteria at the A89 (SP) was 22.8 mg C m^{-2}·h^{-1} (Figure 9a). The average rates of picocyanobacterial carbon uptake at each light gradient were significantly higher in the SP (Table 3; *t*-test, $p < 0.05$). Integrated hourly picocyanobacterial nitrogen uptake rates were 6.32–16.16 mg N m^{-2}·h^{-1} with an average of 9.10 mg N m^{-2}·h^{-1} in the TP and 4.12 mg N m^{-2}·h^{-1} at the A89 in the SP (Figures 7b and 9b). The average nitrate uptake rates by picocyanobacterial communities in the TP and A89 (SP) were 0.21 mg N m^{-2}·h^{-1} (S.D. = ±0.20 mg N m^{-2}·h^{-1}) and 0.40 mg N m^{-2}·h^{-1}, respectively, whereas the average ammonium uptake rates of picocyanobacterial communities were 8.89 mg N m^{-2}·h^{-1} (S.D. = ±3.18 mg N m^{-2}·h^{-1}) and 3.72 mg N m^{-2}·h^{-1}, respectively (Table 3). Picocyanobacterial ammonium uptake rates were more than the nitrate uptake rates in the NPO (Figure 9c,d).

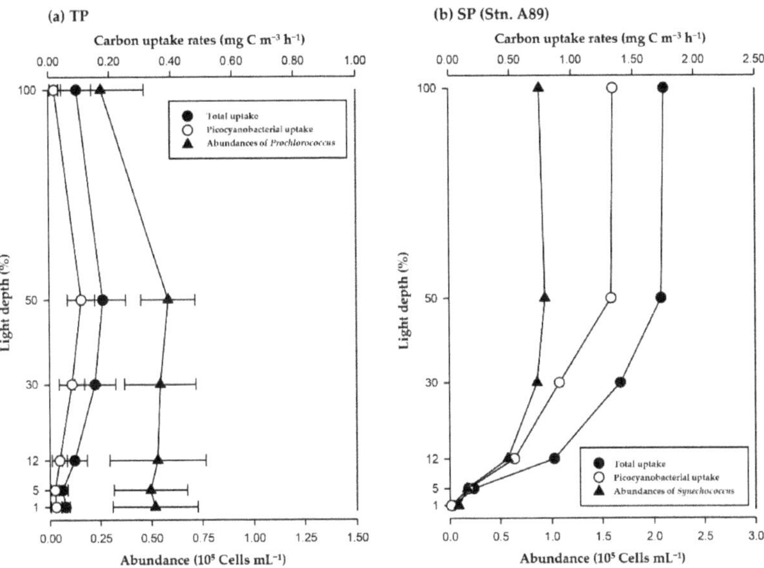

Figure 8. Comparison of the uptake rates for total (closed circles) and bacterial (open circles) carbon uptake with the abundances of the predominant species (closed triangles) in the northwestern Pacific Ocean. TP (**a**) and SP (**b**). SDs are shown by bars.

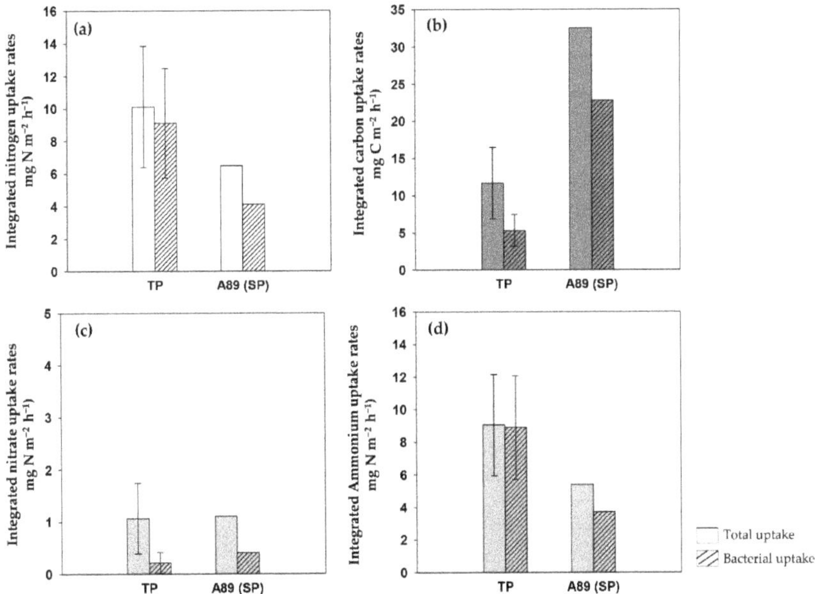

Figure 9. Picocyanobacterial contribution to total carbon and nitrogen uptake rates (primary productivity) in the TP and SP regions of the northwestern Pacific Ocean. Unicolor bars indicate total uptake of each uptake rate, whereas other bars with diagonal stripes indicate picocyanobacterial uptake rates. SDs are shown by bars. Integrated nitrogen uptake rates (**a**), Integrated carbon uptake rates (**b**), Integrated nitrate uptake rates (**c**), and Integrated ammonium uptake rates (**d**).

Table 3. Carbon and nitrogen (nitrate and ammonium) uptake rates by picocyanobacterial communities in the TP and SP regions of the northwestern Pacific Ocean.

	TP			SP (A89)		
	Mean	**SD**	*n*	**Mean**	**SD**	*n*
Picocyanobacterial carbon specific uptake (h^{-1})	0.000695	0.000548	42	0.004404	0.003065	6
Picocyanobacterial carbon absolute uptake (mg C $m^{-3}\cdot h^{-1}$)	0.044	0.043	42	0.708	0.573	6
Integrated picocyanobacterial carbon uptake (mg C $m^{-2}\cdot h^{-1}$)	5.31	2.16	7	22.83	-	1
Picocyanobacterial nitrate specific uptake (h^{-1})	0.000104	0.000174	42	0.00047	0.000348	6
Picocyanobacterial nitrate absolute uptake (mg NO_3^- $m^{-3}\cdot h^{-1}$)	0.001577	0.003138	42	0.012296	0.00997	6
Integrated picocyanobacterial nitrate uptake (mg NO_3^- $m^{-2}\cdot h^{-1}$)	0.21	0.2	7	0.4	-	1
Picocyanobacterial ammonium specific uptake (h^{-1})	0.005073	0.002693	42	0.004497	0.002518	6
Picocyanobacterial ammonium absolute uptake (mg NH_4^+ $m^{-3}\cdot h^{-1}$)	0.070531	0.046278	42	0.104054	0.068611	6
Integrated picocyanobacterial ammonium uptake (mg NH_4^+ $m^{-2}\cdot h^{-1}$)	8.89	3.18	7	3.72	-	1
Integrated picocyanobacterial nitrogen uptake (mg N $m^{-2}\cdot h^{-1}$)	9.1	3.37	7	4.12	-	1
Picocyanobacterial *f*-ratio	0.02	0.01	7	0.1	-	1

4. Discussion

In this study, the abundance of picophytoplankton was different between the TP and the SP (Figure 5). *Prochlorococcus* were not found but *Synechococcus* and picoeukaryotes co-occurred in the SP, whereas *Prochlorococcus* were the dominant picophytoplankton population in the TP. The difference in abundance of dominant population observed in the TP and the SP might be due to different physico-chemical properties as the result of the major currents. Because distribution and abundance of phytoplankton in the euphotic zone can be altered by the hydrological conditions of the seawater, these physiochemical properties are determined by the major currents [47–50]. In fact, the TP is directly influenced by North Equatorial Current, whereas the SP is influenced mainly by the Kuroshio Current, Tsushima Warm Current, and coastal fresh water, respectively [16,51]. According to Choi et al. [24], the picocyanobacterial distribution in the NPO was distinguished along the physical and chemical properties of the water masses. In this study, the water depth in the SP was shallow and had lower temperature and salinity than the TP (Figure 2), whereas the TP was a typical high-temperature oligotrophic water. Since *Prochlorococcus* have been found to be more abundant in the oligotrophic conditions because of their ecological plasticity with respect to requirements of nutrients and light [18,52–55], *Prochlorococcus* could be dominant under temperature and oligotrophic TP. According to previous studies [12,55], *Synechococcus* are usually dominant in the mesotrophic seawater or shallow waters. Thus, *Synechococcus* and picoeukaryotes could be abundant in relatively mesotrophic and shallow SP, which is consistent with previous study from the western Pacific Ocean [54].

In terms of carbon biomasses estimated from the average carbon contents [56,57], *Prochlorococcus* contributed 66.1% to the total phytoplankton in the TP (Figure 5d). In the SP, *Synechococcus* were 76.4% at A89 and picoeukaryotes were 84.0% at A50, respectively. Especially, the carbon biomass contribution of picoeukaryotes was higher than that of *Synechococcus* at the A50, although picoeukaryotes had lower cell abundances than *Synechococcus*, because picoeukaryotes have higher carbon contents compared to *Synechococcus*.

Based on the hourly carbon uptake rates by total phytoplankton, which were estimated in this study, the average daily primary productivities were 0.15 g C $m^{-2}\cdot d^{-1}$ (S.D. = ±0.06 g C $m^{-2}\cdot d^{-1}$) and 0.29 g C $m^{-2}\cdot d^{-1}$ in the TP and SP, respectively (Table 4). Our daily primary productivities fell within the range of previous studies in both regions [4,5,51]. In the TP, Taniguchi [4] reported 0.09 g C m^{-2}·d^{-1} in the North Equatorial Current (Table 4). Kwak et al. [5] observed a relatively higher range of daily primary productivity from 0.17 to 0.23 g C m^{-2}·d^{-1} in the western Pacific Ocean (Table 4). For the SP, the average daily primary productivity obtained from this study is comparable with those from other previous studies [5,51]. Gong et al. [51] reported 0.31 $\pm$ 0.16 g C m^{-2}·d^{-1} and 0.52 $\pm$ 0.32 g C m^{-2}·d^{-1} during early spring and summer, respectively (Table 4). Our daily primary productivity is nearly identical to the daily production (0.28 g C m^{-2}·d^{-1}) reported by Kwak et al. [5] (Table 4).

Table 4. Comparison of daily primary productivity with previous studies in the northwestern Pacific Ocean.

Region	References	Carbon Uptake Rates Average $\pm$ SD (g C m^{-2}·d^{-1})	Nitrate Uptake Rates Average $\pm$ SD (g N m^{-2}·d^{-1})	Ammonium Uptake Rates Average $\pm$ SD (g N m^{-2}·d^{-1})	Season
TP	Taniguchi (1972)	0.09	-	-	Winter
	Kwak et al. (2013)	0.2	-	-	Summer
	In this study	0.15 $\pm$ 0.06	0.01 $\pm$ 0.01	0.16 $\pm$ 0.01	Late spring
SP	Gong et al. (2003)	0.31 $\pm$ 0.16	-	-	Early spring
		0.52 $\pm$ 0.32	-	-	Summer
	Kwak et al. (2013)	0.28	-	-	Summer
	In this study	0.45 (A89)	0.02 (A89)	0.10 (A89)	Late spring
		0.13 (A50)	0.01 (A50)	0.05 (A50)	

Daily total ammonium uptake rates were calculated by multiplying hourly nitrogen uptake rates and each photoperiod [58] in this study. The average daily total ammonium uptake rates were higher than total nitrate uptakes in the euphotic zone of both regions. The average daily total ammonium and nitrate uptake rates were 0.16 g N m^{-2}·d^{-1} (S.D. = $\pm$0.06 g N m^{-2}·d^{-1}) and 0.01 g N m^{-2}·d^{-1} (S.D. = $\pm$0.01 g N m^{-2}·d^{-1}) in the TP, respectively (Table 4). In the SP, the daily total ammonium and nitrate uptake rates were 0.07 g N·m^{-2} d^{-1} and 0.01 g N m^{-2}·d^{-1} at A89, respectively (Table 4). Accordingly, average f-ratios ([nitrate uptake rate]/[nitrate+ammonium uptake rate], [59]) were 0.10 (S.D. = $\pm$0.03) and 0.13 in the TP and SP (Table 2), respectively, as a result of prominent ammonium uptakes. This indicates that a main nitrogen source for growth of total autotrophic plankton was mainly supported by regenerated ammonium in this region during our observation period.

In this study, the average picocyanobacterial contributions to the total primary productivity accounted for 45.2% (S.D. = $\pm$4.8%) in the TP and 70.2% in the A89 (SP) (Figure 9a). Glover et al. [12] reported that contribution of *Synechococcus* to the total primary production, which varies from 6% to 46% in different water masses in the northwestern Atlantic Ocean. In contrast, Liu et al. [15] observed a high contribution of *Prochlorococcus* up to 82% to the primary production at Station ALOHA in the subtropical North Pacific Ocean.

Based on each nitrate and ammonium uptake rate, the average picocyanobacterial f-ratios were 0.02 (S.D. = $\pm$0.01) and 0.10 in the TP and A89 (SP), respectively (Table 3). This result suggests that picocyanobacterial communities strongly depended on a regenerated nitrogen source (i.e., ammonium) or N$_2$ fixation in our study area during the observation period. From the comparison of f-ratios between the total phytoplankton and picocyanobacterial communities, we found that picocyanobacterial f-ratios were substantially lower compared to those of the total phytoplankton communities in the two regions (Tables 2 and 3).

Depth integrated hourly nitrogen uptake rates of picocyanobacterial communities were 9.10 mg N m^{-2}·h^{-1} (S.D. = $\pm$3.73 mg N m^{-2}·h^{-1}) and 4.12 mg N m^{-2}·h^{-1} in the TP and the A89 (SP), respectively (Figure 9). The total nitrogen uptake rates at the same regions were 10.11 mg N m^{-2}·h^{-1} and 6.50 mg N m^{-2}·h^{-1}, respectively. Given the nitrogen uptake rates, the average picocyanobacterial contributions to the total nitrogen uptake rates were 90.2% (S.D. = $\pm$5.3%) and 63.5% in the TP and the A89 (SP), respectively, in this study. These picocyanobacterial contributions to the total nitrogen uptake rates are substantially higher than those to the total carbon uptake rates of the total plankton communities in TP. However, the nitrogen utilization by heterotrophic bacteria can be important since the heterotrophic bacteria account for a large fraction of nitrogen uptake in the global ocean including the Arctic Ocean [32,60,61]. Although we are incapable of distinguishing each contribution for nitrogen uptake between heterotrophic bacteria and picocyanobacteria from this study using a metabolic inhibitor (cycloheximide) blocking only photosynthetic eukaryotes, we need to consider the heterotrophic bacterial nitrogen utilization from the nitrogen contributions in future studies. Apart from this, the potential N2 fixation by cyanobacteria can vary with environmental conditions, particularly nutrient stoichiometry [62]. When the NH_4^+ concentration is relatively higher than phosphorous, the nitrogenase activity can be stopped and photosynthesis can be activated. On the other hand, if the NH_4^+:P ratio is lower than the Redfield's ratio, N2 fixation can be a more major process than primary production. So, the contribution of picocyanobacteria towards the total primary production can be underestimated in that case. Furthermore, when autotrophic primary production is stopped by the inhibitor, the competition for nutrients in the samples may be lesser than one with autotrophic activity and, hence, the primary production rates by picocyanobacteria could be overestimated. Currently, there are some uncertainties for estimating picocyanobacterial contributions to the primary production and nitrogen uptake rates. Therefore, the combined approaches using several different applications are highly important for further future studies on cyanobacterial ecological roles in various oceans.

5. Summary and Conclusions

In this study, we measured picocyanobacterial contribution to the carbon and nitrogen uptake rates by total phytoplankton in the regions of the NPO. There are different abundances and biomasses of dominant species in the TP and the SP regions. *Prochlorococcus* and *Synechococcus* were abundant in the TP and the SP regions, respectively. The picocyanobacterial contributed 45.2% (S.D. = ±4.8%) to primary production by total picophytoplankton in the TP, whereas the picocyanobacterial contribution was about 70.2% in the SP. The picocyanobacterial community is believed to be more important in terms of nitrogen uptake rates since they could contribute about 90.2% (S.D. = ±5.3%) to the total nitrogen uptake rates by picophytoplankton in both regions.

The importance of picoplankton including cyanobacteria has been raised continuously in research regarding the global ocean [25,63,64]. In particular, the picocyanobacterial *Prochlorococcus* and *Synechococcus* have significant ecological positions in the biomass and production in the ocean, but the relative contributions of these organisms to primary productivity are different under various environmental conditions [22]. Under the global warming scenario, picoplankton contribution relative to large plankton will increase in the strongly stratified upper ocean [3]. This climate change will result in increasing distribution, abundance, and contributions to primary production of picocyanobacteria, especially in tropical and subtropical oceans and, consequently, will cause large impacts on the global ocean ecosystem and biogeochemical cycles [26]. Therefore, more measurements under various environmental conditions are needed to better understand the role of picocyanobacterial in the ecosystem.

Author Contributions: Conceptualization, H.-W.L., J.-H.N. and S.-H.L.; methodology, H.-W.L., P.S.B. and S.-H.L.; validation, H.-W.L., J.-H.N. and S.-H.L.; formal analysis, H.-W.L., J.-H.N., D.-H.C., J.-J.K., J.-H.L. and K.-W.K.; investigation, H.-W.L., J.-H.N. and D.-H.C.; writing—original draft preparation, H.-W.L.; writing—review and editing, M.Y., P.S.B. and S.-H.L.; visualization, J.-J.K., J.-H.L., K.-W.K. and H.-K.J.; supervision, S.-H.L. All authors have read and agreed to the published version of the manuscript.

Funding: This research was supported by the National Research Foundation of Korea (NRF) grant funded by the Korean government (MSIT; NRF-2019R1A2C1003515) and partly by the Korea Institute of Ocean Science and Technology (KIOST; PE99923).

Institutional Review Board Statement: Not applicable.

Informed Consent Statement: Not applicable.

Data Availability Statement: Not applicable.

Acknowledgments: We thank the captain and crew members of the R/V *Onnuri* for their outstanding assistance. Especially, we very much appreciate the KIOST for providing CTD and nutrient data. Finally, we thank the anonymous reviewers who greatly improved an earlier version of manuscript.

Conflicts of Interest: The authors declare no conflict of interest.

References

1. Behrenfeld, M.J.; Randerson, J.T.; McClain, C.R.; Feldman, G.C.; Los, S.O.; Tucker, C.J.; Falkowski, P.G.; Field, C.B.; Frouin, R.; Esaias, W.E.; et al. Biospheric primary production during an ENSO transition. *Science* **2001**, *291*, 2594–2597. [CrossRef]
2. Howarth, R.W. Nutrient Limitation of Net Primary Production in Marine Ecosystems. *Annu. Rev. Ecol. Syst.* **1988**, *19*, 89–110. [CrossRef]
3. Morán, X.A.G.; López-Urrutia, Á.; Calvo-Díaz, A.; LI, W.K.W. Increasing importance of small phytoplankton in a warmer ocean. *J. Glob. Chang. Biol.* **2010**, *16*, 1137–1144. [CrossRef]
4. Taniguchi, A. Geographical variation of primary production in the Western Pacific Ocean and adjacent seas with reference to the inter-relations between various parameters of primary production. *Mem. Fac. Fish. Hokkaido Univ.* **1972**, *19*, 1–33.
5. Kwak, J.H.; Hwang, J.; Choy, E.J.; Park, H.J.; Kang, D.J.; Lee, T.; Chang, K.I.; Kim, K.R.; Kang, C.K. High primary productivity and *f*-ratio in summer in the Ulleung basin of the East/Japan Sea. *Deep-Sea Res. I* **2013**, *79*, 74–785. [CrossRef]
6. Lee, S.H.; Yun, M.S.; Kim, B.K.; Saitoh, S.; Kang, C.K.; Kang, S.H.; Whitledge, T. Latitudinal carbon productivity in the Bering and Chukchi Seas during the summer in 2007. *Cont. Shelf Res.* **2013**, *59*, 28–36.
7. Lee, S.H.; Yun, M.S.; Kim, B.K.; Joo, H.T.; Kang, S.H.; Kang, C.K.; Whitledge, T.E. Contribution of small phytoplankton to total primary production in the Chukchi Sea. *Cont. Shelf Res.* **2013**, *68*, 43–50. [CrossRef]
8. Kim, B.K.; Joo, H.T.; Song, H.J.; Yang, E.J.; Lee, S.H.; Hahm, D.; Rhee, T.S.; Lee, S.H. Large seasonal variation in phytoplankton production in the Amundsen Sea. *Polar Biol.* **2015**, *38*, 319–331. [CrossRef]
9. Fuhrman, J.A.; Hagström, Å. Bacterial and Archaeal Community Structure and its Patterns. In *Microbial Ecology of the Oceans*, 2nd ed.; Kirchman, D.L., Ed.; Wiley-Blackwell: Hoboken, NJ, USA, 2008; pp. 45–90.
10. Lee, C.R.; Choi, K.H.; Kang, H.K.; Yang, E.J.; Noh, J.H.; Choi, D.H. Biomass and trophic structure of the plankton community in subtropical and temperate waters of the northwestern Pacific Ocean. *J. Oceanogr.* **2012**, *68*, 473–482. [CrossRef]
11. Li, W.K.W.; Rao, D.V.S.; Harrison, W.G.; Smith, J.C.; Cullen, J.J.; Irwin, B.; Platt, T. Autotrophic picoplankton in the tropical ocean. *Science* **1983**, *219*, 292–295. [CrossRef]
12. Glover, H.E.; Campbell, L.; Prezelin, B.B. Contribution of *Synechococcus* spp. to size-fractioned primary productivity in three water masses in the Northwest Atlantic Ocean. *Mar. Biol.* **1986**, *91*, 193–203. [CrossRef]
13. Vaulot, D. The Cell Cycle of Phytoplankton: Coupling Cell Growth to Population Growth. In *Molecular Ecology of Aquatic Microbes*; NATO ASI Series (Series G: Ecological Sciences); Joint, I., Ed.; Springer: Berlin/Heidelberg, Germany, 1995. [CrossRef]
14. Chavez, F.P.; Buck, K.R.; Service, S.K.; Newton, J.; Barber, R.T. Phytoplankton variability in the central and eastern tropical pacific. *Deep Sea Res. II* **1996**, *43*, 835–870. [CrossRef]
15. Liu, H.; Nolla, H.A.; Campbell, L. *Prochlorococcus* growth rate and contribution to primary production in the equatorial and subtropical North Pacific Ocean. *Aquat. Microb. Ecol.* **1997**, *12*, 39–47. [CrossRef]
16. Choi, D.H.; Selph, K.E.; Noh, J.H. Niche partitioning of picocyanobacterial lineages in the oligotrophic northwestern Pacific Ocean. *Algae* **2015**, *30*, 223–232. [CrossRef]
17. Bhavya, P.S.; Min, J.O.; Kim, M.S.; Jang, H.K.; Kim, K.; Kang, J.J.; Lee, J.H.; Lee, D.; Jo, N.; Kim, M.J.; et al. A Review on Marine N_2 Fixation: Mechanism, Evolution of Methodologies, Rates, and Future Concerns. *Ocean Sci. J.* **2019**, *54*, 515–528. [CrossRef]
18. Vaulot, D.; Partensky, F.; Neveux, J.; Mantoura, R.F.C.; Llewellyn, C.A. Winter presence of prochlorophytes in surface waters of the northwestern Mediterranean Sea. *Limnol. Oceanogr.* **1990**, *35*, 1156–1164. [CrossRef]
19. Chisholm, S.W.; Frankel, S.L.; Goericke, R.; Olson, R.J.; Palenik, B.; Waterbury, J.B.; West-Johnsrud, L.; Zettler, E.R. *Prochlorococcus marinus* nov. gen. nov. sp.: An oxyphototrophic marine prokaryote containing divinyl chlorophyll *a* and *b*. *Arch. Microbiol.* **1992**, *157*, 297–300. [CrossRef]
20. Goericke, R.; Welschmeyer, N.A. The marine prochlorophyte *Prochlorococcus* contributes significantly to phytoplankton biomass and primary production in the Sargasso Sea. *Deep-Sea Res. I* **1993**, *40*, 2283–2294. [CrossRef]
21. Landry, M.R.; Kirshtein, J.; Constantinou, J. Abundances and distributions of picoplankton populations in the central equatorial pacific from 12° N to 12° S, 140° W. *Deep-Sea Res. II* **1996**, *43*, 871–890. [CrossRef]
22. Partensky, F.; Blanchot, J.; Vaulot, D. Differential distribution of *Prochlorococcus* and *Synechococcus* in oceanic waters: A review. Ocean. *Monaco-Numero Spec.* **1999**, *19*, 457–476.
23. Zinser, E.R.; Johnson, Z.I.; Coe, A.; Karaca, E.; Veneziano, D.; Chisholm, S.W. Influence of light and temperature on *Prochlorococcus* ecotype distributions in the Atlantic Ocean. *Limnol. Oceanogr.* **2007**, *52*, 2205–2220. [CrossRef]

24. Choi, D.H.; Noh, J.H.; An, S.M.; Choi, Y.R.; Lee, H.; Ra, K.; Kim, D.; Rho, T.; Lee, S.H.; Kim, K.T.; et al. Spatial distribution of cold-adapted *Synechococcus* during spring in seas adjacent to Korea. *Algae* **2016**, *31*, 231–241. [CrossRef]

25. Barber, R.T. Picoplankton do some heavy lifting. *Science* **2007**, *315*, 777–778. [CrossRef]

26. Flombaum, P.; Gallegos, J.L.; Gordillo, R.A.; Rincón, J.; Zabala, L.L.; Jiao, N.; Karl, D.M.; Li, W.K.W.; Lomas, M.W.; Veneziano, D.; et al. Present and future global distributions of the marine Cyanobacteria *Prochlorococcus* and *Synechococcus*. *Proc. Natl. Acad. Sci. USA* **2013**, *110*, 9824–9829. [CrossRef]

27. Lee, S.H.; Joo, H.T.; Lee, J.H.; Lee, J.H.; Kang, J.J.; Lee, H.W.; Lee, D.; Kang, C.K. Seasonal carbon uptake rates of phytoplankton in the northern East/Japan Sea. *Deep-Sea Res. II* **2017**, *143*, 45–53. [CrossRef]

28. Lee, S.H.; Kim, B.K.; Lim, Y.J.; Joo, H.T.; Kang, J.J.; Lee, D.; Park, J.; Ha, S.Y.; Lee, S.H. Small phytoplankton contribution to the standing stocks and the total primary production in the Amundsen Sea. *Biogeosciences* **2017**, *14*, 3705–3713. [CrossRef]

29. Jang, H.K.; Kang, J.J.; Lee, J.H.; Kim, M.; Ahn, S.H.; Jeong, J.Y.; Yun, M.S.; Han, I.S.; Lee, S.H. Recent Primary Production and Small Phytoplankton Contribution in the Yellow Sea during the Summer in 2016. *Ocean Sci. J.* **2018**, *53*, 509–519. [CrossRef]

30. Kang, J.J.; Jang, H.K.; Lim, J.H.; Lee, D.; Lee, J.H.; Bae, H.; Lee, C.H.; Kang, C.K.; Lee, S.H. Characteristics of Different Size Phytoplankton for Primary Production and Biochemical Compositions in the Western East/Japan Sea. *Front. Microbiol.* **2020**, *11*, 1–16. [CrossRef]

31. Sarma, V.V.S.S.; Chopra, M.; Rao, D.N.; Priya, M.M.R.; Rajula, G.R.; Lakshmi, D.S.R.; Rao, V.D. Role of eddies on controlling total and size-fractionated primary production in the Bay of Bengal. *Cont. Shelf Res.* **2020**, *204*, 104186. [CrossRef]

32. Fouilland, E.; Gosselin, M.; Rivkin, R.B.; Vasseur, C.; Mostajir, B. Nitrogen uptake by heterotrophic bacteria and phytoplankton in Arctic surface waters. *J. Plankton Res.* **2007**, *29*, 369–376. [CrossRef]

33. Wheeler, P.A.; Kirchman, D.L. Utilization of inorganic and organic nitrogen by bacteria in marine systems. *Limnol. Oceanogr.* **1986**, *31*, 998–1009. [CrossRef]

34. Middelburg, J.J.; Nieuwenhuize, J. Nitrogen uptake by heterotrophic bacteria and phytoplankton in the nitrate-rich Thames estuary. *Mar. Ecol. Prog. Ser.* **2000**, *203*, 13–21. [CrossRef]

35. Middelburg, J.J.; Nieuwenhuize, J. Uptake of dissolved inorganic nitrogen in turbid, tidal estuaries. *Mar. Ecol. Prog. Ser.* **2000**, *192*, 79–88. [CrossRef]

36. Vazquez, D. Inhibitors of protein synthesis. *FEBS Lett.* **1974**, *40*, S63–S84. [CrossRef]

37. Jacoby, G.A.; Gorini, L. The Effect of Streptomycin and Other Aminoglycoside Antibiotics on Protein Synthesis. In *Antibiotics*; Gottlieb, D., Shaw, P.D., Eds.; Springer: Berlin/Heidelberg, Germany, 1967. [CrossRef]

38. Parsons, T.R.; Maita, Y.; Lalli, C.M. *A Manual of Chemical and Biological Methods for Seawater Analysis*; Pergamon Press: New York, NY, USA, 1984.

39. Olson, R.J.; Shalapyonok, A.; Sosik, H.M. An automated submersible flow cytometer for analyzing pico- and nanophytoplankton: FlowCytobot. *Deep-Sea Res. I* **2003**, *50*, 301–315. [CrossRef]

40. Lee, S.H.; Whitledge, T.E.; Kang, S.H. Recent carbon and nitrogen uptake rates of phytoplankton in Bering Strait and the Chukchi Sea. *Cont. Shelf Res.* **2007**, *27*, 2231–2249. [CrossRef]

41. Song, H.J.; Kim, K.; Lee, J.H.; Ahn, S.H.; Joo, H.M.; Jeong, J.Y.; Yang, E.J.; Kang, S.H.; Yun, M.S.; Lee, S.H. In-situ Measured Carbon and Nitrogen Uptake Rates of Melt Pond Algae in the Western Arctic Ocean, 2014. *Ocean Sci. J.* **2018**, *53*, 107–117. [CrossRef]

42. Lim, Y.J.; Kim, T.W.; Lee, S.H.; Lee, D.; Park, J.; Kim, B.K.; Kim, K.; Jang, H.K.; Bhavya, P.S.; Lee, S.H. Seasonal Variations in the Small Phytoplankton Contribution to the Total Primary Production in the Amundsen Sea, Antarctica. *J. Geophys. Res. Oceans.* **2019**, *124*, 8324–8341. [CrossRef]

43. Yun, M.S.; Kim, Y.; Jeong, Y.; Joo, H.T.; Jo, Y.H.; Lee, C.H.; Bae, H.; Lee, D.; Bhavya, P.S.; Kim, D.; et al. Weak Response of Biological Productivity and Community Structure of Phytoplankton to Mesoscale Eddies in the Oligotrophic Philippine Sea. *J. Geophys. Res. Oceans* **2020**, *125*, e2020JC016436. [CrossRef]

44. Hama, T.; Miyazaki, T.; Ogawa, Y.; Iwakuma, T.; Takahashi, M.; Otsuki, A.; Ichimura, S. Measurement of photosynthetic production of a marine phytoplankton population using a stable ^{13}C isotope. *Mar. Biol.* **1983**, *73*, 31–36. [CrossRef]

45. Dugdale, R.C.; Goering, J.J. Uptake of New and Regenerated Forms of Nitrogen in Primary Productivity. *Limnol. Oceanogr.* **1967**, *12*, 196–206. [CrossRef]

46. Gosselin, M.; Levasseur, M.; Wheeler, P.A.; Horner, R.A.; Booth, B.C. New measurements of phytoplankton and ice algal production in the Arctic Ocean. *Deep Sea Res. Part II Top. Stud. Ocean.* **1997**, *44*, 1623–1644. [CrossRef]

47. Furuya, K.; Kurita, K.; Odate, T. Distribution of phytoplankton in the East China sea in the winter of 1993. *J. Oceanogr.* **1996**, *52*, 323–333. [CrossRef]

48. Gibb, S.W.; Barlow, R.G.; Cummings, D.G.; Rees, N.W.; Trees, C.C.; Holligan, P.; Suggett, D. Surface phytoplankton pigment distributions in the Atlantic Ocean: An assessment of basin scale variability between 50° N and 50° S. *Prog. Oceanogr.* **2000**, *45*, 339–368. [CrossRef]

49. Platt, T.; Bouman, H.; Devred, E.; Fuentes-Yaco, C.; Sathyendranath, S. Physical forcing and phytoplankton distributions. *Sci. Mar.* **2005**, *69*, 55–73. [CrossRef]

50. Liu, X.; Huang, B.; Huang, Q.; Wang, L.; Ni, X.; Tang, Q.; Sun, S.; Wei, H.; Liu, S.; Li, C.; et al. Seasonal phytoplankton response to physical processes in the southern Yellow Sea. *J. Sea Res.* **2015**, *95*, 45–55. [CrossRef]

51. Gong, G.C.; Wen, Y.H.; Wang, B.W.; Liu, G.J. Seasonal variation of chlorophyll a concentration, primary production and environmental conditions in the subtropical East China Sea. *Deep-Sea Res. II* **2003**, *50*, 1219–1236. [CrossRef]

52. Moore, L.R.; Chisholm, S.W. Photophysiology of the marine cyanobacterium *Prochlorococcus*: Ecotypic differences among cultured isolates. *Limnol. Oceanogr.* **1999**, *44*, 628–638. [CrossRef]
53. Moore, L.R.; Post, A.F.; Rocap, G.; Chisholm, S.W. Utilization of different nitrogen sources by the marine cyanobacteria *Prochlorococcus* and *Synechococcus*. *Limnol. Oceanogr.* **2002**, *47*, 989–996. [CrossRef]
54. Zhao, S.; Wei, J.; Yue, H.; Xiao, T. Picophytoplankton abundance and community structure in the Philippine Sea, western Pacific. *Chin. J. Oceanol. Limnol.* **2010**, *28*, 88–95. [CrossRef]
55. Choi, D.H.; Noh, J.H.; Hahm, M.S.; Lee, C.M. Picocyanobacterial abundances and diversity in surface water of the northwestern Pacific Ocean. *Ocean Sci. J.* **2011**, *46*, 265–271. [CrossRef]
56. Charpy, L.; Blanchot, J. Photosynthetic picoplankton in French Polynesian atoll lagoons: Estimation of taxa contribution to biomass and production by flow cytometry. *Mar. Ecol. Prog. Ser.* **1998**, *162*, 57–70. [CrossRef]
57. Blanchot, J.; André, J.M.; Navarette, C.; Neveux, J.; Radenac, M.H. Picophytoplankton in the equatorial pacific: Vertical distributions in the warm pool and in the high nutrient low chlorophyll conditions. *Deep-Sea Res. I* **2001**, *48*, 297–314. [CrossRef]
58. McCarthy, J.J.; Garside, C.; Nevins, J.L.; Barber, R.T. New production along 140 degrees W in the equatorial Pacific during and following the 1992 El Niño event. *Deep-Sea Res. II* **1996**, *43*, 1065–1093. [CrossRef]
59. Eppley, R.W.; Peterson, B.J. Particulate organic matter flux and planktonic new production in the deep ocean. *Nature* **1979**, *282*, 677–680. [CrossRef]
60. Kirchman, D.L. The uptake of inorganic nutrients by heterotrophic bacteria. *Microb. Ecol.* **1994**, *28*, 225–271. [CrossRef]
61. Rodrigues, R.M.N.V.; Williams, P.J.L.B. Inorganic nitrogen assimilation by picoplankton and whole plankton in a coastal ecosystem. *Limnol. Oceanogr.* **2002**, *47*, 1608–1616. [CrossRef]
62. Bhavya, P.S.; Kumar, S.; Gupta, G.V.M.; Sudheesh, V.; Sudharma, K.V.; Varrier, D.S.; Dhanya, K.R.; Saravanane, N. Nitrogen Uptake Dynamics in a Tropical Eutrophic Estuary (Cochin, India) and Adjacent Coastal Waters. *Estuaries Coasts* **2016**, *39*, 54–67. [CrossRef]
63. Waterbury, J.B.; Watson, S.W.; Guillard, R.R.; Brand, L.E. Widespread occurrence of a unicellular, marine, planktonic, cyanobacterium. *Nature* **1979**, *227*, 293–294. [CrossRef]
64. Chisholm, S.W.; Olson, R.J.; Zettler, E.R.; Goericke, R.; Waterbury, J.B.; Welschmeyer, N.A. A novel free-living prochlorophyte occurs at high cell concentrations in the oceanic euphotic zone. *Nature* **1988**, *334*, 340–343. [CrossRef]

Article

Spatiotemporal Variation in Phytoplankton Community Driven by Environmental Factors in the Northern East China Sea

Yejin Kim [1], Seok-Hyun Youn [2], Hyun Ju Oh [2], Jae Joong Kang [1], Jae Hyung Lee [1], Dabin Lee [1], Kwanwoo Kim [1], Hyo Keun Jang [1], Junbeom Lee [1] and Sang Heon Lee [1,*]

[1]   Department of Oceanography, Pusan National University, Busan 46241, Korea; yejini@pusan.ac.kr (Y.K.); jaejung@pusan.ac.kr (J.J.K.); tlyljh78@pusan.ac.kr (J.H.L.); ldb1370@pusan.ac.kr (D.L.); goanwoo7@pusan.ac.kr (K.K.); janghk@pusan.ac.kr (H.K.J.); june2548@korea.kr (J.L.)
[2]   Oceanic Climate & Ecology Research Division, National Institute of Fisheries Science, Busan 46083, Korea; younsh@korea.kr (S.-H.Y.); hyunjuoh@korea.kr (H.J.O.)
*   Correspondence: sanglee@pusan.ac.kr

Received: 13 July 2020; Accepted: 24 September 2020; Published: 26 September 2020

Abstract: The East China Sea (ECS) is the largest marginal sea in the northern western Pacific Ocean. In comparison to various physical studies, little information on the seasonal patterns in community structure of phytoplankton is currently available. Based on high performance liquid chromatography (HPLC) pigment analysis, spatiotemporal variations in phytoplankton community compositions were investigated in the northern ECS. Water temperature and salinity generally decreased toward the western part of the study area but warmer conditions in August led to strong vertical stratification of the water column. In general, major inorganic nutrient concentrations were considerably higher in the western part with a shallow water depth, and consistent with previous results, had no discernable vertical pattern during our observation period except in August. This study also revealed PO_4-limited environmental conditions in May and August. The monthly averaged integral chlorophyll-a concentration varied seasonally, highest (35.2 ± 20.22 mg m^{-2}) in May and lowest (5.2 ± 2.54 mg m^{-2}) in February. No distinct vertical differences in phytoplankton community compositions were observed for all the sampling seasons except in August when cyanobacteria predominated in the nutrient-deficient surface layer and diatoms prevailed at deep layer. Canonical correlation analysis results revealed that nutrient distribution and the water temperature were the major drivers of the vertical distribution of phytoplankton communities in August. Spatially, a noticeable difference in phytoplankton community structure between the eastern and western parts was observed in November with diatom domination in the western part and cyanobacteria domination in the eastern part, which were significantly ($p < 0.01$) correlated with water temperature, salinity, light conditions, and nutrient concentrations. Overall, the two major phytoplankton groups were diatoms (32.0%) and cyanobacteria (20.6%) in the northern ECS and the two groups were negatively correlated, which holds a significant ecological meaning under expected warming ocean conditions.

Keywords: East China Sea; phytoplankton; HPLC; diatoms; cyanobacteria

---

1. Introduction

Phytoplankton communities play an important role in marine ecosystems, affecting carbon and nutrient cycling, the structure and efficiency of the food web, and the flux of particles to deep waters [1–3]. Phytoplankton show a clear variation in community structure and abundance in response to environmental changes, so the phytoplankton community structure can be used as a useful indicator of ecosystem and water quality characteristics [4–6]. Therefore, in order to understand the structure and function of the ecosystem, it is necessary to monitor the spatiotemporal changes in the phytoplankton community [7]. Various methods such as microscopy, flow cytometry, and pigment analysis have been used to quantitatively analyze phytoplankton community structure. Traditionally, microscopic methods have been the most commonly used to assess biomass and community structure [8]. Microscopes can provide detailed information on species and size, but this method requires taxonomic expertise and very considerable time. Furthermore, microscopic methods fall short when identifying small organisms such as some of picophytoplankton and nano flagellates [9], and the structure of fragile cells of many species can be altered during the process of fixation in Lugol's solution, formaldehyde, glutar-aldehyde, and similar fixatives [10,11]. Flow-cytometric analysis has been developed for providing more rapid and automated method for identification of communities of smaller phytoplankton. Flow-cytometric analysis requires a full understanding of the optical characteristics of the species and can mainly separate phytoplankton communities into picoplanktonic prokaryotes, picoeukaryotes, and nanoeukaryotes [12–14]. High performance liquid chromatography (HPLC) was used for this study because HPLC method can be used to measure the concentration of each pigment separately, and possible to determine the clustering of phytoplankton using the extracted marker pigments [15]. In particular, this method can provide useful information on nano- and pico-sized phytoplankton communities that are difficult to distinguish based on microscopic observations [16].

The East China Sea (ECS) is the largest marginal sea in the northern west Pacific and approximately 70% of the area is made up of a wide continental shelf. ECS is one of the most productive areas and possible sinks of carbon dioxide [17]. Furthermore, it is considered one of the most important marine fishing grounds in China [18]. Various water masses affect in the ECS, such as the Yellow Sea bottom cold water (YSCW) from the north, Changiiang diluted water (CDW) from the world's largest Yangtze river from the west, Kuroshio water (KW) from the east and Taiwan current warm water (TCWW) from the south [19–22]. Generally, the environmental conditions vary from the eastern part to the western part in the ECS. This complex topography and various water masses cause show heterogeneous and complex environmental characteristics seasonally and spatially [23]. Previous studies for phytoplankton community in the northern ECS are quite limited and most of the studies have focused on the Yangtze River estuary and adjacent waters [24–26]. Three different phytoplankton communities in the Yangtze River estuary have been identified according to water mass [27–29]. Diatoms are generally the most dominant groups in this area [27–31]. In the northern ECS near Korea, several previous studies focused on the phytoplankton community were carried out in spring [32] and summer [33,34] and mostly conducted over one season. In addition, most of the studies have focused on the spatial distribution of diatoms and dinoflagellates which can be identified under the microscope. To date, little information on the seasonal patterns in community structure that is inclusive of all phytoplankton is currently available in the northern ECS. Therefore, the present study aimed to investigate spatiotemporal changes in composition and distribution of phytoplankton community structure in the northern ECS that is possible using pigment analysis through HPLC.

2. Materials and Methods

2.1. Sampling Site and Water Sampling

Four cruises were carried out in the northern ECS from 1–9 February, 30 April–10 May, 2–10 August, and 7–17 November in 2018, as representatives for winter, spring, summer, and autumn, respectively (Figure 1; Table 1). Water samples were collected from three light depths (100%, 30%, and 1% penetration of surface irradiance, PAR) using a CTD/rosette sampler fitted with Niskin bottles. The light depths were determined by a Secchi disk. Phytoplankton pigments and physicochemical parameters (temperature, salinity, and major nutrients; N, P, and Si) were analyzed in samples drawn from the three light depths. The vertical temperature and salinity were measured by SBE9/11 CTD (Sea-Bird Electronics, Bellevue, WA, USA) sensors.

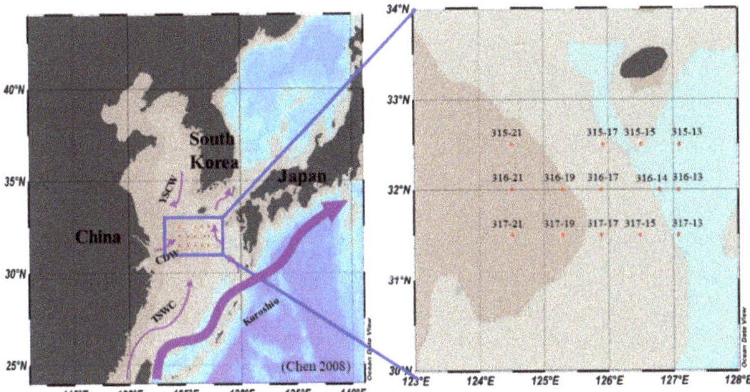

Figure 1. Sampling stations in the northern East China Sea, 2018. The major currents in the northern East China Sea are based on [19].

Table 1. Description of sampling stations in the northern East China Sea for each cruise period, 2018.

Date	Station	Latitude	Longitude	Euphotic Depth (m)
February	315-13	32.5	127.0	19
	315-17	32.5	125.9	5
	316-14	32.0	126.8	19
	316-17	32.0	125.9	14
	317-17	31.5	125.9	19
	317-21	31.5	124.5	5
May	315-13	32.5	127.0	27
	315-15	32.5	126.5	38
	315-21	32.5	124.5	11
	316-13	32	127.0	27
	316-17	32	125.9	27
	316-21	32	124.5	5
	317-13	31.5	127.0	16
	317-15	31.5	126.5	33
	317-21	31.5	124.5	14
August	315-13	32.5	127.0	41
	315-17	32.5	125.9	46
	315-21	32.5	124.5	22
	316-13	32	127.0	41
	316-17	32	125.9	27
	316-21	32	124.5	30
	317-13	31.5	127.0	35
	317-15	31.5	126.5	54
	317-19	31.5	125.3	35
	317-21	31.5	124.5	19
October	315-13	32.5	127.0	41
	315-15	32.5	126.5	41
	315-21	32.5	124.5	8
	316-13	32	127.0	33
	316-17	32	125.9	27
	316-19	32	125.3	8
	316-21	32	124.5	5
	317-13	31.5	127.0	49
	317-15	31.5	126.5	49
	317-21	31.5	124.5	5

2.2. Phytoplankton Pigment Analysis

Water samples for photosynthetic pigment analysis were filtered through 47 mm GF/F filters (Whatman, Maidstone, UK; 07 µm), and then stored in a freezer at −80 °C to avoid degradation. Pigments were extracted in 100% acetone (5 mL) with cantaxanthin (100 µL) as an internal standard for 24 h in the dark at 4 °C and placed in an ultra-sonic bath to disrupt a cell [35,36]. An aliquot water of 1 mL was passed through a 0.45 µm PTFE syringe filter to rid the samples of particles. After the extracts were centrifuged for 10 min at 3500 rpm to remove cellular debris and glass fibers. All procedures were carried out under low light conditions to minimize pigment degradation. Pigments were analyzed using a HPLC (Agilent Infinite 1260, Santa Clara, CA, USA), and the separation of pigments was performed using a slightly modified method of [37] and [38]. The peaks were identified based on their retention time compared with those of pure standards (chlorophyll a, chlorophyll b, β-carotene, fucoxanthin, prasinoxanthin, 19'-hexanoyloxyfucoxanthin, diadinoxanthin, 19'-butanoyloxy-fucoxanthin, peridinin, alloxanthin, neoxanthin, violaxanthin, prasinoxanthin, lutein, and zeaxanthin obtained from DHI, Denmark). The concentrations of pigments in samples were calculated as following equation. Standard response factor (Rf) was calculated based on the standard pigment and dividing the concentration of the standard by the measured peak area [38].

$$\text{Concentration} = \text{Area} \times \text{Rf} \times (\text{Ve/Vs})\ [\text{ngL}^{-1}] \tag{1}$$

Area = area of the peak in the sample [area]
Rf = standard response factor [ngL^{-1} area^{-1}]
Ve = AIS/(peak area of IS added to sample) × (Volume of IS added to sample) [L]
Vs = volume of filtered water sample [L]
AIS = peak area of IS when 1 mL IS is mixed with 300 µL of H$_2$O
IS = Internal Standard

The CHEMTAX program was used to estimate the contribution of the different phytoplankton community structure to the total chlorophyll a [15,16]. The contribution of diatoms, dinoflagellates, prymnesiophytes, chlorophytes, chrysophytes, cryptophytes, cyanobacteria and prasinophytes were calculated based on the program. Twelve pigments and initial pigment ratios for around the Korean peninsula were used for this study [38]. In the following CHEMTAX, to derive the most accurate phytoplankton groups, data was binned according to sampling month and three light depths (100%, 30% and 1% penetration of surface irradiance, PAR) [39,40].

2.3. Dissolved Inorganic Nutrient Concentration

An aliquot of water (100 mL) was filtered onboard through GF/F filters (Whatman, Maidstone, UK; 07 µm) for dissolved inorganic nutrient concentrations (NH$_4$, NO$_2$, NO$_3$, PO$_4$, and SiO$_2$) and kept frozen (−20 °C) until further analysis. Concentrations of nutrients were determined in an automatic analyzer (Quaatro, Bran + Luebbe, Germany) belonging to the National Institute of Fisheries Science (NIFS), Korea. Dissolved inorganic nitrogen (DIN) concentrations were calculated as the sum of NH$_4$, NO$_2$ and NO$_3$.

For verifying P-limited water conditions, Excess Nitrate (ExN), which is calculated as ExN = DIN-(R*PO$_4$) (R = Redfield N:P ratio of 16), was used in this study [41–43]. ExN values of <0 indicate PO$_4$-enriched condition, while ExN > 0 indicates the converse condition [41–43].

2.4. Statistical Analysis

Canonical correspondence analysis (CCA) was performed using "past 3" software to explain the relationship between environmental parameters and phytoplankton community structure [44]. Temperature, salinity, depth, DIN, PO$_4$, SiO$_2$, and ExN were include for the environmental parameters.

3. Results

3.1. Physical Environments

Seasonal distribution patterns of temperature and salinity during the four cruises are summarized in Table 1. The average temperature was lowest in February (winter) at 13.7 ± 2.9 °C and gradually increased to highest in August (summer) at 24.2 ± 4.7 °C. The average salinity was highest at 34.1 ± 0.6 in February and lowest at 32.3 ± 0.7 in August. In February, the water temperature decreased toward the western part from the eastern part in the study area and the salinity showed the same trend as the water temperature (Table 1). The water temperature and salinity in May (spring) were also relatively higher in the eastern part and lower in the western but the difference in water temperature was smaller in May compared to that in February. On the other hand, the water temperature and salinity were inversely spatially distributed in August with low in the eastern and high in the western parts and the differences were smallest during the observation period. Vertically, the temperature increased with depth in August, which resulted in a strong stratification (Figure 2). In November (autumn), the patterns in water temperature and salinity were similar to those in February and May.

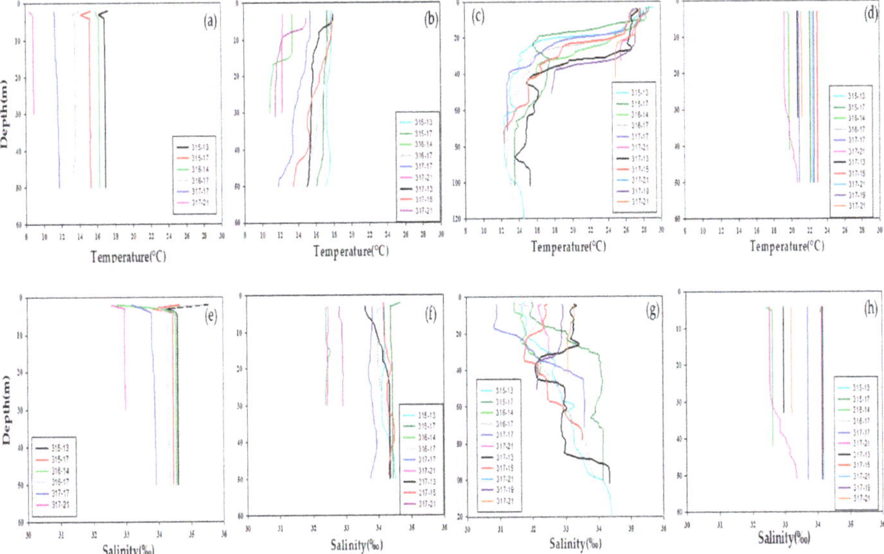

Figure 2. Vertical profiles of temperature and salinity in the northern East China Sea, 2018. (**a**) February, (**b**) May, (**c**) August, and (**d**) Nov for temperature; and (**e**) February, (**f**) May, (**g**) August, and (**h**) November for salinity.

3.2. Dissolved Inorganic Nutrient Concentrations

Inorganic nutrient concentrations at the three light depths for each cruise are summarized in Table 2. DIN and PO_4 concentrations were highest in February and remained low in other seasons, whereas SiO_2 tended to increase from May to August and November. In February, the ranges of DIN, PO_4 and SiO_2 concentrations from surface to 1% light depths were 5.3–14.1 μM, 0.3–0.6 μM, and 6.0–16.8 μM, respectively. There were no distinct vertical patterns, but in the horizontal direction, DIN, PO_4, and SiO_2 tended to increase from the northeast to the southwest stations. In May, the ranges of DIN, PO_4, and SiO_2 concentrations were 2.5–12.3 μM, <0.1–0.3 μM, and 3.4–12.4 μM, respectively. No marked vertical patterns in the concentrations were observed but horizontally, DIN and SiO_2 showed relatively higher in the western part compared to the eastern part in May. Generally, PO_4 concentrations in May were very low at all the stations with an average of 0.1 μM except for St. 316-21. The ranges of DIN, PO_4, and SiO_2 concentrations were 1.6–16.9 μM, <0.1–0.5 μM, and 2.1–14.7 μM, respectively, in August. Unlike other seasons, noticeable vertical distributions of nutrients were observed in August with low concentrations at surface but increasing with depth. In November, the ranges of DIN, PO_4, and SiO_2 concentrations were 2.1–15.9 μM, 0.1–0.6 μM, and 2.0–15.8 μM, respectively. Nutrient concentrations were relatively higher in the western part compared to the eastern part and the differences in the concentrations between the eastern and western parts were largest in November among the four cruises but no vertically distinct distributions were found.

Table 2. The dissolved inorganic nutrient concentrations (μM) at the euphotic depths (100%, 30%, and 1%) of water column in the northern East China Sea, 2018.

	Station	Light (%)	NH_4	NO_2	NO_3	DIN	PO_4	SiO_2
February	315-13	100%	0.84	0.40	4.81	6.04	0.33	7.15
		30%	0.92	0.36	4.01	5.29	0.29	6.27
		1%	1.01	0.39	5.07	6.48	0.32	7.11
	315-17	100%	0.80	0.10	10.66	11.56	0.60	13.67
		30%	0.93	0.08	5.07	6.08	0.33	6.79
		1%	0.90	0.08	4.92	5.90	0.35	8.07
	316-14	100%	0.93	0.09	8.31	9.33	0.33	8.76
		30%	0.91	0.09	7.95	8.95	0.35	8.65
		1%	0.91	0.10	8.06	9.07	0.33	8.50
	316-17	100%	0.85	0.28	5.10	6.23	0.29	6.96
		30%	0.73	0.27	4.79	5.78	0.29	6.83
		1%	1.05	0.24	4.66	5.95	0.29	6.60
	317-17	100%	1.01	0.18	7.87	9.07	0.41	9.82
		30%	0.80	0.15	5.24	6.19	0.34	6.04
		1%	0.79	0.11	7.80	8.70	0.41	9.97
	317-21	100%	0.81	0.09	11.55	12.46	0.53	14.46
		30%	0.78	0.10	12.34	13.22	0.61	15.37
		1%	0.82	0.10	13.19	14.11	0.60	16.77
	average		0.88	0.18	7.30	8.36	0.39	9.32
	S.D		0.09	0.11	2.94	2.84	0.11	3.40
May	315-13	100%	1.58	0.09	1.98	3.66	0.02	3.45
		30%	1.51	0.05	1.17	2.74	0.02	3.44
		1%	1.56	0.06	1.31	2.93	0.02	4.68
	315-15	100%	1.59	0.06	3.23	4.88	0.03	3.99
		30%	1.49	0.04	1.43	2.95	0.02	4.11
		1%	1.56	0.05	1.29	2.89	0.04	4.59
	315-21	100%	1.64	0.23	4.47	6.34	0.11	11.36
		30%	1.65	0.21	3.68	5.54	0.13	11.54
		1%	1.52	0.16	3.51	5.19	0.08	10.68
	316-13	100%	1.96	0.07	3.43	5.46	0.03	5.78
		30%	1.60	0.05	1.04	2.68	0.02	4.36
		1%	1.77	0.19	4.51	6.47	0.18	6.91
	316-17	100%	2.14	0.06	1.41	3.61	0.03	6.64
		30%	1.61	0.07	1.11	2.79	0.04	7.15
		1%	1.54	0.04	1.30	2.88	0.03	6.59
	316-21	100%	1.53	0.51	10.23	12.26	0.33	9.86
		30%	1.44	0.48	8.90	10.82	0.30	9.50
		1%	1.47	0.44	7.51	9.43	0.30	7.83
	317-13	100%	2.65	0.30	1.69	4.64	0.03	3.72
		30%	1.55	0.06	1.06	2.66	0.01	5.95
		1%	1.60	0.06	0.85	2.51	0.01	5.52
	317-15	100%	1.55	0.05	1.27	2.86	0.01	3.89
		30%	1.57	0.05	1.01	2.63	0.01	3.97
		1%	1.57	0.26	2.61	4.44	0.10	6.03
	317-21	100%	2.06	0.10	2.61	4.78	0.01	8.87
		30%	1.72	0.08	1.45	3.26	0.01	9.58
		1%	1.58	0.45	8.12	10.16	0.31	12.41
	average		1.67	0.16	3.04	4.87	0.08	6.75
	S.D		0.26	0.15	2.66	2.76	0.10	2.79

Table 2. *Cont.*

	Station	Light (%)	NH$_4$	NO$_2$	NO$_3$	DIN	PO$_4$	SiO$_2$
August	315-13	100%	0.41	0.06	1.49	1.97	0.06	8.31
		30%	0.41	0.09	1.99	2.50	0.06	7.94
		1%	0.44	0.09	14.14	14.67	0.53	13.75
	315-17	100%	1.46	0.09	1.41	2.97	0.07	8.48
		30%	0.40	0.05	1.16	1.61	0.05	8.62
		1%	0.36	0.09	8.45	8.90	0.49	10.75
	315-21	100%	0.80	0.14	2.18	3.13	0.07	7.07
		30%	0.80	0.11	2.00	2.92	0.07	6.91
		1%	1.17	1.66	5.68	8.51	0.11	11.79
	316-13	100%	0.54	0.06	2.04	2.63	0.03	8.29
		30%	2.76	0.07	0.96	3.78	0.03	3.31
		1%	0.43	0.06	9.11	9.59	0.42	8.88
	316-17	100%	0.72	0.20	2.96	3.88	0.08	2.61
		30%	0.61	0.17	3.05	3.83	0.05	2.70
		1%	0.74	0.29	13.84	14.86	0.34	14.71
	316-21	100%	1.29	0.05	1.54	2.88	0.06	2.10
		30%	0.56	0.06	1.68	2.31	0.07	2.74
		1%	0.74	1.26	4.06	6.06	0.18	7.80
	317-13	100%	0.58	0.04	3.35	3.96	0.06	5.45
		30%	3.15	0.04	0.79	3.98	0.03	2.11
		1%	0.49	0.45	10.73	11.67	0.28	11.89
	317-15	100%	1.30	0.06	1.56	2.93	0.06	3.59
		30%	1.02	0.06	1.96	3.04	0.09	4.12
		1%	3.71	0.06	9.46	13.23	0.43	9.97
	317-19	100%	0.67	0.05	1.63	2.35	0.05	2.12
		30%	0.64	0.05	1.85	2.54	0.05	2.59
		1%	1.11	0.69	6.50	8.30	0.15	8.53
	317-21	100%	1.54	0.04	1.16	2.74	0.06	2.38
		30%	0.40	0.04	1.68	2.11	0.06	3.73
		1%	0.56	0.10	2.30	2.96	0.08	3.81
	average		0.99	0.21	4.02	5.23	0.14	6.57
	S.D		0.83	0.37	3.88	4.00	0.15	3.76
October	315-13	100%	0.66	0.15	1.79	2.61	0.18	2.72
		30%	0.74	0.10	1.38	2.22	0.13	1.99
		1%	0.79	0.14	1.18	2.11	0.17	2.32
	315-15	100%	0.49	0.48	2.66	3.62	0.22	3.71
		30%	0.63	0.40	1.88	2.90	0.17	3.09
		1%	0.73	0.47	2.32	3.51	0.23	3.63
	315-21	100%	2.07	0.13	13.73	15.93	0.56	13.36
		30%	0.70	0.11	10.35	11.16	0.57	15.78
		1%	0.62	0.12	10.91	11.65	0.56	14.81
	316-13	100%	1.04	0.21	2.42	3.67	0.19	2.66
		30%	0.60	0.07	1.93	2.60	0.10	2.79
		1%	0.74	0.11	1.41	2.27	0.17	2.35
	316-17	100%	0.60	0.49	5.38	6.47	0.36	7.99
		30%	0.54	0.52	4.62	5.68	0.39	7.44
		1%	0.60	0.54	4.34	5.49	0.41	6.93
	316-19	100%	1.24	0.14	8.70	10.07	0.49	11.69
		30%	0.78	0.12	9.28	10.18	0.47	12.39
		1%	0.56	0.11	7.79	8.47	0.50	10.50
	316-21	100%	0.87	0.07	6.30	7.24	0.47	8.25
		30%	0.75	0.11	8.29	9.15	0.59	10.99
		1%	0.89	0.11	9.28	10.28	0.55	12.24
	317-13	100%	1.10	0.18	4.74	6.02	0.22	2.84
		30%	0.66	0.09	1.81	2.55	0.17	2.35
		1%	0.71	0.11	1.47	2.29	0.22	2.13
	317-15	100%	0.88	0.16	2.27	3.30	0.22	2.99
		30%	0.85	0.15	1.90	2.90	0.21	2.91
		1%	1.39	0.21	2.00	3.60	0.23	3.04
	317-21	100%	1.05	0.14	9.23	10.41	0.51	11.52
		30%	0.65	0.12	8.16	8.93	0.53	10.68
		1%	0.65	0.12	8.70	9.48	0.51	11.18
	average		0.82	0.20	5.21	6.22	0.34	6.91
	S.D		0.32	0.15	3.67	3.75	0.17	4.59

3.3. Phytoplankton Biomass and Community Structure

The monthly averaged chlorophyll-a concentration integrated from surface to 1% light depth was highest (35.2 ± 20.22 mg m^{-2}) in May and lowest (5.2 ± 2.54 mg m^{-2}) in February (Figure 3). In February, the integral chlorophyll-a concentration was relatively lower in the western part (2.8 mg m^{-2}) compared to that in the eastern part (6.7 ± 2.46 mg m^{-2}) of our study area, which is similar to the temperature distribution. In May, the integral chlorophyll-a concentration was highly variable across the study area with the range of 8.2–70.0 mg m^{-2} and the chlorophyll-a concentration was relatively higher in the southern part than in the northern part. In August, no distinct spatial distribution in the chlorophyll-a concentration was observed. In November, the spatial distribution in the integral chlorophyll-a concentration was opposite to that in August, which is similar to the nutrient distribution patterns (Table 2).

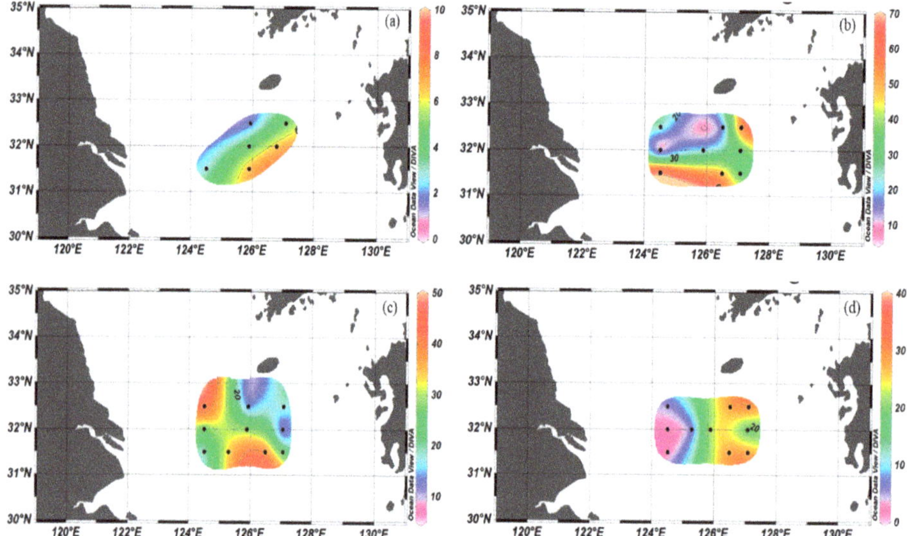

Figure 3. Horizontal distributions of water column-integrated chlorophyll-a concentration from surface to 1% light depth in the northern East China Sea (**a**) February, (**b**) May, (**c**) August, and (**d**) November.

Generally, no distinct vertical differences in phytoplankton community compositions were observed at 100%, 30%, and 1% light depths for all the sampling seasons except August (Figure 4). The phytoplankton community compositions in August were conspicuously different between 30–100% light depths and 1% light depths. Cyanobacteria predominated, contributing 63.3% to the total phytoplankton biomass and diatoms were the second most abundant group (15.5%) at 100% light depths, whereas diatoms contributed 58.2% followed by dinoflagellates (13.0%) and other classes (<10%) at 1% light depths (Figure 4). Spatially, noticeable differences in phytoplankton community between the eastern and western parts were observed season, especially in November. Diatoms predominated in the western part, contributing 58.6% to the total phytoplankton biomass and cryptophytes were the second most abundant group (27.4%), whereas cyanobacteria predominated (45.0%) in the eastern part followed by cryptophytes (31.0%) in November. These two dominant groups were significantly ($p < 0.01$) correlated with water temperature (Figure 5). The contribution of diatoms was negatively related with water temperature (y = −0.0227x + 0.8061, r^2 = 0.7207), whereas the contribution of cyanobacteria had a positive relationship with water temperature (y = 0.0309x − 0.3506, r^2 = 0.824).

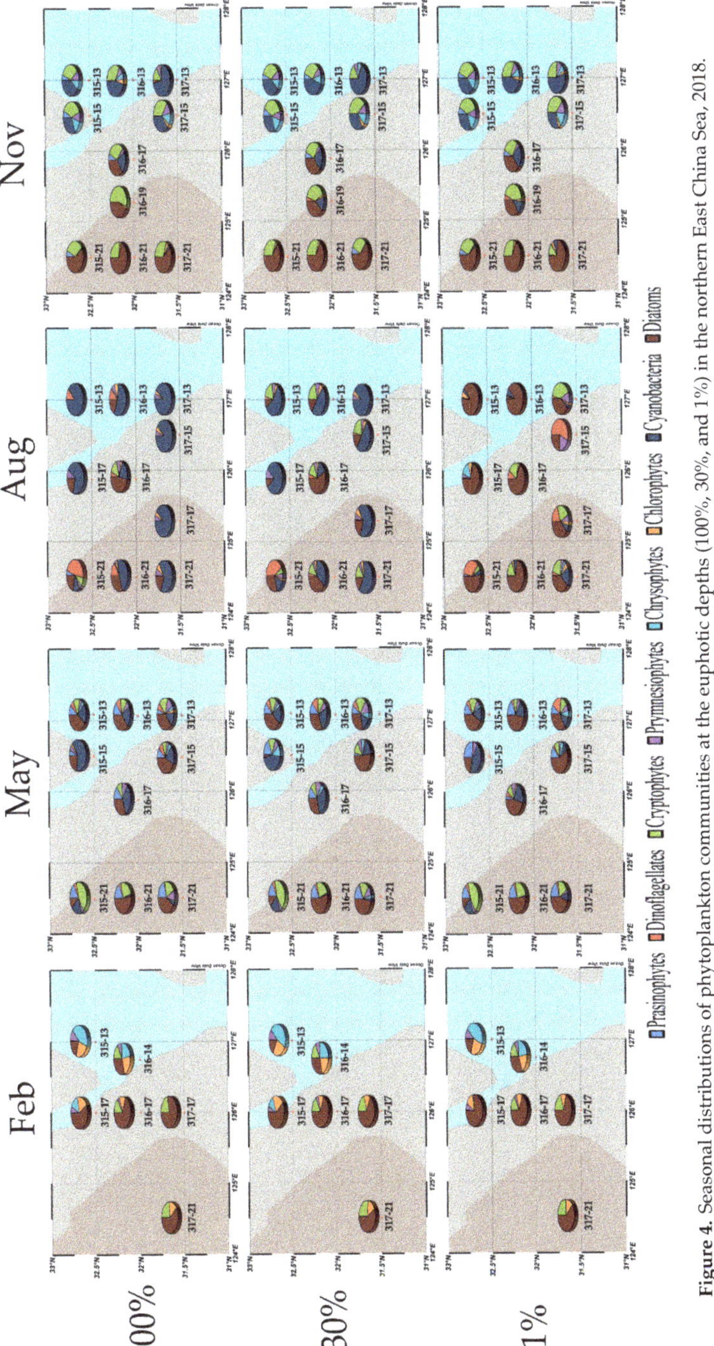

Figure 4. Seasonal distributions of phytoplankton communities at the euphotic depths (100%, 30%, and 1%) in the northern East China Sea, 2018.

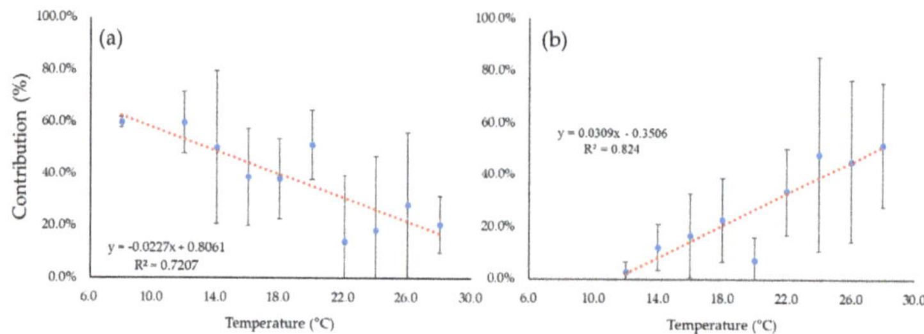

Figure 5. Relationships between contributions of two major phytoplankton communities and water temperature for all the cruise period, 2018. (**a**) Diatom (**b**) Cyanobacteria.

Overall, the major phytoplankton community in the study site was diatoms with a contribution more than 30% although it varied seasonally from 9.8% (November) and 50.0% (February) (Figure 6). Cyanobacteria were the second highest contributors ranging from 0% to 38.3% during our study period. Cyanobacteria were not appeared in February but their contribution increased steadily from May to November. The contributions of cryptophytes ranged from 7.8% to 30.7%. The contributions of prymnesiophytes were 5.4–7.6%, with a similar contribution for each cruise. Chlorophytes contributed 0.5–16.1%, with the highest contribution in February and were hardly observed in May and November (0.5% and 0.8%, respectively). Chrysophytes had the contributions of 0.6–14.0%, showing their highest contribution in February. Dinoflagellates showed their contributions of 0–17.4% and their highest contribution was in August. In the case of prasinophytes, they showed the contributions ranging from 0% to 14.2% and the highest contribution was in May (Figure 6).

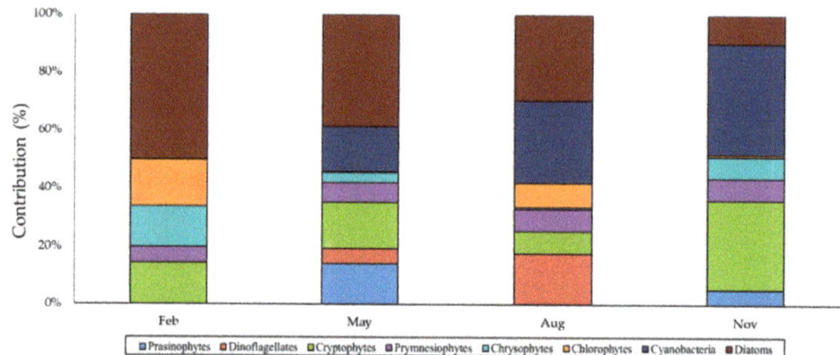

Figure 6. Seasonal contributions of phytoplankton communities averaged from all the stations for each cruise period in the northern East China Sea, 2018.

3.4. Canonical Correspondence Analysis (CCA)

CCA results between phytoplankton community and environmental parameters for each season are presented in Figure 7. In February, diatoms and chryptophytes showed negative correlations with temperature and salinity and positive correlations with nutrients, whereas chrysophytes and chlorophytes had positive correlations with temperature and salinity. In May, diatoms had no significant correlation with any environment parameter, whereas cyanobacteria and chrysophytes had negative correlations with nutrients and cyanobacteria had positive correlations with temperature and salinity. In August, cyanobacteria had a positive correlation with temperature and negative correlations with nutrients and depth. In comparison, diatoms had significantly positive correlations with nutrients and depth in August. Similarly, cyanobacteria showed a strong positive correlation with temperature and negative correlations with depth and nutrients in November. In comparison, diatoms had negative correlations with temperature and depth, and positive correlations with nutrients in November.

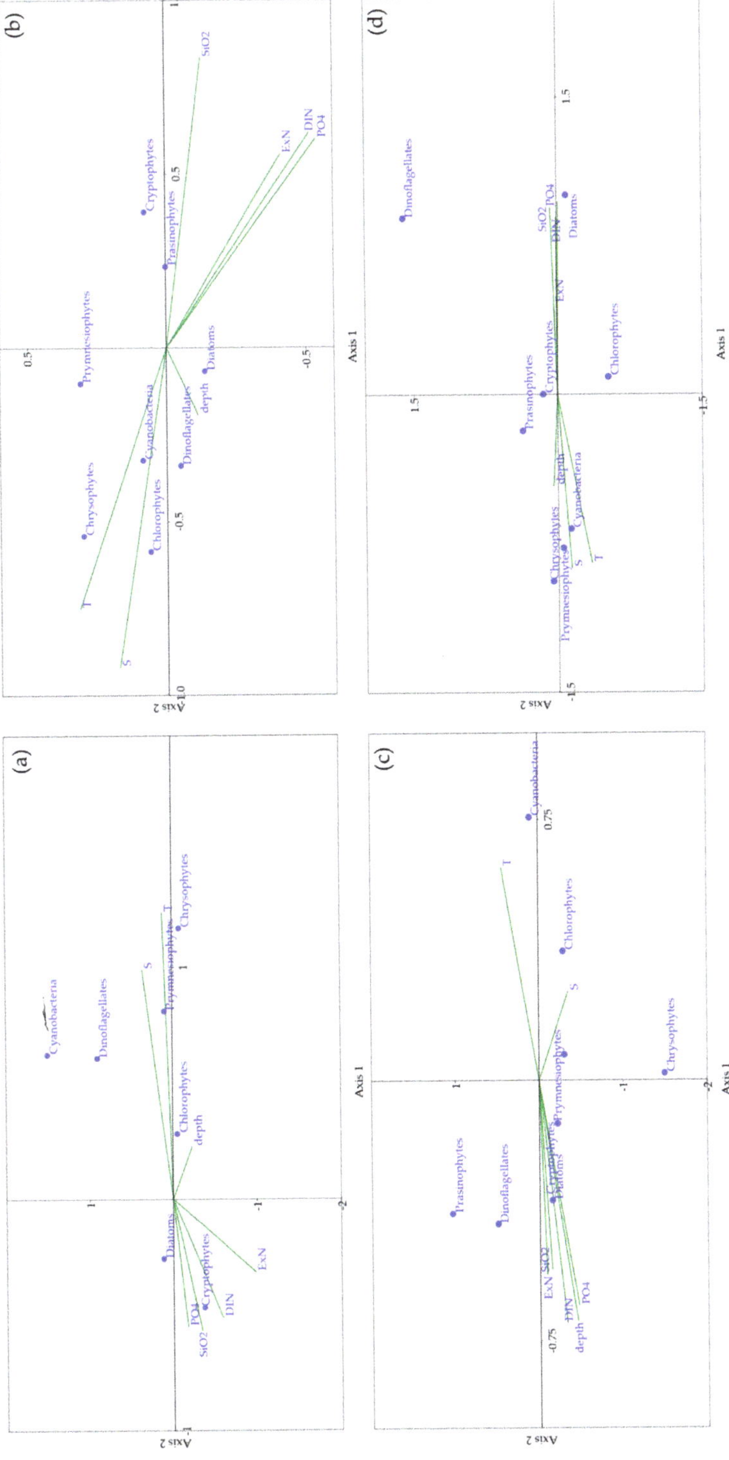

Figure 7. Canonical correspondence analysis of seasonal phytoplankton communities and environmental parameters measured at each cruise period in the northern East China Sea (**a**) February (**b**) May (**c**) August (**d**) November.

4. Discussion

The northern ECS is a typical temperate water seasonally affected mainly by four different water masses. The CDW, TCWW, KW, and YSCW, but their influence can vary seasonally [45,46]. Mixed waters were mainly distributed in our study area in February but YSCW was found at the most western stations (Sts. 315-21, 316-21, and 317-21) in May based on T-S diagrams. Low temperature, strong winds, and vigorous vertical mixing are generally observed in February during the Northeast Monsoon [47,48]. Weak surface stratification begins May and the water column was well stratified in August with TCWW mainly distributed at surface layer whereas the YSCW is mainly distributed at bottom layer (73 m). Normally, the surface layer in summer has a low density due to a high temperature and low salinity water from the CDW and the lower layer forms a strong stratification due to the distribution of the low temperature water from the YSCW and high salinity water from the TCWW [49]. In the northern ECS, the runoff from the Changjiang river is maximum in summer and minimum in winter [50]. According to a previous study, CDW is a main source of fresh water input in the ECS, increasing from spring to summer [51]. In November, the water masses were relatively well mixed (Figure 2).

In this study, we found that major inorganic nutrient concentrations were considerably higher in the western part compared to those in the eastern part in February and November during this study (Table 2), which is consistent with previous results [52]. The waters in the western part of the ECS are fully mixed from surface to the bottom because of the shallow water depth (<50 m), but in the eastern part vertical mixings occur only in the upper layer [35]. The noticeable vertical difference in nutrient concentrations were observed in August (Table 2) due to a strong stratified water column (Figure 2) which suppressed the upwelling of nutrients from the bottom layer. In addition, the seasonal average N:P ratios in the study area ranged from 10.5 to 422.9 (55.1 ± 64.6) which are higher than the Redfield ratio of 16 generally found in various oceans. [53] defined nutrient limitations following as; PO_4 limitation when Si:P > 22 and DIN:P > 22; N limitation when DIN:P < 10 and Si:DIN > 1; Si limitation when Si:P < 10 and Si:DIN < 1. Various studies suggested that PO_4 is a limiting nutrient to phytoplankton growth in the ECS [41,54–56]. This study also verified PO_4-limited environmental conditions in May (124.5 ± 91.1) and August (50.3 ± 29.3) (Figure 8). According to [55] a high N:P ratio is related to very low PO_4 concentration. Indeed, low PO_4 concentrations (approximately 0.1 µM) were observed in May and August. These PO_4-limited conditions could have caused the seasonal variation in phytoplankton community in the ECS. According to [57], Diatoms would have a higher phosphorus demand relative to other phytoplankton groups which may be reflected by lower N:P ratios in diatoms compared to those in other groups. Indeed, [43] showed the phytoplankton community in mid-shelf ECS in summer and identified 2 distinct phytoplankton communities under two major water masses with different nutrient conditions: PO_4-rich Kuroshio intermediate water (KW) indicated by a low ExN value leading to diatom domination and PO_4-limited CDW indicated by high ExN leading to small phytoplankton domination such as chlorophytes and cyanobacteria. However, we did not find the relationships in this study. In spring, diatoms were mostly dominant despite of PO_4-limited water conditions. The PO_4- limited condition in spring could be due to the spring bloom of diatoms, which is consistent with the results in Chesapeake Bay [58]. Even in our summer cruise period, opposite relationship between diatoms and ExN was observed. This discrepancy between this and previous studies could be caused by several factors. [42] observed pronounced effects of KW and CDW and a large range of PO_4 concentration, whereas in this study, TCWW current was largely dominant rather than the KW and CDW and narrow range of PO_4 concentration. In addition, the analysis in [42] was performed only in the surface layer, whereas this study was performed within the euphotic layer (surface to 1% light depth). Indeed, we also found a dominance of cyanobacteria in the surface layer with a lack of PO_4 and diatoms dominant at the nutrient-rich depths as discussed in detail below.

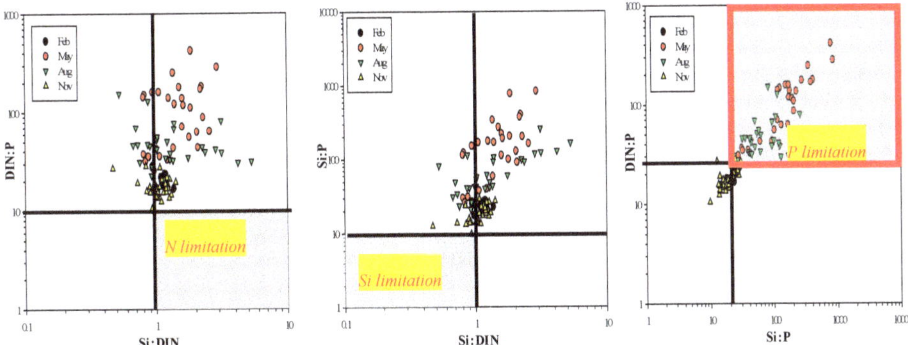

Figure 8. Scatter diagrams of atomic nutrient ratios at the euphotic depths (100%, 30%, and 1%) in the northern East China Sea, 2018.

The distinct vertical difference in dominant phytoplankton communities was observed in our study area in August with cyanobacteria predominated at surface layer and diatoms prevailed at deep layer. A strong water stratification appeared in the study area could have caused the vertical pattern of phytoplankton community [59]. A stratified water column restricts the upward supply of major inorganic nutrients to the upper euphotic surface layer. According to the resource competition theory [60–62], pico-phytoplankton are favored over larger phytoplankton in nutrient-limited conditions because of their higher nutrient affinity associated to their small size [63–65]. Thus, small size cyanobacteria are predominant in the nutrient-deficient surface layer in August. Since zeaxanthin is a marker pigment in cyanobacteria and plays an important role in protecting cyanobacteria against photoinhibition [66], the high concentration of zeaxanthin at surface might be due to much higher photosensitivity than that at the deep water column [67,68]. In addition, the water temperature was approximately 7 °C higher at surface than the deep layer in August. Indeed, CCA revealed a positive correlation between cyanobacteria and temperature in this study (Figure 7c). As water temperature exceeds 20 °C, the growth rates of eukaryotic phytoplankton usually stabilize or decrease whereas those of many cyanobacteria species increase because of their competitive advantage over high temperature [69–71]. Therefore, the water temperature and nutrient distribution in August had a great influence on the vertical distribution of phytoplankton communities.

During the four research periods, spatial difference in phytoplankton community was not significantly high, but in November, there was a clear difference in phytoplankton community between the western and eastern parts. The most predominant phytoplankton communities were diatoms in the western part and cyanobacteria in the eastern part. The cryptophytes were the third dominant species in both western and eastern parts. CCA result showed that cyanobacteria are associated with high temperature, high salinity, low nutrient concentrations, and depth, whereas diatoms are associated with low temperature and high nutrient concentrations in November (Figure 7d). Nutrient concentrations were also horizontally different, gradually decreasing toward the west. In relation to the distribution of these nutrients, zeaxanthin (major pigments of cyanobacteria) showed a negative correlation with nutrients ($p < 0.01$, t-test), whereas fucoxanthin (major pigments of diatoms) showed a positive correlation with nutrients ($p < 0.01$, t-test). These correlations with the nutrient concentrations indicate that nutrients are a major driver of the spatial difference in phytoplankton community distribution in November. Moreover, there was a significant difference in light condition based on the euphotic depths between in the western and eastern parts. The euphotic layer up to 1% depth was 43 m on average in the eastern part, whereas it was 7 m in the western part. Light can be a limiting factor largely influencing the spatial distribution of picophytoplankton, probably because the decreasing light in water is mostly variable in the water column [72,73]. According to [74], diatoms have high growth efficiency under a low light condition. In comparison to the phytoplankton community in the ECS, several studies in other oceanic basins influenced by large rivers were compared. Similar to the East China Sea, the Gulf of Mexico is a phosphate-limited environment during summer period [75,76]. These studies showed that diatoms generally predominate and cryptophytes are the second most abundant group in the winter and spring periods and cyanobacteria are most dominant during PO_4-limited summer time compared to other seasons in the Gulf of Mexico. The spatiotemporal variations in the region are controlled mainly by river flow runoff, along with other environmental variables such as wind pressure and stratification [75,76]. In contrast, Western Tropical North Atlantic, which is a region largely affected by the Amazon River, is mainly dominated by the diatom-diazotroph associations (DDAs) [43]. In this region, the phytoplankton community structure and distributions are controlled by low concentrations of inorganic nitrite and nitrate (NO_2 + NO_3) [43].

Based on the four different seasonal observations in this study, the yearly average contributions of different phytoplankton communities were 32.0%, 20.6%, 17.2%, 6.9%, 6.4%, 6.4%, 5.7%, and 5.0% for diatoms, cyanobacteria, cryptophytes, prymnesiophytes, chlorophytes, chrysophytes, dinoflagellates, and prasinophytes, respectively, in the northern ECS. Chlorophyll-a concentrations were highest in May and lowest in February in this study which is consistent with previous results in the ECS [56,77]. Previous studies reported that diatoms are associated with phytoplankton blooms in early spring and that the dominant species in the ECS are mostly chain-forming diatoms such as *Pseudonitzschia delicatissima*, *Thalassionema nitzschioides*, and *Paralia sulcate* [28]. Consistent with previous observations, this study also verified that the dominant species were diatoms during the spring bloom in May.

5. Summary and Conclusions

There are multiple factors including light intensity, stability of water column, temperature, and nutrient conditions [78] that can cause variations in phytoplankton compositions and spatial distributions. The seasonal variations in the phytoplankton community were distinct in our study area although spatial and vertical variations were observed along the seasons. Diatoms appeared to be dominant in the northern ECS throughout the year in this study. Normally, diatoms are known to be competitive over other species at low water temperatures [79]. Therefore, in February with a low water temperature (Figure 2) and high nutrient concentrations (Table 2), diatoms were most predominated among our study periods. Moreover, diatoms are more efficient at high nutrient concentrations than small phytoplankton [80] and they can quickly respond to nutrient inputs [81]. Contrary to diatoms, cyanobacteria, as the next dominant species, started to appear in May and showed their contribution gradually increased from May to November in this study. According to previous research, water temperature is the main control factor for the distribution of cyanobacteria [82]. In this study, we also found that water temperature is a main factor driving the seasonal variation in the cyanobacteria contribution in the northern ECS throughout the year based on CCA result (Figure 7). Overall, the cyanobacteria contribution was strongly negatively correlated with the diatom contribution in the northern ECS during our study period in 2018 (Figure 6). This result implies an ecologically significant meaning for the marine ecosystem in the northern ECS. Under expected warming ocean scenarios, the potential change in dominant phytoplankton groups from diatoms to cyanobacteria could cause substantial differences in quantity and qualitative aspects of primary marine food sources in the northern ECS. Comprehensive monitoring for qualitative and quantitative characteristics of different phytoplankton communities is warrant for a better understanding their potential consequences on the entire marine ecosystem in the ECS.

Author Contributions: Conceptualization, Y.K., S.-H.Y., H.J.O., and S.H.L.; methodology, Y.K. and J.J.K.; validation, Y.K., S.-H.Y., and S.H.L.; formal analysis, Y.K.; investigation, J.J.K., J.H.L., D.L., K.K., H.K.J., and J.L.; data curation, Y.K.; writing—original draft preparation, Y.K.; writing—review and editing, Y.K., J.J.K., and S.H.L.; and visualization, Y.K. All authors have read and agreed to the published version of the manuscript.

Funding: This research was supported by the "Development of marine ecological forecasting system for Korean waters (R2018067)" funded by the National Institute of Fisheries Science (NIFS), Korea. This research also partly supported by a part of the project entitled "Construction of Ocean Research Station and their application" funded by the Ministry of Oceans and Fisheries, Korea.

Acknowledgments: We thank the anonymous reviewers who greatly improved an earlier version of manuscript.

Conflicts of Interest: The authors declare no conflict of interest.

References

1. Smith, W., Jr.; Sakshaug, E. Polar Phytoplankton. In *Polar Oceanography*; Academic: San Diego, CA, USA, 1990; pp. 477–525.
2. Sommer, U. The impact of light intensity and daylength on silicate and nitrate competition among marine phytoplankton. *Limnol. Oceanogr.* **1994**, *39*, 1680–1688. [CrossRef]
3. Hilligsøe, K.M.; Richardson, K.; Bendtsen, J.; Sørensen, L.-L.; Nielsen, T.G.; Lyngsgaard, M.M. Linking phytoplankton community size composition with temperature, plankton food web structure and sea–air CO_2 flux. *Deep Sea Res. Part I* **2011**, *58*, 826–838.
4. Kilham, P.; Hecky, R.E. Comparative ecology of marine and freshwater phytoplankton 1. *Limnol. Oceanogr.* **1988**, *33*, 776–795. [CrossRef]
5. Domingues, R.B.; Galvao, H. Phytoplankton and environmental variability in a dam regulated temperate estuary. *Hydrobiologia* **2007**, *586*, 117–134. [CrossRef]
6. Wu, N.; Schmalz, B.; Fohrer, N. Development and testing of a phytoplankton index of biotic integrity (P-IBI) for a German lowland river. *Ecol. Indic.* **2012**, *13*, 158–167. [CrossRef]
7. Smayda, T. Biogeographical meaning; indicators. In *Phytoplankton Manual*; UNESCO: Paris, France, 1978; pp. 225–229.
8. Soares, M.C.S.; Lobão, L.M.; Vidal, L.O.; Oyma, N.P.; Barros, N.O.; Cardoso, S.J.; Roland, F. Light microscopy in aquatic ecology: Methods for plankton communities studies. *Methods Mol. Biol.* **2011**, *689*, 215–227.
9. Chisholm, S.W.; Olson, R.J.; Zettler, E.R.; Goericke, R.; Waterbury, J.B.; Welschmeyer, N.A. A novel free-living prochlorophyte abundant in the oceanic euphotic zone. *Nature* **1988**, *334*, 340. [CrossRef]
10. Gieskes, W.; Kraay, G. Dominance of Cryptophyceae during the phytoplankton spring bloom in the central North Sea detected by HPLC analysis of pigments. *Mar. Biol.* **1983**, *75*, 179–185. [CrossRef]
11. Simon, N.; Barlow, R.G.; Marie, D.; Partensky, F.; Vaulot, D. Characterization of oceanic photosynthetic picoeukaryotes by flow cytometry. *J. Phycol.* **1994**, *30*, 922–935. [CrossRef]
12. Readman, J.; Devilla, R.; Tarran, G.; Llewellyn, C.; Fileman, T.; Easton, A.; Burkill, P.; Mantoura, R. Flow cytometry and pigment analyses as tools to investigate the toxicity of herbicides to natural phytoplankton communities. *Mar. Environ. Res.* **2004**, *58*, 353–358. [CrossRef]
13. Campbell, L.; Nolla, H.; Vaulot, D. The importance of Prochlorococcus to community structure in the central North Pacific Ocean. *Limnol. Oceanogr.* **1994**, *39*, 954–961. [CrossRef]
14. Partensky, F.; Blanchot, J.; Lantoine, F.; Neveux, J.; Marie, D. Vertical structure of picophytoplankton at different trophic sites of the tropical northeastern Atlantic Ocean. *Deep Sea Res. Part I* **1996**, *43*, 1191–1213. [CrossRef]
15. Wright, S.; Jeffrey, S.; Mantoura, R.; Llewellyn, C.; Bjørnland, T.; Repeta, D.; Welschmeyer, N. Improved HPLC method for the analysis of chlorophylls and carotenoids from marine phytoplankton. *Mar. Ecol. Prog. Ser.* **1991**, *77*, 183–196. [CrossRef]
16. Mackey, M.; Mackey, D.; Higgins, H.; Wright, S. CHEMTAX-a program for estimating class abundances from chemical markers: Application to HPLC measurements of phytoplankton. *Mar. Ecol. Prog. Ser.* **1996**, *144*, 265–283. [CrossRef]
17. Furuya, K.; Kurita, K.; Odate, T. Distribution of phytoplankton in the East China Sea in the winter of 1993. *J. Oceanol.* **1996**, *52*, 323–333. [CrossRef]
18. Jiang, Y.-Z.; Cheng, J.-H.; Li, S.-F. Temporal changes in the fish community resulting from a summer fishing moratorium in the northern East China Sea. *Mar. Ecol. Prog. Ser.* **2009**, *387*, 265–273. [CrossRef]
19. Chen, C.-T.A. Distributions of nutrients in the East China Sea and the South China Sea connection. *J. Oceanol.* **2008**, *64*, 737–751. [CrossRef]
20. Moriyasu, S. The Tsushima Current. In *Kuroshio, Its Physical Aspects*; Stommel, H., Yoshida, K., Eds.; University of Tokyo Press: Tokyo, Japan, 1972; pp. 353–369.
21. Nitani, H. Beginning of the Kuroshio. In *Kuroshio Physical Aspect of the Japan Current*; University of Washington Press: Seattle, WA, USA, 1972; pp. 129–163.
22. Seung, Y. Water masses and circulations around Korean Peninsula. *J. Oceanol. Soc. Korea* **1992**, *27*, 324–331.

23. Qilong, Z.; Xuechuan, W.; Yuling, Y. Analysis of water masses in the south Yellow Sea in spring. *Oceanol. Limnol. Sin.* **1996**, *27*, 42–428. (In Chinese)

24. Zhou, M.-J.; Shen, Z.-L.; Yu, R.-C. Responses of a coastal phytoplankton community to increased nutrient input from the Changjiang (Yangtze) River. *Cont. Shelf Res.* **2008**, *28*, 1483–1489. [CrossRef]

25. Guo, C.; Liu, H.; Zheng, L.; Song, S.; Chen, B.; Huang, B. Seasonal and spatial patterns of picophytoplankton growth, grazing and distribution in the East China Sea. *Biogeosciences* **2014**, *11*, 1847–1862. [CrossRef]

26. Furuya, K.; Hayashi, M.; Yabushita, Y.; Ishikawa, A. Phytoplankton dynamics in the East China Sea in spring and summer as revealed by HPLC-derived pigment signatures. *Deep Sea Res. Part II* **2003**, *50*, 367–387. [CrossRef]

27. Liu, X.; Xiao, W.; Landry, M.R.; Chiang, K.-P.; Wang, L.; Huang, B. Responses of phytoplankton communities to environmental variability in the East China Sea. *Ecosystems* **2016**, *19*, 832–849. [CrossRef]

28. Guo, S.; Feng, Y.; Wang, L.; Dai, M.; Liu, Z.; Bai, Y.; Sun, J. Seasonal variation in the phytoplankton community of a continental-shelf sea: The East China Sea. *Mar. Ecol. Prog. Ser.* **2014**, *516*, 103–126. [CrossRef]

29. Jiang, Z.; Chen, J.; Zhou, F.; Shou, L.; Chen, Q.; Tao, B.; Yan, X.; Wang, K. Controlling factors of summer phytoplankton community in the Changjiang (Yangtze River) Estuary and adjacent East China Sea shelf. *Cont. Shelf Res.* **2015**, *101*, 71–84. [CrossRef]

30. Zhao, R.; Sun, J.; Bai, J. Phytoplankton assemblages in Yangtze River Estuary and its adjacent water in autumn 2006. *Mar. Sci.* **2010**, *34*, 32–39.

31. He, Q.; Sun, J.; Luan, Q.-S.; Yu, Z.-M. Phytoplankton in Changjiang Estuary and adjacent waters in winter. *Mar. Environ. Sci.* **2009**, *28*, 360–365.

32. Yoon, Y. Spatial distribution of phytoplankton community and red tide of dinoflagellate, Prorocentrum donghaience in the East China Sea during early summer. *Korean J. Environ. Biol.* **2003**, *21*, 132–141.

33. Park, M.-O.; Kang, S.-W.; Lee, C.-I.; Choi, T.-S.; Lantoine, F. Structure of the phytoplanktonic communities in Jeju Strait and northern East China Sea and dinoflagellate blooms in spring 2004: Analysis of photosynthetic pigments. *Sea* **2008**, *13*, 27–41.

34. Yoon, Y.-H.; Park, J.-S.; Soh, H.-Y.; Hwang, D.-J. On the marine environment and distribution of phytoplankton community in the northern East China Sea in early summer 2004. *J. Korean Soc. Mar. Environ. Energy* **2005**, *8*, 100–110.

35. kim, D.; Choi, S.H.; Kim, K.H.; Shim, J.; Yoo, S.; Kim, C.H. Spatial and temporal variations in nutrient and chlorophyll-a concentrations in the northern East China Sea surrounding Cheju Island. *Cont. Shelf Res.* **2009**, *29*, 1426–1436. [CrossRef]

36. Kang, J.J.; Lee, J.H.; Kim, H.C.; Lee, W.C.; Lee, D.; Jo, N.; Min, J.-O.; Lee, S.H. Monthly Variations of Phytoplankton Community in Geoje-Hansan Bay of the Southern Part of Korea Based on HPLC Pigment Analysis. *J. Coast. Res.* **2018**, *85*, 356–360. [CrossRef]

37. Zapata, M.; Rodríguez, F.; Garrido, J.L. Separation of chlorophylls and carotenoids from marine phytoplankton: A new HPLC method using a reversed phase C8 column and pyridine-containing mobile phases. *Mar. Ecol. Prog. Ser.* **2000**, *195*, 29–45. [CrossRef]

38. Lee, Y.-W.; Park, M.-O.; Im, Y.-S.; Kim, S.-S.; Kang, C.-K. Application of photosynthetic pigment analysis using a HPLC and CHEMTAX program to studies of phytoplankton community composition. *Sea* **2011**, *16*, 117–124. [CrossRef]

39. Wolf, C.; Frickenhaus, S.; Kilias, E.S.; Peeken, I.; Metfies, K. Protist community composition in the Pacific sector of the Southern Ocean during austral summer 2010. *Polar Biol.* **2014**, *37*, 375–389. [CrossRef]

40. Torrecilla, E.; Stramski, D.; Reynolds, R.A.; Millán-Núñez, E.; Piera, J. Cluster analysis of hyperspectral optical data for discriminating phytoplankton pigment assemblages in the open ocean. *Remote Sens. Environ.* **2011**, *115*, 2578–2593. [CrossRef]

41. Wong, G.; Gong, G.; Liu, K.; Pai, S. 'Excess nitrate'in the East China Sea. *Estuar. Coast. Shelf Sci.* **1998**, *46*, 411–418. [CrossRef]

42. Xu, Q.; Sukigara, C.; Goes, J.I.; do Rosario Gomes, H.; Zhu, Y.; Wang, S.; Shen, A.; de Raús Maúre, E.; Matsuno, T.; Yuji, W. Interannual changes in summer phytoplankton community composition in relation to water mass variability in the East China Sea. *J. Oceanol.* **2019**, *75*, 61–79. [CrossRef]

43.   Goes, J.I.; do Rosario Gomes, H.; Chekalyuk, A.M.; Carpenter, E.J.; Montoya, J.P.; Coles, V.J.; Yager, P.L.; Berelson, W.M.; Capone, D.G.; Foster, R.A. Influence of the Amazon River discharge on the biogeography of phytoplankton communities in the western tropical north Atlantic. *Prog. Oceanogr.* **2014**, *120*, 29–40. [CrossRef]

44.   Ter Braak, C.J.; Smilauer, P. *CANOCO Reference Manual and CanoDraw for Windows User's Guide: Software for Canonical Community Ordination (Version 4.5)*; Microcomputer Power: Ithaca, NY, USA, 2002.

45.   Gong, G.-C.; Chen, Y.-L.L.; Liu, K.-K. Chemical hydrography and chlorophyll a distribution in the East China Sea in summer: Implications in nutrient dynamics. *Cont. Shelf Res.* **1996**, *16*, 1561–1590. [CrossRef]

46.   Chen, C.-T.A. Chemical and physical fronts in the Bohai, Yellow and East China seas. *J. Mar. Sci.* **2009**, *78*, 394–410. [CrossRef]

47.   Chen, C.; Beardsley, R.C.; Limeburner, R.; Kim, K. Comparison of winter and summer hydrographic observations in the Yellow and East China Seas and adjacent Kuroshio during 1986. *Cont. Shelf Res.* **1994**, *14*, 909–929. [CrossRef]

48.   Kusakabe, M. Hydrographic feature of the East China Sea. *Bull. Coast. Oceanogr.* **1998**, *36*, 5–17.

49.   Jang, S.-T.; Lee, J.-H.; Hong, C.-S. Mixing of Sea Waters in the Northern Part of the East China Sea in Summer. *Sea* **2007**, *12*, 390–399.

50.   Beardsley, R.; Limeburner, R.; Yu, H.; Cannon, G. Discharge of the Changjiang (Yangtze river) into the East China sea. *Cont. Shelf Res.* **1985**, *4*, 57–76. [CrossRef]

51.   Zuo-sheng, Y.; Milliman, J.D.; Fitzgerald, M.G. Transfer of water and sediment from the Yangtze River to the East China Sea, June 1980. *Can. J. Fish. Aqua. Sci.* **1983**, *40*, s72–s82. [CrossRef]

52.   Kim, D.; Shim, J.; Yoo, S. Seasonal variations in nutrients and chlorophyll-a concentrations in the northern East China Sea. *Ocean Sci. J.* **2006**, *41*, 125–137. [CrossRef]

53.   Justić, D.; Rabalais, N.N.; Turner, R.E.; Dortch, Q. Changes in nutrient structure of river-dominated coastal waters: Stoichiometric nutrient balance and its consequences. *Estuar. Coast. Shelf Sci.* **1995**, *40*, 339–356. [CrossRef]

54.   Harrison, P.; Hu, M.; Yang, Y.; Lu, X. Phosphate limitation in estuarine and coastal waters of China. *J. Exp. Mar. Biol. Ecol.* **1990**, *140*, 79–87. [CrossRef]

55.   Chen, Y.-L.L.; Lu, H.-B.; Shiah, F.-K.; Gong, G.; Liu, K.-K.; Kanda, J. New production and f-ratio on the continental shelf of the East China Sea: Comparisons between nitrate inputs from the subsurface Kuroshio Current and the Changjiang River. *Estuar. Coast. Shelf Sci.* **1999**, *48*, 59–75. [CrossRef]

56.   Gong, G.-C.; Wen, Y.-H.; Wang, B.-W.; Liu, G.-J. Seasonal variation of chlorophyll a concentration, primary production and environmental conditions in the subtropical East China Sea. *Deep Sea Res. Part II* **2003**, *50*, 1219–1236. [CrossRef]

57.   Hillebrand, H.; Steinert, G.; Boersma, M.; Malzahn, A.; Meunier, C.L.; Plum, C.; Ptacnik, R. Goldman revisited: Faster-growing phytoplankton has lower N: P and lower stoichiometric flexibility. *Limnol. Oceanogr.* **2013**, *58*, 2076–2088. [CrossRef]

58.   Li, J.; Glibert, P.M.; Zhou, M.; Lu, S.; Lu, D. Relationships between nitrogen and phosphorus forms and ratios and the development of dinoflagellate blooms in the East China Sea. *Mar. Ecol. Prog. Ser.* **2009**, *383*, 11–26. [CrossRef]

59.   Hampton, S.E.; Gray, D.K.; Izmest'eva, L.R.; Moore, M.V.; Ozersky, T. The rise and fall of plankton: Long-term changes in the vertical distribution of algae and grazers in Lake Baikal, Siberia. *PLoS ONE* **2014**, *9*, e88920. [CrossRef] [PubMed]

60.   Tilman, D. Resource competition between plankton algae: An experimental and theoretical approach. *Ecology* **1977**, *58*, 338–348. [CrossRef]

61.   Tilman, D. *Resource Competition and Community Structure*; Princeton University Press: Princeton, NJ, USA, 1982.

62.   Tilman, D.; Kilham, S.S.; Kilham, P. Phytoplankton community ecology: The role of limiting nutrients. *Annu. Rev. Ecol. Syst.* **1982**, *13*, 349–372. [CrossRef]

63.   Smith, R.E.; Kalff, J. Size dependence of growth rate, respiratory electron transport system activity and chemical composition in marine diatoms in the laboratory. *J. Phycol.* **1982**, *18*, 275–284. [CrossRef]

64.   Fogg, G.E. Review Lecture-Picoplankton. *Proc. R. Soc. Lond. B Biol. Sci.* **1986**, *228*, 1–30.

65.   Raven, J. The twelfth Tansley Lecture. Small is beautiful: The picophytoplankton. *Funct. Ecol.* **1998**, *12*, 503–513. [CrossRef]

66. Demmig-Adams, B.; Adams, W.W.; Czygan, F.-C.; Schreiber, U.; Lange, O.L. Differences in the capacity for radiationless energy dissipation in the photochemical apparatus of green and blue-green algal lichens associated with differences in carotenoid composition. *Planta* **1990**, *180*, 582–589. [CrossRef]

67. Descy, J.-P.; Sarmento, H.; Higgins, H.W. Variability of phytoplankton pigment ratios across aquatic environments. *Eur. J. Phycol.* **2009**, *44*, 319–330. [CrossRef]

68. Claustre, H.; Kerhervé, P.; Marty, J.-C.; Prieur, L. Phytoplankton photoadaptation related to some frontal physical processes. *J. Mar. Syst.* **1994**, *5*, 251–265. [CrossRef]

69. Mur, R.; Skulberg, O.; Utkilen, H. Cyanobacteria in the environment. In *Toxic Cyanobacteria in Water*; Chorus, I., Batram, J., Eds.; E&FN Spon: London, UK, 1999; pp. 15–40.

70. Peperzak, L. Climate change and harmful algal blooms in the North Sea. *Acta Oecol.* **2003**, *24*, S139–S144. [CrossRef]

71. Paerl, H.W.; Huisman, J. Climate change: A catalyst for global expansion of harmful cyanobacterial blooms. *Environ. Microbiol. Rep.* **2009**, *1*, 27–37. [CrossRef] [PubMed]

72. Dusenberry, J.; Olson, R.; Chisholm, S. Field observations of oceanic mixed layer dynamics and picophytoplankton photoacclimation. *J. Mar. Syst.* **2000**, *24*, 221–232. [CrossRef]

73. Moore, L.R.; Goericke, R.; Chisholm, S.W. Comparative physiology of Synechococcus and Prochlorococcus: Influence of light and temperature on growth, pigments, fluorescence and absorptive properties. *Mar. Ecol. Prog. Ser.* **1995**, *116*, 259–275. [CrossRef]

74. Goldman, J.C.; McGillicuddy, D.J.J. Effect of large marine diatoms growing at low light on episodic new production. *Limnol. Oceanogr.* **2003**, *48*, 1176–1182. [CrossRef]

75. Chakraborty, S.; Lohrenz, S.E. Phytoplankton community structure in the river-influenced continental margin of the northern Gulf of Mexico. *Mar. Ecol. Prog. Ser.* **2015**, *521*, 31–47. [CrossRef]

76. Anglès, S.; Jordi, A.; Henrichs, D.W.; Campbell, L. Influence of coastal upwelling and river discharge on the phytoplankton community composition in the northwestern Gulf of Mexico. *Prog. Oceanogr.* **2019**, *173*, 26–36. [CrossRef]

77. Shi, W.; Wang, M. Satellite views of the bohai sea, yellow sea, and East China Sea. *Prog. Oceanogr.* **2012**, *104*, 30–45. [CrossRef]

78. Reid, P.; Lancelot, C.; Gieskes, W.; Hagmeier, E.; Weichart, G. Phytoplankton of the North Sea and its dynamics: A review. *Neth. J. Sea Res.* **1990**, *26*, 295–331. [CrossRef]

79. Johnson, F.A. Green, bluegreen and diatom algae: Taxonomic differences in competitive ability for phosphorus, silicon and nitrogen. *Arch. Hydrobiol.* **1986**, *106*, 473485.

80. Hassen, M.B.; Drira, Z.; Hamza, A.; Ayadi, H.; Akrout, F.; Issaoui, H. Summer phytoplankton pigments and community composition related to water mass properties in the Gulf of Gabes. *Estuar. Coast. Shelf Sci.* **2008**, *77*, 645–656. [CrossRef]

81. Lagus, A.; Suomela, J.; Weithoff, G.; Heikkilä, K.; Helminen, H.; Sipura, J. Species-specific differences in phytoplankton responses to N and P enrichments and the N: P ratio in the Archipelago Sea, northern Baltic Sea. *J. Plankton Res.* **2004**, *26*, 779–798. [CrossRef]

82. Kanoshina, I.; Lips, U.; Leppänen, J.-M. The influence of weather conditions (temperature and wind) on cyanobacterial bloom development in the Gulf of Finland (Baltic Sea). *Harmful Algae* **2003**, *2*, 29–41. [CrossRef]

Article

Phytoplankton Community in the Western South China Sea in Winter and Summer

Changling Ding [1,2], Jun Sun [2,3,*], Dhiraj Dhondiram Narale [2] and Haijiao Liu [2]

1   College of Biotechnology, Tianjin University of Science and Technology, Tianjin 300457, China; dingcl0405@163.com
2   Research Centre for Indian Ocean Ecosystem, Tianjin University of Science and Technology, Tianjin 300457, China; dhirajnarale2000@gmail.com (D.D.N.); coccolith@126.com (H.L.)
3   College of Marine Science and Technology, China University of Geosciences (Wuhan), Wuhan 430074, China
*   Correspondence: phytoplankton@163.com; Tel.: +86-022-60601116

Abstract: Phytoplankton are known as important harbingers of climate change in aquatic ecosystems. Here, the influence of the oceanographic settings on the phytoplankton community structure in the western South China Sea (SCS) was investigated during two seasons, i.e., the winter (December 2006) and summer (August–September, 2007). The phytoplankton community was mainly composed of diatoms (192 taxa), dinoflagellates (109 taxa), and cyanobacteria (4 taxa). The chain-forming diatoms and cyanobacteria *Trichodesmium* were the dominants throughout the study period. The phytoplankton community structure displayed distinct variation between two seasons, shifting from a diatom-dominated regime in winter to a cyanobacteria-dominated system in summer. The increased abundance of overall phytoplankton and cyanobacteria in the water column during the summer signifies the impact of nutrient advection due to upwelling and enriched eddy activity. That the symbiotic cyanobacteria–diatom (*Rhizosolenia–Richelia*) association was abundant during the winter signifies the influence of cool temperature. On the contrary, *Trichodesmium* dominance during the summer implies its tolerance to increased temperature. Overall, the two seasonal variations within the local phytoplankton community in the western SCS could simulate their community shift over the forthcoming climatic conditions.

Keywords: South China Sea; upwelling; eddy; diatom; *Trichodesmium*; *Rhizosolenia–Richelia*

Citation: Ding, C.; Sun, J.; Narale, D.D.; Liu, H. Phytoplankton Community in the Western South China Sea in Winter and Summer. *Water* **2021**, *13*, 1209. https://doi.org/10.3390/w13091209

Academic Editor: Bo Kyung Kim

Received: 9 February 2021
Accepted: 9 April 2021
Published: 27 April 2021

Publisher's Note: MDPI stays neutral with regard to jurisdictional claims in published maps and institutional affiliations.

1. Introduction

The South China Sea (SCS), a typically oligotrophic area, is the largest marginal sea in the tropical Pacific Ocean. The upper SCS is characterized by the monsoon-induced circulation and mesoscale eddies which predominantly impact biogeochemical progress; concurrently, the riverine input from the Pearl and Mekong Rivers dramatically affects nutrient exchange in the SCS [1–4]. Despite receiving large amounts of terrestrial nutrient input through the riverine discharge, the SCS only utilizes a small portion to support productivity [5,6]. Nutrient concentrations in the SCS are often below the detectable limits [7]. The ratios of nitrogen to soluble reactive phosphorus (N/P) were much lower than 16 (the Redfield N/P Ratio), suggesting nitrogen limitations in the SCS [8]. Nutrient deficiency causes relatively low chlorophyll concentrations [9–11], and low phytoplankton stock compared with other adjacent marginal seas [12–14]. The western SCS is located towards the east of the Vietnam coast, where the deep basin is extended by steep slopes, with a maximum depth reaching 4000 m. Vietnamese upwelling is one of the typical features in the western SCS, and the Vietnam offshore flowing to the north in summer causes a local enhancement of Vietnamese upwelling intensity [6]. In the western SCS, cyclonic eddies form frequently with a raised thermocline in winter, and anticyclonic eddies form with a depressed thermocline in summer [15]. Besides, summer circulation often has a dipole structure associated with an eastward jet, appearing off central Vietnam [16].

These physical processes control nutrient flux from the deep water into the euphotic zone and subsequently affect the ocean's ecological status [17].

Marine phytoplankton, as the most important primary producer at the base of the marine food chain, are responsible for generating roughly half of the global net primary production, and play a key role in the elements cycle and energy flow in a marine ecosystem as the primary producers [18]. Physical processes such as upwelling and eddies are particularly relevant to phytoplankton productivity [19]. The instabilities of these processes help to create and maintain localized environments that favor the growth of phytoplankton [17]. The coupling between these physical and biological processes influences phytoplankton biomass and seasonal succession [20]. In the western SCS, a series of physical processes, controlling nutrient flux into the euphotic zone, play a profound role in supporting the phytoplankton growth and their spatio-temporal distribution [21–26]. High chlorophyll *a* concentration often occurs in the western SCS, where phytoplankton blooms even appear in summer when southwest monsoons are parallel to the Vietnamese coast [21,22]. Wang and Tang (2014) observed that the patchiness in spatial and vertical phytoplankton distribution was controlled by the vertical flux of nutrients caused by curl-driven upwelling in the western SCS [23]. Liang et al. (2018) found that the high chlorophyll *a* belt was determined by the advection of coastal upwelling water by the northeastward jet and the resultant cyclonic/anticyclonic eddies, which were defined as a 'jet-eddy system' [27]. Wang et al. (2016) calculated that the contribution of phytoplankton groups to the total chlorophyll *a* biomass changed along with cyclonic eddy dimensional structure [25].

Many studies have investigated phytoplankton biomass and the coupling of biological–physical processes in the western SCS. These existing related studies were focused mainly on pigments and remote sensing observations. However, yet, quantitative measures of phytoplankton diversity, a comprehensive interpretation of phytoplankton successions, and knowledge of interactions with diverse hydrodynamic settings are still meager. Knowledge of phytoplankton species and their response within the marine environment is essential to understand the responses of ocean biota to a dynamic ecosystem and changing global climate [28]. Here, we carried out a series of biogeochemical investigations during two seasons (winter and summer) in the western SCS. In this study, the cold-core cyclonic eddies and warm-core anticyclonic eddy were observed during the summer investigation [29]. The main objectives of this study were to evaluate the spatial and temporal difference of the phytoplankton community structure in different seasons, aiming to supply a cue of how physicochemical influence on the phytoplankton community shifts, to provide insights into the acclimation and adaptation of the phytoplankton community to a changing marine environment.

2. Materials and Methods

2.1. Study Area

The sampling was carried out from the western SCS extending eastwards from the Vietnamese shelf region towards the eastern deep basin (10–15° N, 110–112.5° E) (Figure 1). In this region, the seasonal reversal of monsoon winds mainly controls the upper-ocean circulations (Shaw and Chao, 1994). During the northeast (or winter) monsoon (November to March) a stronger cyclonic gyre exists in the western portion of the southern SCS [30]. A strong coastal jet occurs in the western boundary of the SCS, southward along the continental shelf from the Chinese coast to southern Vietnam [31], causing the basin-scale circulation. On the contrary, during the southwest (or summer) monsoon (April to August), the weaker anticyclonic gyre dominates upper layer circulations in the southwestern SCS. The northward jet separates from the Vietnamese coast at about 12° N in summer [32] and eddy pairs associate with the jet forms [33]. The upwelling takes place off the Vietnamese coast, which flows northeastward and carries the cold continental water into the open basin [31]. Two cruises were conducted on the R.V. 'Dongfanghong 2' during the southwest monsoon (December 2006) and northeast monsoon (August–September, 2007) periods to assess the phytoplankton community structure in the western SCS. Two cyclonic mesoscale cold eddies were monitored in August and September, which were named as cold eddy 1 (CE1) and cold eddy 2 (CE2), respectively, during the cruise using in situ current, hydrographic measurements as well as concurrent satellite altimeter observations [29,34]. With a relatively steady intensity and radius, the CE2 endured for two weeks after its swift formation in late August and prior to its quick dissipation in mid-September. The anticyclonic warm eddies, marked WE, were also observed in the survey area [29]. During this study, a total of 15 and 36 stations were investigated in winter and summer, respectively. The sampling stations marked with dotted circles were located within the eddy area (Figure 1).

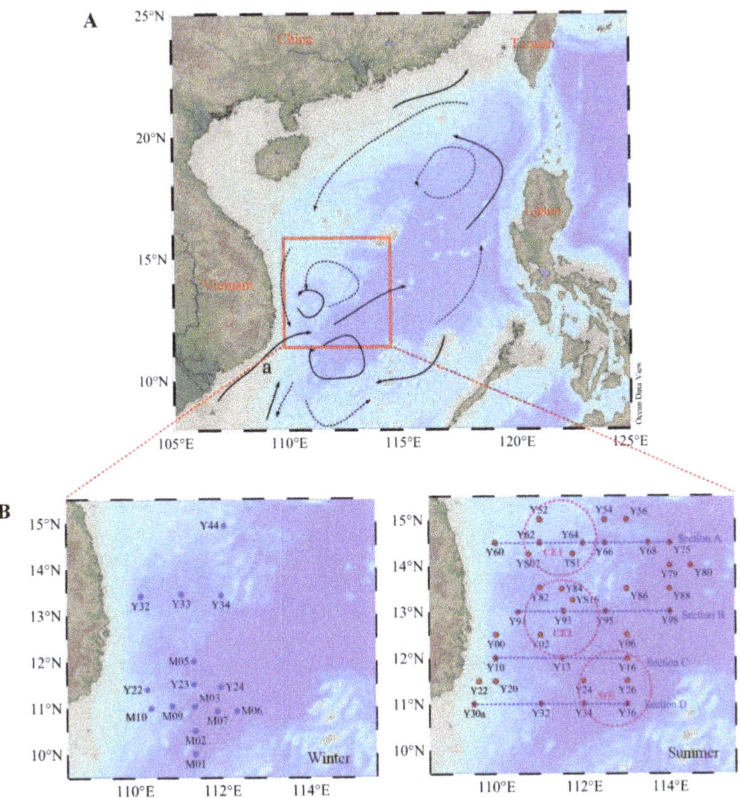

Figure 1. Map indicating the sampling stations along the southwestern region of the South China Sea (SCS) (Vietnamese upwelling region) (**A**). The arrows indicate the general surface current patterns in the SCS during the winter (black dotted arrows) and summer (black solid arrows) [35]. Map indicating the sampling locations during the winter (**B**) and summer seasons. The red dotted circle shows the eddy area, i.e., CE1: cold eddy 1, CE2: cold eddy 2, and WE: warm eddy [29,35]. The blue dotted lines show sampling sections defined as Section A, Section B, Section C, and Section D.

2.2. Sample Collection and Analysis

Seawater samples were collected from seven depths (0, 25, 50, 75, 100, 150, and 200 m) at 51 sampling stations using the Niskin bottles attached to a Rosette water sampler fitted with a Seabird 917 Plus site CTD system. A total of 40 and 230 samples for phytoplankton analysis were collected in winter and summer respectively. Temperature and salinity data were derived from the Seabird CTD. For enumeration of the phytoplankton community, a 3 L seawater sample was concentrated to 1 L by using 10 μm mesh and taken into polyethylene (PE) bottles, then fixed with 2% buffered formaldehyde solution and stored in darkness until completing the voyage.

After returning to the laboratory, the Utermöhl method was applied for phytoplankton water sample analysis [36]. A 1 L subsample was stood for 48 h, then 800 mL supernatant was removed carefully by siphoning through a catheter; it was important to note that the position of the catheter avoided touching the bottom of the bottle. After that, the remaining 200 mL liquid was well mixed gently, half of which was further concentrated with a 100 mL sedimentation column (Utermöhl method) for 48 h sedimentation [37]. Then, the phytoplankton species were identified and enumerated under an inverted microscope (AE2000, Motic, Xiamen, China) at 400× (or 200×) magnification, and five enumerations were performed under the non-overlapping field (529 field in total). The size limit of resolution for this analysis was ~5 μm. The phytoplankton species were identified using published standard literature [38] and the World Register of Marine species (http://www. marinespecies.org, Updated: 12 April 2021). The species identification was as close as possible to the species level.

For nutrient estimation, 100 mL of seawater was collected in the clean plastic bottles and stored at −20 °C till further analysis. Nutrition data were supplied by Dr. Min Han Dai's lab, Xiamen University. In detail, dissolved inorganic nitrogen NO_x (NO_3^- + NO_2^-) was analyzed by reducing NO_3^- to NO NO_2^- with a Cd column and then determining NO_2^- using the standard pink azo dye method, and a flow injection analyzer [39]. The dissolved inorganic phosphorus (PO_4^{3-}) concentrations were measured using two independent methods. For PO_4^{3-} concentrations > 500 nM, the concentration was measured by the standard molybdenum blue procedure [40], and for PO_4^{3-} concentrations < 500 nM, measurements were taken with a home-made ship-board C18 enrichment-flow injection analysis system [24,41]. Silicate concentrations were estimated using the standard silica aluminum blue spectrophotometry method [39].

2.3. Data Analysis

Horizontal and depth-integrated distribution of phytoplankton and physiochemical parameters were projected using Ocean Data View 4.7.6 (https://odv.awi.de/en/software/, released on 2 March 2018). The histogram, scatter diagram, and box-whisker plots were plotted with Origin (Version 8.5) [42]. The Spearman's correlation analysis and canonical correspondence analysis (CCA) between assemblages and physicochemical parameters were performed using Past3 software (http://www.canadiancontent.net/tech/download/PAST.html, released on June 2013).

The phytoplankton community diversity was evaluated mainly using the Shannon–Wiener diversity index (H'), Pielou evenness index (J), and dominance index (Y) [43]. The dominant species of phytoplankton was determined by dominance index (Y).

The Shannon–Wiener (S–W) diversity index (H') was calculated by the equation below:

$$H' = -\sum_{i=1}^{S} P_i \log_2 P_i \rightarrow H_{\max} = \log_2 S \qquad (1)$$

where P_i is the relative cell abundance of a species, i is the numbers of the i-th species, and S is the numbers of total species in a sample. The evenness index (J) was calculated from H' using the following formula:

$$J = \frac{H'}{\log_2 S} \qquad (2)$$

where H' is the S–W diversity index, and S is the number of the total species in a sample.

The phytoplankton dominance index (Y) was calculated as follows:

$$Y = \frac{n_i}{N} \cdot f_i \qquad (3)$$

where n_i is the number of the individual species, N is the total number of all species, and f_i is the occurrence frequency of the species in a sample.

Community alpha diversity indices (Shannon–Weiner index *H'*, and Pielou evenness index *J*, Species Richness, Simpson, Chao1) were calculated and performed using the 'vegan' package by R version 3.6.1. (https://www.r-project.org/, released on 5 July 2019) [44]. The Kruskal–Wallis test was used to compare the abundance differences of phytoplankton groups and diversity indices among defined groups. The two-tailed *t*-test was used to compare the abundance of phytoplankton groups between different defined groups.

3. Results

3.1. Seasonality in the Environmental Variables

The surface temperature and salinity in the winter ranged from 27.06 to 28.69 °C and from 33.15 to 33.72, respectively. In summer the surface temperature varied from 26.53 to 29.78 °C, whereas surface salinity ranged from 28.86 to 34.14. During this period, two cyclonic eddies were accurately captured throughout the cruise using in situ current and hydrographic measurements as well as the concurrent satellite altimeter observations [29,35]. One cold eddy, CE1, lay in the north region (112° E, 14° N), which lasted from 15 to 31 August. The other cold eddy, CE2, located in the south region (111° E, 12° N), endured for one week (1–8 September). Meanwhile, a warm eddy (WE) was observed near the CE2 (112° E, 10° N) (6–8 September) (Figure 1). During both seasons, the entire study region was divided into the following eddy stages (as adopted by [35]): no eddy stage (NE, in winter (December 2006)), CE1 stage (15–24 August 2007), CE1 relaxation stage (CE1-r, 25 to 31 August 2007), CE2 (1–8 September 2007), and WE (6–8 September 2007). The sampling stations marked with dotted circles were located within the eddy area (Figure 1). The data of various environmental factors during both seasons are given in Table A1. During all stages, the temperature decreased with water depth. The temperature at 50–100 m was relatively high in the WE stage compared with other stages (Table A1). The salinity increased with water depth. The nutrition concentrations also increased with water depth, and they were relatively high in the winter compared with that in the summer (the eddy stages). Among the different cyclonic stages in the summer, the average concentration of inorganic nitrogen (nitrate and nitrite) was almost below 0.2 μmol/L (except in the CE1 stage) in the upper water (0–50 m). However, in the middle water column (50–100 m), the average concentration of inorganic nitrogen, phosphate, and silicate were relatively higher during the CE1 and CE2 stages than during the CE1-r and WE stages.

3.2. Phytoplankton Species Composition in the Study Region

During the winter, a total of 112 phytoplankton taxa belonging to six phyla (Bacillariophyta, Dinophyta, Cyanophyta, Chlorophyta, Haptophyta, and Chrysophyta) and 39 genera were identified in the study region. The phytoplankton community was mainly composed of diatoms with 99 taxa belonging to 29 genera. Among diatoms, *Chaetoceros* and *Rhizosolenia* emerged frequently. Species in the genera *Bacteriastrum* and *Coscinodiscus* declined evidently. Nine dinoflagellate taxa belonging to six genera were reported during the present study. The frequency of most dinoflagellate species, especially the species belonging to the genera *Protoperidinium*, *Ceratium*, *Oxytoxum*, *Amphisolenia*, *Ornithocercus*, *Podolampas*, and *Dinophysis*, was decreased in the winter compared to summer. During the winter season, the cyanobacteria group was mainly comprised of *Trichodesmium thiebautii*, *Trichodesmium erythraeum*, and the symbiotic cyanobacteria *Richelia intracellularis*. Among them, *T. thiebautii* was the most abundant (0.01), although with an extremely low occurrence frequency (0.50) (Table 1). Besides *T. thiebautii*, the diatom species *Thalassionema nitzschioides*, *Nitzschia* spp., *Thalassiosira rotula*, *Navicula* spp., and *Chaetoceros* spp. dominated the species assemblage (Table 1). Species such as *Dictyocha fibula* and *Scenedesmus quadricauda* were also observed during the winter (Table 1). The symbiont cyanobacteria *R. intracellularis* was mainly associated with diatom species such as *Guinardia cylindrus*, *Rhizosolenia styliformis*, and *Rhizosolenia hebetata*. Notably, its association with *R. hebetata* was more dominant during the winter.

During the summer season, 320 taxa belonging to 148 genera and six phyla (Bacillariophyta, Dinophyta, Cyanophyta, Chlorophyta, Haptophyta, and Chrysophyta) were identified in the southwestern SCS. Among them, diatoms represented 187 taxa belonging to 54 genera and they were more dominant than the dinoflagellates (109 taxa from 22 genera). The phytoplankton community was more diverse in the summer than in the winter. The number of taxa and genera almost increased by a factor of two and three, respectively. *Trichodesmium* and *Chaetoceros* were the fpredominant genera in the phytoplankton community. The chain-forming species, including *T. thiebautii*, *T. nitzschioides*, *T. erythraeum*, *Chaetoceros dichaeta*, *Chaetoceros affinis*, *Chaetoceros lorenzianus*, *Thalassionema frauenfeldii*, *Pseudo-nitzschia delicatissima*, *Pseudo-nitzschia pungens*, *Leptocylindrus danicus*, *Hemiaulus hauckii*, and *Bacteriastrum comosum*, dominated the phytoplankton assemblage during the summer season (Table 1). In addition, the small-sized diatoms *Nitzschia* spp. and *Navicula* spp. were widely distributed in the study area. The cyanobacteria species *T. thiebautii*, *T. erythraeum*, and symbiotic cyanobacteria *R. intracellularis* and *Calothrix rhizosoleniae* were also reported during the summer. *R. intracellularis* was mainly associated with the diatom hosts such as *R. styliformis*, *G. cylindrus*, *R. hebetata*, and *Hemiaulus membranaceus* in the intercellular location. However, *C. rhizosoleniae* was attached externally to species like *Chaetoceros subsecumdus*, *C. affinis*, *Chaetoceros compressus*, *Chaetoceros glandazii*, and *Chaetoceros tortissimus*.

Table 1. List of the dominant phytoplankton species (with their occurrence frequency *f* and dominance index *Y*) observed during the winter and summer seasons in the southwestern South China Sea.

Winter			Summer		
Species	*f*	*Y*	**Species**	*f*	*Y*
Thalassionema nitzschioides	0.23	0.0231	*Trichodesmium thiebautii*	0.29	0.1842
Nitzschia spp.	0.33	0.0136	*Thalassionema nitzschioides*	0.57	0.0171
Trichodesmium thiebautii	0.05	0.0120	*Trichodesmium erythraeum*	0.13	0.0071
Thalassiosira rotula	0.30	0.0102	*Chaetoceros dichaeta*	0.28	0.0066
Navicula spp.	0.28	0.0089	*Chaetoceros affinis*	0.40	0.0054
Chaetoceros spp.	0.23	0.0081	*Thalassionema frauenfeldii*	0.54	0.0052
Bacteriastrum spp.	0.13	0.0066	*Chaetoceros lorenzianus*	0.30	0.0043
Dictyocha fibula	0.25	0.0057	*Pseudo-nitzschia delicatissima*	0.35	0.0041
Thalassiosira subtilis	0.20	0.0049	*Pseudo-nitzschia pungens*	0.28	0.0030
Chaetoceros affinis	0.08	0.0019	*Leptocylindrus danicus*	0.30	0.0026
Chaetoceros coarctatus	0.05	0.0016	*Hemiaulus hauckii*	0.37	0.0025
Chaetoceros lorenzianus	0.03	0.0014	*Nitzschia* spp.	0.65	0.0025
Corethron hystrix	0.05	0.0013	*Navicula* spp.	0.68	0.0022
Chaetoceros atlanticus	0.10	0.0008	*Bacteriastrum comosum*	0.23	0.0020
Chaetoceros laciniosus	0.05	0.0007	*Bacteriastrum hyalinum*	0.20	0.0019
Thalassiothrix frauenfeldii	0.05	0.0006	*Chaetoceros messanense*	0.23	0.0015
Rizizosoleniu Rhizosolenia hebetata-Richelia	0.08	0.0005	*Chaetoceros curvisetus*	0.14	0.0013
Rhizosolenia hebetata	0.03	0.0005	*Chaetoceros tortissimus*	0.16	0.0013
Octactis octonaria	0.08	0.0004	*Bacteriastrum elongatum*	0.21	0.0012
Leptocylindrus mediterraneus	0.03	0.0004	*Dactyliosolen blavyanus*	0.36	0.0010

3.3. Seasonal Distribution of Phytoplankton Community

The phytoplankton abundance during the winter ranged from 0.08×10^3 to 9.52×10^3 cells L^{-1}, with an average of 2.74×10^3 cells L^{-1}. Diatom abundance ranged from 0.08×10^3 to 3.36×10^3 cells L^{-1} (average 0.67×10^3 cells L^{-1}) and comprised ~63% of the total phytoplankton abundance (Table A2). *Chaetoceros* was a common genus in the diatom group with an average abundance of 0.61×10^3 cells L^{-1}. Diatoms dominated the phytoplankton assemblage at most stations, except St. Y22, where *Trichodesmium* contributed 92% (total 8.8×10^3 cells L^{-1}) to total phytoplankton abundance in the surface water (Figure 2a). Cyanobacteria were observed only in three stations, with the abundance of 0.56×10^3 cells L^{-1} at St. Y23, 2.16×10^3 cells L^{-1} at St. Y34, and 8.80×10^3 cells L^{-1} at St. Y22 (continental region) (Figure A1). Moreover, the total abundance of *T. thiebautii* was 10.16×10^3 cells L^{-1}, which accounted for a 24% proportion of the whole community. The symbiotic cyanobacteria only consisted of *Richelia intracellularis*, with a total abundance of 1.36×10^3 cells L^{-1}, contributing ~27% to total community abundance. Dinoflagellate abundance ranged from 0.08×10^3 to 0.16×10^3 cells L^{-1} (average 0.10×10^3 cells L^{-1}) and contributed approximately 1% to total phytoplankton abundance. The highest dinoflagellate abundance was observed at open water station M07 (Figure 2a).

On the contrary, the overall phytoplankton abundance ranged from 0.02×10^3 to 128.82×10^3 cells L^{-1}, with an average of 1.05×10^3 cells L^{-1}. Compared to the winter season, the overall proportion of diatoms in the phytoplankton community decreased by 33.78% in summer, although the ratio of diatom/dinoflagellate was comparable. Conversely, the proportion of cyanobacteria increased by 12.11%, where *Trichodesmium* contributed up to 44.84% (Figure 2b). The diatom–diazotrophic associations decreased by 2.70%. Cyanobacteria was the most abundant group, as the abundance ranged from 0.02×10^3 to 123.15×10^3 cells L^{-1} (average 1.89×10^3 cells L^{-1}) and contributed to 69% of the total phytoplankton abundance. The abundance of *T. thiebautii* reached up to 121.55×10^3 cells L^{-1}. The total abundance of symbiotic cyanobacteria in the region was 3.36×10^3 cells L^{-1}, with a *Richelia intracellularis* and *Calothrix rhizosoleniae* abundance of 2.20×10^3 and 1.16×10^3 cells L^{-1}, respectively. The cyanobacterial population was even distributed in deeper waters in the summer than that in the winter. The average abundance (0.83×10^3 cells L^{-1}) and the proportion of diatoms were about half that of cyanobacteria (Table A2). Among the diatoms, *Chaetoceros* was the most dominant species and contributed ~42% to diatom abundance. Dinoflagellates, together with other groups, contributed a small proportion (below 10%) of the phytoplankton community (Table A2). Compared with historical data of phytoplankton composition and abundance in similar regions and seasons, we detected a relatively higher species number in the summer, and the average abundance of phytoplankton was in accordance with previous data (Table 2).

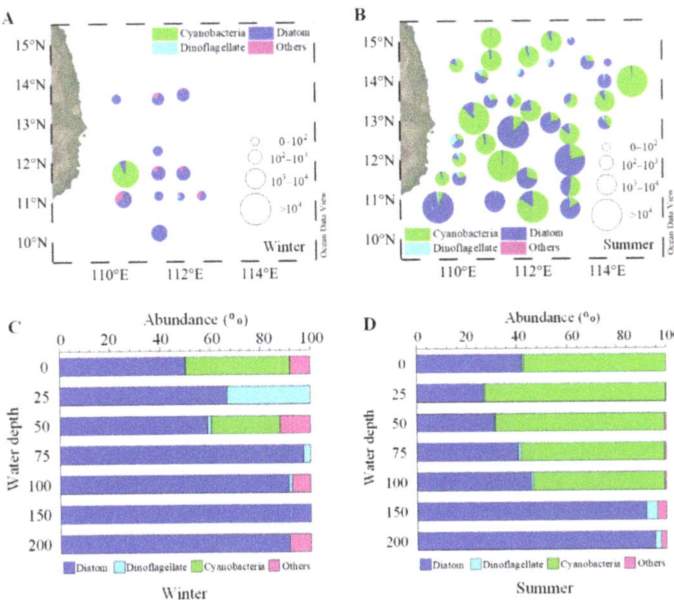

Figure 2. Composition and abundance (cell L^{-1}) of phytoplankton community in the surface water in the winter (**A**) and summer (**B**), and vertical distribution (station average) patterns of phytoplankton group relative abundance (%) in the winter (**C**) and summer (**D**).

Table 2. Comparison of historical data of phytoplankton with average cell abundance in the South China Sea.

Sampling Date	Region	Water Depth (m)	Species Number	Abundance (10^3 cell L^{-1})	Reference
1998–06–07	6–23° N,108–120° E	0–150 m	88	0.84	[17]
1998.06	5–25° N,105–120° E	Surface	63	0.83	[45]
1998.08	18–22° N, 105–117° E	Surface	58	181.00	[45]
1998.11–12	18–22° N, 105–117° E	0–150 m	85	8.46	[17]
2006.12	10–15° N, 110–112.5° E	0–200 m	117	2.74	This study
2007.08–09	10–15° N, 110–112.5° E	0–200 m	314	1.05	This study
2007.08	18–22° N, 110–120° E	0–200 m	216	11.22	[46]
2009.08	18–22° N, 110–117° E	0–200 m	109	8.20	[47]
2009.07–08	18–23.5° N, 109–120° E	0–200 m	150	26.49	[48]
2014.08	18–22° N, 114–116° E	0–200 m	229	16.32	[14]
2014.08	18–22° N, 114–116° E	Surface	98	0.23	[14]
2015.07–08	21–23.5° N, 111–117° E	0–200 m	212	45.61	[49]
2017.07–08	14–23° N, 114–124° E	0–200 m	287	2.14	[50]

3.4. Vertical Distribution of Phytoplankton Community at Different Eddy Stages

The phytoplankton composition and abundance during the summer and winter seasons were not only different in the surface layer but also the water column (Figure 2). In the winter, the abundances and proportions of cyanobacteria were comparable to that of diatoms, whereas in the summer, the cyanobacterial population was close to double that of diatoms. During the winter, the relative abundance of diatoms was greater below 50 m (peak abundance 2.56×10^3 cells L^{-1} at Stn 23), whereas the dinoflagellates (25 m) and cyanobacteria (0 and 25 m) abundance increased above 50 m (Figures 2c and A1). The highest abundance of cyanobacteria (8.8×10^3 cells L^{-1}) was observed in the surface water at Stn 22. Dinoflagellates contributed more to total phytoplankton at M07. Other phytoplankton species, mainly *Dictyocha fibula*, were relatively abundant and contributed 7.95% to total phytoplankton abundance in the surface water. During the summer, phytoplankton were mainly distributed towards the south of 13° N. Among them, the cyanobacteria mainly flourished in the area with eddy existence (especially around the sites of 14.5° N, 112° E and 12° N, 114° E), while diatoms were distributed in the southern area and the open basin (Figure 2b, Figure A2). The abundances of diatoms and dinoflagellates in the eddy mature stage (CE2) rose to 10 times that in the eddy relaxation stage (Figure 2b). Dinoflagellates accounted for a relatively high proportion of total phytoplankton at Y00 and stations along 14° N (Figure 2a). The abundance of *Dictyocha fibula* was very low, with the proportion of 0.06% of total phytoplankton abundance. Overall, depth-wise the cyanobacterial population was relatively abundant (>50%) until 100 m, whereas diatoms and (to some extent) dinoflagellates dominated deeper layers (Table A3). Moreover, the vertical distribution of cyanobacteria revealed their dominance on the edge of cold eddies and the open basin (Figure 3b, Section A, and Section B), as well as in the area influenced by the warm eddy (Figure 3b, Section C, and Section D). Differently, other phytoplankton (excluding cyanobacteria) mainly emerged in the center of cold eddies (Figure 3a, Section A, and Section B), followed by the continental area impacted by coastal upwelling (Figure 3a, Section D).

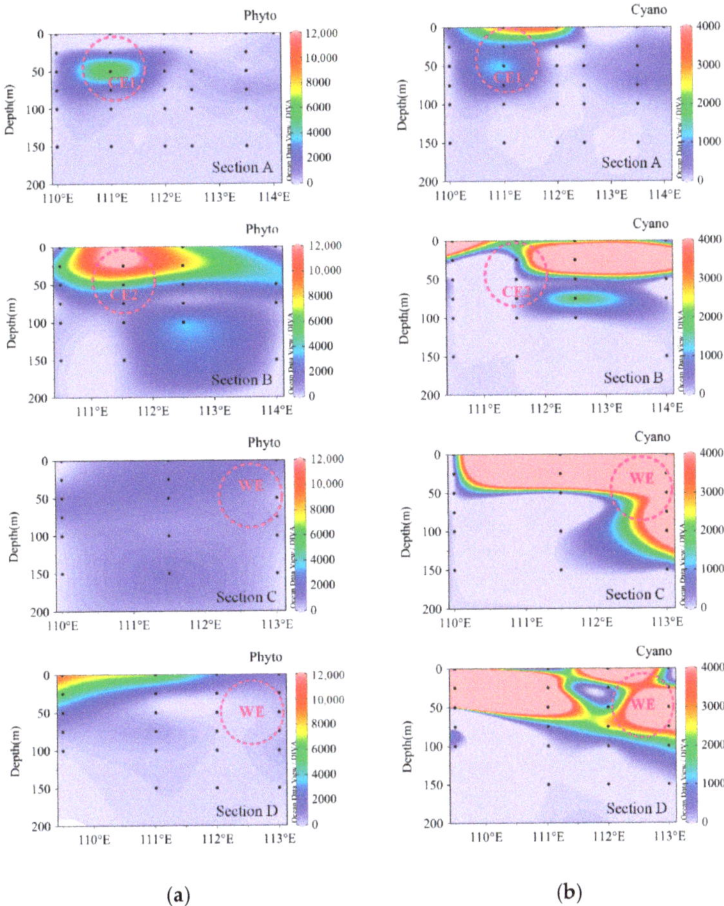

(a)   **(b)**

Figure 3. Vertical distribution of phytoplankton abundance during the summer. (**a**) Phytoplankton (Phyto) excluding cyanobacteria; (**b**) cyanobacteria (Cyano). CE1: cold eddy 1, CE2: cold eddy 2, and WE: warm eddy.

The proportions of the main phytoplankton groups changed remarkably with different stages (Figure 4a). The abundance of *Trichodesmium* in the phytoplankton community increased in the summer. The ratio of diatom to dinoflagellate (dia/din) was comparable during the winter and summer, whereas it was lower in CE2 (34.10) than in WE (Table A4). The diatom to cyanobacteria ratio (dia/cya) in the winter was five times more than in the summer. Furthermore, the dia/cya ratio was lower in the eddy periods than the no eddy and eddy relaxation stages. The relative contribution of algae groups including diatoms, dinoflagellates, and symbionts decreased dramatically during cold eddy mature stages but increased significantly during the later stages (Figure 4a). The dominant genus *Trichodesmium* presented a discrepant occurrence; however, its relative contribution to the phytoplankton community was much higher than that in non-eddy stages. The proportion of *Trichodesmium* (80.92%) and the dia/cya ratio (0.23) in the cold eddy mature period (CE2) were comparable to that in the warm eddy period (WE) (80.06%, 0.24, respectively) (Figure 4a). The phytoplankton abundance varied significantly above 50 m in the water column, while no significant difference was observed below the 50 m layer (Figure 4b). The phytoplankton community significantly varied seasonally (two-tailed *t*-test, $p < 0.05$)

(Table A5). Similarly, the variation of diatoms and dinoflagellates other than cyanobacteria showed a significant difference between eddy stages ($p < 0.01$). Moreover, the abundance of different phytoplankton groups, excluding cyanobacteria, had statistical differences among different periods (Kruskal–Wallis test, $p < 0.01$) (Figure A3). Thus, the overall results indicate that the phytoplankton community composition and structure changed with season and eddy development.

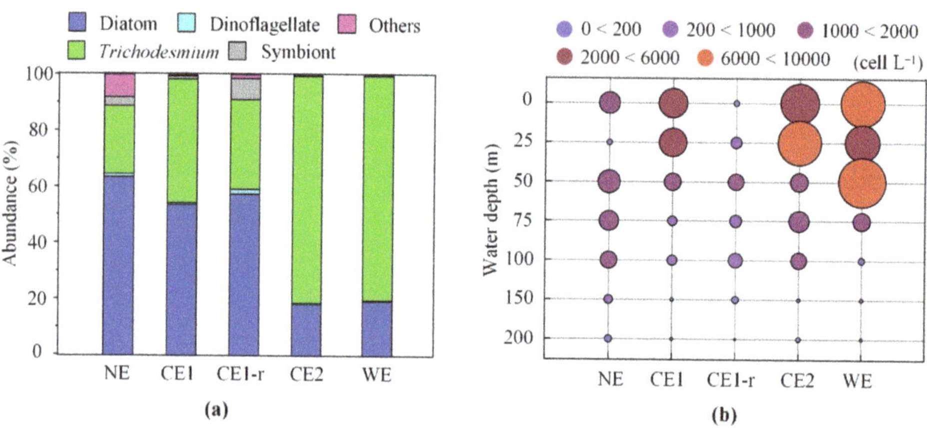

Figure 4. The relative abundance (%) contribution of different groups to the phytoplankton community (**a**), and the vertical variation of phytoplankton abundance (cell L^{-1}) at different eddy stages (**b**). (NE: winter, CE1: cold eddy 1, CE1-r: cold eddy 1 relaxation, CE2: cold eddy 2, and WE: warm eddy).

3.5. Diversity of Phytoplankton Community

The Shannon–Weiner index (H') and Pielou evenness index (J) were used for analyzing phytoplankton community diversity in this study. Our results show that the Shannon–Weiner index had a similar distribution pattern to the Pielou evenness index in both seasons (Figure A1). During the winter, the Shannon–Weiner index and Pielou evenness index ranged from 0.50 to 3.35 (avg. 1.28) and 0.09 to 0.57 (avg. 0.22), respectively, in the surface water (Figure 5a). High phytoplankton community diversity was observed in the open ocean region (Figure 5b). In the summer, the Shannon–Weiner and Pielou evenness indexes ranged from 0.01 to 5.14 (avg. 2.70) and 0.01 to 0.62 (avg. 0.32), respectively, in the surface water. Diversity indices were significantly higher in the summer than winter ($p < 0.001$), whereas the Pielou evenness index was significantly lower ($p < 0.05$) during the summer (Figures A1 and A4). This emphasizes that the surface water phytoplankton community in the summer was significantly more diverse than in the winter. Moreover, the phytoplankton community diversity in the summer was high in the water column (from 25 to 75 m depth) and also in eddy-controlled areas. Then, phytoplankton diversity was also higher towards the southern part around 13° N (Figure 5). In the study region, the diatoms and cyanobacteria controlled the phytoplankton community structure, whereas dinoflagellates and other groups contributed significantly to the transformation of the phytoplankton community and diversity.

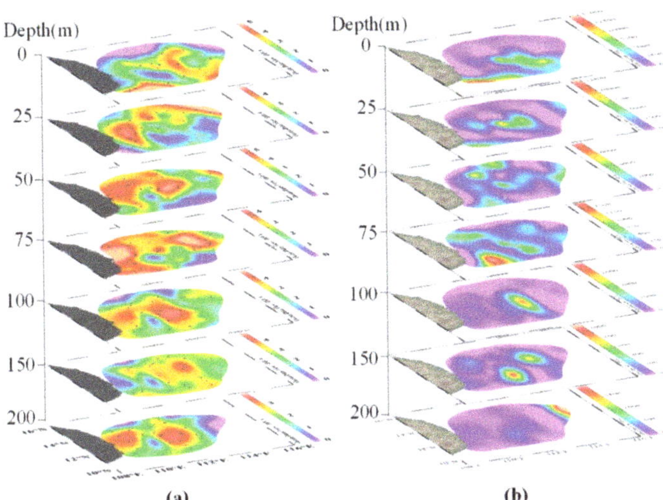

Figure 5. Shannon–Wiener diversity index (*H'*) (**a**) and cell abundance (**b**) of the phytoplankton community at different layers in summer.

3.6. Effect of the Environmental Cues on the Phytoplankton Community

The influence of the environmental factors on shaping the phytoplankton community structure in the western SCS was assessed using Spearman's correlation and CCA analysis (Figure 6). The phytoplankton community in the region was significantly influenced by the seasonality in the environmental characteristics. During the winter season, phytoplankton was positively correlated with temperature and Si/N ratio, and was mainly influenced by nitrogen (nitrate and nitrite) (Figure 6a,b). The various diatom groups had a different response to the aquatic environment. *Bacteriastrum* and *Chaetoceros*, which belong to the class of Centricae, exhibited significant relationships with environmental parameters, whereas diatoms that belonged to the class Pennatae had no significant relationship with any environmental parameter (Figure 6a,b). In detail, Centricae was positively influenced by temperature but negatively by water depth. However, Pennatae showed a discrepant relationship with the environment as compared to Centricae. These disparate responses evidence that the Centricae thrives in upper warm water, whereas Pennatae prefers the cool lower water. The Chrysophyte member *Dictyocha* was significantly associated with the various environmental parameters, which could eventually fuel its growth during the winter (Figure 6a). Cyanobacteria were significantly influenced by N/P and Si/N ratios, while dinoflagellates did not reveal an obvious correlation with environmental parameters.

The abundance of phytoplankton groups, excluding cyanobacteria, was significantly different during both the summer and winter seasons (Kruskal–Wallis test, $p < 0.01$) (Table A3). During the summer, the various phytoplankton groups (except Pennatae diatoms) were significantly influenced by the changing environmental factors (such as temperature, salinity, and nutrients at the respective depths) (Figure 6c,d)). However, the Pennatae diatom species did not show a significant relationship with environmental parameters, similar to that in winter (Figure 6c,d). Unexpectedly, *Dictyocha* also was not influenced by the changing water characteristics in the summer season. Cyanobacteria, including *Trichodesmium* and symbionts, were significantly influenced by the spatially and temporally changing environmental parameters. In summer, temperature, salinity, and nutrients were the important factors significantly controlling the phytoplankton growth ($p < 0.05$). This could evidence that the dynamic change in the environment (due to eddies and upwelling) results in temperature and nutrient variations, which in turn influence profoundly the phytoplankton community structure.

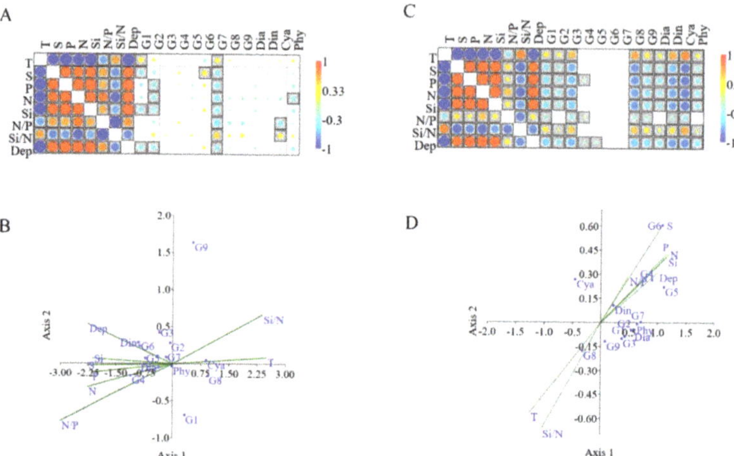

Figure 6. The influence of the environmental factors on the phytoplankton community shown with Spearman's correlation and CCA value, respectively. Notes: (**A,B**) represents Spearman's correlation and CCA for the winter samples, and (**C,D**) represents Spearman's correlation and CCA for the summer samples. Environmental parameters include temperature (T), salinity (S), depth (Dep), PO_4^{3-} (P), N: NO_x (P), SiO_3^{2-} (Si), N/P ratio, and Si/N ratio. Phytoplankton are listed as the following groups: diatoms G1 to G3 represent *Bacteriastrum*, *Chaetoceros*, and *Rhizosolenia*, belonging to the class of Centricae, G4 to G6 represent Fragilariaceae, Naviculaceae, and Nitzschiaceae, belonging to the class of Pennatae, and G7 to G9 represent *Dictyocha*, *Trichodesmium*, and symbionts. Dia: diatom, Din: dinoflagellate, Cya: cyanobacteria, and Phy: phytoplankton.

4. Discussion

4.1. Influence of Hydrological Processes on the Phytoplankton Community

The East Asian monsoon system has a strong bearing on the oceanographic and resultant biological features of the SCS. During the winter monsoons, the circulation in the southern SCS forms a cyclonic gyre, and an anticyclonic gyre during the summer monsoon. The surface water in most areas of the SCS is impoverished of nutrients due to a strong pycnocline, leading to a paucity of phytoplankton stock and production [19]. In winter, towards the western boundary of the SCS, the Vietnam offshore flow (which exists between 11° and 16° N) drifts northwards (along the coast) in the summer and southwards during the winter [31]. This flow pattern forms an offshore jet between 12°–13° N, resulting in a local enhancement of the upwelling intensity during the summer. The peculiarity of stretching deformation separates the Vietnamese upwelling from the offshore area and water masses [6,36]. Simultaneously, a strong coastal jet forms a dipole recirculation pattern and flows northeastward between a cyclonic cold eddy (CE2) and an anticyclonic warm eddy (WE) [29]. In this study, the phytoplankton community, especially diatoms, showed relatively high diversity in the continental margin influenced by the Vietnamese upwelling. Previously, the high Chlorophyll *a* concentrations were observed along the Vietnamese coast [21,27,51]. Loick-Wilde et al. (2017) estimated that the diatoms dominated the cell-carbon biomass in the Vietnamese upwelling area [11]. The nutrient advection during the coastal upwelling stimulates the phytoplankton growth in the upper layers [27].

During the summer season, eddies were persistent in the upper ocean layers in the SCS. The phytoplankton community structure changed with eddy developments. Statistical analysis revealed that the growth of various algae groups (except cyanobacteria) significantly varied with different eddy development phases, suggesting a significant influence of eddies on the phytoplankton community in the study region. The high phytoplankton abundance observed between the 25 and 100 m depth layers was mainly influenced by eddies. Earlier studies in the SCS pointed out that the maximum chlorophyll *a* concentration often appeared from 50 to 100 m in the non-eddy region, and appeared at 75 m in the eddy [23,27]. The cold eddy occurrence resulted in a continuous increase in diatom abundance compared to the non-eddy period, as observed previously [20]. The nutrient advection due to variable vertical motion could support the difference in phytoplankton abundance variation in the subsurface water [27]. Mesoscale eddies were proved to supply 20–40% of the nutrient requirements of phytoplankton [52,53]. The enhanced productivity in eddies could be even comparable to the productivity supported by upwelled subsurface nitrate driven during the prevailing monsoon [9]. The increased chlorophyll *a* concentrations due to nutrient enrichment during cyclonic eddies were also observed elsewhere [54]. Nutrient supplements derived from eddy occurrence, resulting in phytoplankton development, reflected the inter-coupling between physical and biochemical processes in the SCS region.

Different phytoplankton functional groups have a varied response to seasonal and spatial fluctuations of environmental factors [55]. Diatoms are a major starter of food chains and food webs, and important contributors to marine primary production and the ocean carbon cycle [56,57]. Cyanobacteria could maintain the balance of the global ocean nitrogen budget by biological nitrogen fixation [58]. In this study, the diversity and abundance of phytoplankton were much higher in the summer than that in the winter. Diatoms contributed more to phytoplankton abundance in the winter, but cyanobacteria (*Trichodesmium* dominance) contributed more in the summer. This seasonal variation clearly explains the shifts within the phytoplankton community from diatoms (in winter) to cyanobacteria (in summer). Earlier, in the SCS, a higher proportion of diatoms in the phytoplankton assemblage was reported during the winter than in summer [19]. However, discrepantly, the dominance of *Trichodesmium* was not reported earlier [19]. Moreover, high phytoplankton abundance was reported at a deeper layer due to the deepened thermocline in the summer, compared with that in the winter. The variations of major phytoplankton groups were explained by different adaptive strategies to overcome the constraints imposed by temperature and nutrient concentration variations in the SCS [59]. Here, wind pumping also played a significant role in inducing high biological productivity during the summer monsoon. The upwelling and cold eddies both fueled the nutrient enrichment and eventually the phytoplankton diversity during the summer season. Overall, in the western SCS, the seasonality of the phytoplankton community and growth dynamics could be significantly influenced by the coupled physical processes mostly driven by the East Asian Monsoon. Changes in compositions of phytoplankton in this study provide clues in understanding the mechanisms that regulate their acclimation and adaptation to changing environments.

4.2. Significance of Diazotrophic Cyanobacteria in the Western SCS

Nitrogen acted as an essential but limiting factor for phytoplankton. Nitrogen concentration in the surface water was mostly below the detection limit in the western SCS. Although frequent eddies, driven by upwelling and monsoon, replenish nutrients from the deeper water, nitrogen lost through denitrification (leading to Redfield ratios below 16) in the water column becomes a major limiting factor for phytoplankton growth [60]. Therefore, diazotrophic cyanobacteria essentially alleviate nitrogen limitation and are involved in regulating marine productivity [1,61,62]. In this study, diazotrophic cyanobacteria containing *Trichodesmium* and the diatom-associated symbionts *Richelia* and *Calothrix* were the highest in abundance. *Trichodesmium* was the most dominant species in the phytoplankton community in the continental margin and the oligotrophic basin during both seasons. The dominance of *Trichodesmium* was also recorded in the SCS earlier [62]. The high abundance of *Trichodesmium* that appeared in the subsurface water could be controlled by upwelling and eddies. Together with eddy perturbations, the abundance of *Trichodesmium* was more than 12 times higher in the eddy mature period than that in the degenerating stage. Compared to the previous study [19], the *Trichodesmium* abundance was above one order of magnitude higher in our study. Here, temperature and nutrient concentration were significant influencing factors for the *Trichodesmium* population. Earlier studies in this region indicated that *Trichodesmium* regulated the higher N_2 fixation and primary production rates in the oligotrophic offshore waters [11]. The nitrogen-fixing by *Trichodesmium* was quickly converted to plankton biomass and, in particular, the abundance of the diatoms (increased by 1.4–15 factor) in the Pacific Ocean [63]. Thus, here, it can be speculated that the thriving *Trichodesmium* population potentially contributes the bioavailable nitrogen into the oligotrophic waters in the western SCS. In addition, the symbionts *Calothrix* and *Richelia*, and their host diatoms, were relatively abundant in the summer, whereas *Rhizosolenia–Richelia* dominated in the winter. Symbionts often formed blooms in the low-nutrient water of the Pacific and the Atlantic Oceans [64–67]. Foster et al. (2011) estimated that the diatom partners influenced the growth and metabolism of their cyanobacterial symbionts *Richelia* and *Calothrix*, and the export of diazotroph-derived nitrogen supported the growth of the diatom partners [68]. Thus, diazotrophic symbioses and *Trichodesmium* would potentially play an important role in the nitrogen supplementation and phytoplankton growth of the oligotrophic ocean.

4.3. Phytoplankton Thermal Adaptations Inferred from Seasonal Successions

Global warming has increased steadily and increasingly involved deeper layers of the ocean since 1990 [69]. The warming ocean temperature would cause an alteration in the succession of the phytoplankton community [70–72]. Rising temperatures this century will cause poleward shifts in species' thermal niches [73]. Concomitantly, the ongoing global climate change is also linked to prolonged periods of anomalously high sea surface temperatures, which are defined as marine heatwaves [74]. From 1925 to 2016, the global average marine heatwave frequency and duration increased by 34 and 17%, respectively, resulting in a 54% increase in annual marine heatwave days globally [75]. Marine heatwaves have been accompanied by a large-scale change in surface chlorophyll levels, shifts in marine species location, and the reshaping of community structure [76,77]. Evidence from the field indicates temperature changes may lead to changes in diatom biogeography [59,78], and each species, even within the same genus, has its own characteristic temperature performance curve [79]. In this study, the seasonal succession of phytoplankton showed a predominance of diatoms in the phytoplankton community in the cool winter, which further shifted to cyanobacterial prevalence during the warm summer. Furthermore, in this study diatoms belonging to Centricae (represented by *Chaetoceros*, *Rhizosolenia*, and *Bacteriastrum*) were significantly related to temperature, as compared to Pennatae groups (such as Fragilariaceae, Naviculaceae, and Nitzschiaceae). The dominance of Centricae diatom species is often observed in the tropical ocean [14,80,81]. In the future, Centricae will become a potentially more sensitive group in the succession of the phytoplankton community as a consequence of the ocean temperature rise. On the other side, ocean stratification caused by rising temperature could result in nutrient deficiency [82]. Cyanobacteria prefer such a warm habitat with the low-nutrient oligotrophic condition. Here, in this study, the declining diatom predominance in the phytoplankton community during the warm condition could reveal their vulnerability to increasing temperature. On the contrary, flourishing cyanobacterial populations, mainly *Trichodesmium*, in warm conditions reflected their preference and adaptability in response to environmental change. The overall findings of our study could provide insight into phytoplankton community succession in future global temperature rise, and its further influence on biogeochemical cycles.

5. Conclusions

Here, we addressed the seasonal variability of the phytoplankton population in the western SCS, during the summer and winter monsoon periods. The seasonal changes of the phytoplankton community shifted from a diatom-dominated regime in winter to a cyanobacteria-dominated regime in the summer. This community change was controlled by eddies and upwelling activities during this season. Precisely, nutrient advection due to eddy activity triggered phytoplankton abundance, diversity, and *Trichodesmium* proliferation in summer. However, elevated temperature adversely influenced the diatom–diazotrophic association during the summer. The phytoplankton community succession responses to local oceanographic forces provide insights into forecasting biotic community evolution in the future global climate change.

Author Contributions: Conceptualization, J.S. and C.D.; methodology, J.S.; software, C.D.; validation, J.S. and C.D.; formal analysis, C.D.; investigation, C.D.; resources, J.S.; data curation, J.S.; writing—original draft preparation, C.D.; writing—review and editing, J.S., D.D.N., and H.L.; visualization, C.D.; supervision, J.S.; project administration, J.S.; funding acquisition, J.S. All authors have read and agreed to the published version of the manuscript.

Funding: This research was funded by the National Key Research and Development Project of China (2019YFC1407805), the National Natural Science Foundation of China (41876134, 41676112, 41276124, and 41406155), the Key Project of Key Laboratory of Integrated Marine Monitoring and Applied Technologies for Harmful Algal Blooms (MATHAB201805), the Tianjin 131 Innovation Team Program (20180314), and the Changjiang Scholar Program of Chinese Ministry of Education (T2014253) to Jun Sun.

Institutional Review Board Statement: Not applicable.

Informed Consent Statement: Not applicable.

Data Availability Statement: The data presented in this study are available in insert article.

Acknowledgments: We would like to thank Minhan Dai from Xiamen University for the nutrients data and CTD data supply. Special thanks to Qingshan Luan and Qing He for their hard work in sample collection.

Conflicts of Interest: The authors declare no conflict of interest.

Appendix A

Table A1. The average of environmental factors during different eddy developmental stages (-: missing data).

Stage	Layer	Temperature (°C)	Salinity	NO_x (μmol/L)	PO_4^{3-} (μmol/L)	SiO_3^{2-} (μmol/L)
Winter/NE	0–50	27.61	33.48	<0.2	0.03	2.39
	50–100	21.11	34.35	7.2	0.44	8.52
	100–150	16.92	34.54	15.62	1.02	18.06
	150–200	15.20	34.50	18.42	1.28	23.63
CE1	0–50	27.55	33.77	0.50	0.05	2.70
	50–100	22.27	34.28	5.87	0.40	7.75
	100–150	17.71	34.53	14.45	0.97	15.52
	150–200	13.64	34.51	15.67	1.33	24.67
CE1-r	0–50	27.56	33.80	<0.2	0.05	2.20
	50–100	22.38	34.27	4.55	0.33	6.28
	100–150	18.94	34.50	11.05	0.83	13.32
	150–200	17.38	34.57	15.60	1.24	22.71
CE2	0–50	28.36	33.37	<0.2	0.03	2.28
	50–100	22.36	34.29	6.44	0.40	7.34
	100–150	17.31	34.53	14.39	0.94	14.16
	150–200	-	-	18.80	1.27	22.19
WE	0–50	29.18	32.99	<0.2	0.05	2.25
	50–100	24.92	33.93	4.55	0.17	3.96
	100–150	17.74	34.55	14.72	0.86	13.47
	150–200	-	-	17.50	1.11	18.16

Table A2. The abundance of phytoplankton during the winter and summer seasons (cell L^{-1}).

Group		Winter	Summer
Diatom	Range	80–3360	5–13,482
	Average	670	827
	Total	26,800	190,045
	Proportion	63.57%	30.18%
Dinoflagellate	Range	80–160	2–245
	Average	12	16
	Total	480	3585
	Proportion	1.14%	0.57%
Cyanobacteria	Range	560–8800	16–123,150
	Average	288	1891
	Total	11,520	434,912
	Proportion	27.32%	69.07%
Others	Range	80–560	3–60
	Average	84	5
	Total	3360	1166
	Proportion	8.00%	0.19%
Phytoplankton	Range	80–9520	20–128,820
	Average	1054	2738
	Total	42,160	629,708

Table A3. The in-depth average abundances and proportions of phytoplankton groups during the winter and summer seasons.

Season	Depth (m)	Diatom Average (cell L^{-1})	Diatom Proportion	Dinoflagellate Average (cell L^{-1})	Dinoflagellate Proportion	Cyanobacteria Average (cell L^{-1})	Cyanobacteria Proportion	Others Average (cell L^{-1})	Others Proportion	Total Average (cell L^{-1})
Winter	0	704	50.00%	5.33	0.38%	586.67	41.67%	112	7.95%	1408
	25	53.33	66.67%	26.67	33.33%	0	0	0	0	80
	50	880	51.16%	0	0	680	39.53%	160	9.30%	1720
	75	1706.67	86.49%	80	4.05%	0	0	186.67	9.46%	1973.33
	100	1008	91.30%	16	1.45%	0	0	80	7.25%	1104
	150	304	100%	0	0	0	0	0	0	304
	200	176	91.67%	0	0	0	0	16	8.33%	192
Summer	0	1978.19	38.02%	39.14	0.75%	3182.31	61.16%	3.04	0.06%	5202.69
	25	1372.29	17.06%	32.58	0.40%	6633.21	82.45%	6.50	0.08%	8044.56
	50	1295.67	41.45%	13.52	0.43%	1803.94	57.71%	13.10	0.42%	3126.23
	75	414.11	31.61%	10.49	0.80%	880.32	67.20%	5.42	0.41%	1310.35
	100	264.39	43.48%	4.90	0.81%	335.40	55.17%	3.68	0.61%	608.39
	150	125.54	95.84%	3.17	2.42%	0	0	2.08	1.58%	130.79
	200	107.60	97.82%	1.23	1.12%	0	0	1.09	0.99%	109.92

Table A4. TThe abundance of phytoplankton groups at different eddy developmental stages.

	Group	NE	CE1	CE1-r	CE2	WE
Abundance (cell L^{-1})	Diatoms	26,800	104,334	3654	53,827	28,231
	Dinoflagellate	480	1258	116	1579	627
	Cyanobacteria	11,520	87,816	2507	236,753	116,098
	Trichodesmium	10,160	85,635	2025	236,341	115,810
	R. intracellularis	1360	2181	482	411	288
	Other groups	3360	621	82	406	56
	Total	42,160	194,029	6359	292,564	145,012
Proportion	Diatoms	63.57%	53.77%	57.46%	18.40%	19.47%
	Dinoflagellate	1.14%	0.65%	1.82%	0.54%	0.43%
	Cyanobacteria	27.32%	45.26%	39.42%	80.92%	80.06%
	Trichodesmium	24.10%	44.14%	31.84%	80.78%	79.86%
	R. intracellularis	3.23%	1.12%	7.58%	0.14%	0.20%
	Other groups	7.97%	0.32%	1.29%	0.14%	0.04%
Abundance ratio	Diatoms/Dinoflagellate	55.83	82.94	31.50	34.10	44.99
	Diatoms/Cyanobacteris	2.33	1.19	1.46	0.23	0.24

Table A5. The average abundances of phytoplankton groups were analyzed by *t*-test between different eddy developmental stages.

	Winter	CE1	CE1-r	CE2
Summer	a^1, b^1, c^3, d^1			
CE1-r		a^2, b^3, c^3, d^3		
CE2		a^2, b^1, c^3, d^1	a^2, b^2, c^3, d^1	
WE		a^2, b^2, c^3, d^1	a^2, b^2, c^3, d^3	a^1, b^2, c^3, d^1

Note: a: diatom, b: dino, c: cyan, d: others; *t*-test: [1]: $0.01 < p < 0.05$, [2]: $p < 0.01$, [3]: $p > 0.05$.

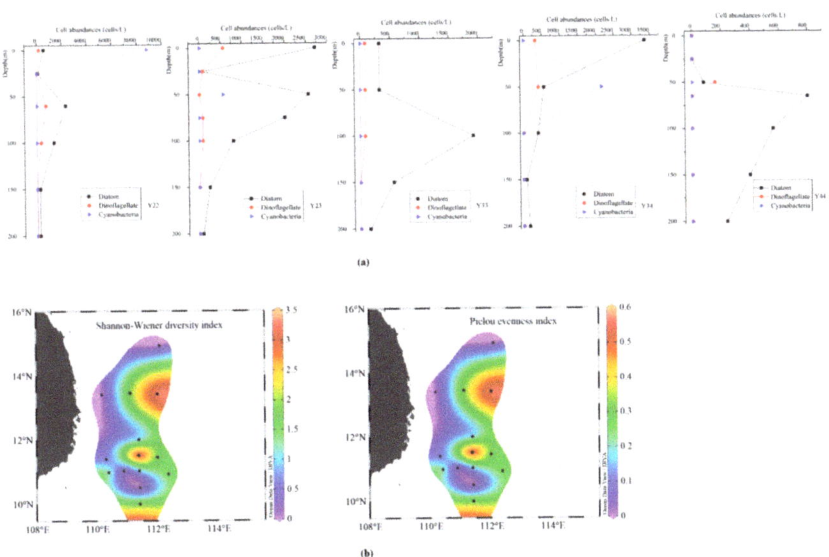

Figure A1. Vertical variation of phytoplankton abundance (**a**) and surface distribution diagram of community diversity (**b**) in winter.

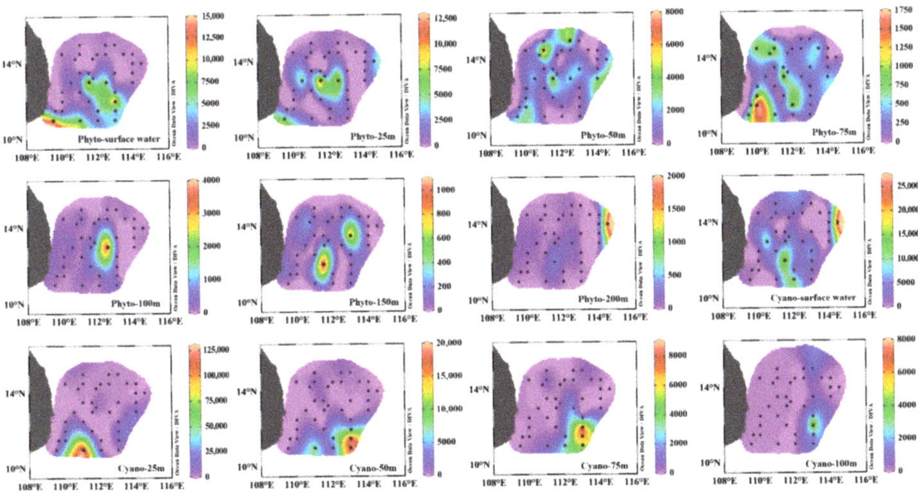

Figure A2. Horizontal distributions of cyanobacteria (Cyano) and other phytoplankton (Phyto) abundances at the different water layers in summer.

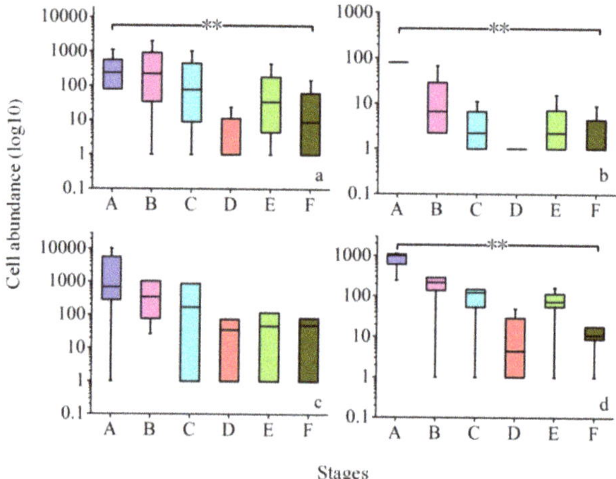

Figure A3. The phytoplankton group difference among defined stages by Kruskal–Wallis test (**: $p < 0.01$, A: winter, B: summer, C: warm eddy, D: Eddy II, E: Eddy I, F: Eddy I relaxation, a: phytoplankton, b: diatom, c: dinoflagellate, and d: others).

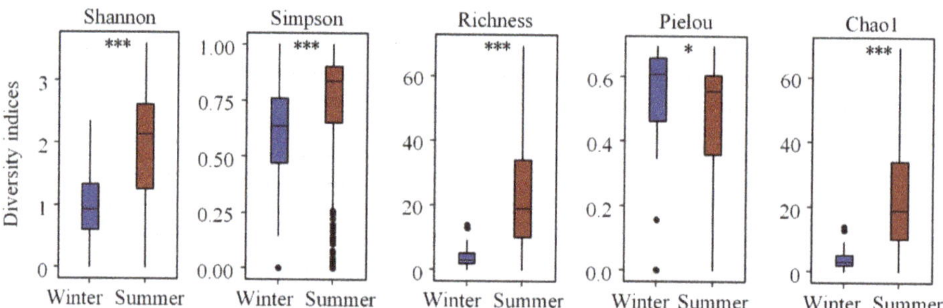

Figure A4. Alpha diversity indices were analyzed between the two seasons (*: $0.01 < p < 0.05$, ***: $p < 0.001$).

References

1. Voss, M.; Bombar, D.; Loick, N.; Dippner, J.W. Riverine influence on nitrogen fixation in the upwelling region off Vietnam, South China Sea. *Geophys. Res. Lett.* **2006**, *33*, L07604. [CrossRef]
2. Cai, W.-J.; Dai, M.; Wang, Y.; Zhai, W.; Huang, T.; Chen, S.; Zhang, F.; Chen, Z.; Wang, Z. The biogeochemistry of inorganic carbon and nutrients in the Pearl River estuary and the adjacent Northern South China Sea. *Cont. Shelf Res.* **2004**, *24*, 1301–1319. [CrossRef]
3. Guo, M.; Chai, F.; Xiu, P.; Li, S.; Rao, S. Impacts of mesoscale eddies in the South China Sea on biogeochemical cycles. *Ocean Dyn.* **2015**, *65*, 1335–1352. [CrossRef]
4. Yang, H.; Liu, Q.; Liu, Z.; Wang, D.; Liu, X. A general circulation model study of the dynamics of the upper ocean circulation of the South China Sea. *J. Geophys. Res.* **2002**, *107*. [CrossRef]
5. Cheng, Z.D. Fluxes of seawater and some dissolved chemicals in the Bashi Strait. *Chin. J. Oceanol. Limnol.* **1995**, *13*, 240–246.
6. Dippner, J.W.; Nguyen, K.V.; Hein, H.; Ohde, T.; Loick, N. Monsoon-induced upwelling off the Vietnamese coast. *Ocean Dyn.* **2007**, *57*, 46–62. [CrossRef]
7. Gong, G.-C.; Liu, K.K.; Liu, C.-T.; Pai, S.-C. The chemical hydrography of the South China Sea west of Luzon and a comparison with the west Philippine Sea. *Terr. Atmos. Ocean. Sci.* **1992**, *3*, 587–602. [CrossRef]
8. Chen, Y.-L.L.; Chen, H.-Y.; Karl, D.M.; Takahashi, M. Nitrogen modulates phytoplankton growth in spring in the South China Sea. *Cont. Shelf Res.* **2004**, *24*, 527–541. [CrossRef]
9. Chen, Y.-L.L.; Chen, H.-Y.; Lin, I.-I.; Lee, M.-A.; Chang, J. Effects of cold eddy on phytoplankton production and assemblages in Luzon strait bordering the South China Sea. *J. Oceanogr.* **2007**, *63*, 671–683. [CrossRef]

10. Liu, K.-K.; Chao, S.-Y.; Shaw, P.-T.; Gong, G.-C.; Chen, C.-C.; Tang, T. Monsoon-forced chlorophyll distribution and primary production in the South China Sea: Observational and a numerical study. *Deep Sea Res. Part I Oceanogr. Res. Pap.* **2002**, *49*, 1387–1412. [CrossRef]

11. Loick-Wilde, N.; Bombar, D.; Doan, H.N.; Nguyen, L.N.; Nguyen-Thi, A.M.; Voss, M.; Dippner, J.W. Microplankton biomass and diversity in the Vietnamese upwelling area during SW monsoon under normal conditions and after an ENSO event. *Prog. Oceanogr.* **2017**, *153*, 1–15. [CrossRef]

12. Sun, J.; Liu, D.Y. Net-phytoplankton community of the Bohai Sea in the autumn of 2000. *Acta Oceanol. Sin.* **2005**, *27*, 124–132.

13. Guo, S.; Feng, Y.; Wang, L.; Dai, M.; Liu, Z.; Bai, Y.; Jun, S. Seasonal variation in the phytoplankton community of a continental-shelf sea: The East China Sea. *Mar. Ecol. Prog. Ser.* **2014**, *516*, 103–126. [CrossRef]

14. Xue, B.; Sun, J.; Li, T.T. Phytoplankton community structure of northern South China Sea in summer of 2014. *Acta Oceanol. Sin.* **2016**, *38*, 54–65.

15. Gan, J.; Li, H.; Curchitser, E.N.; Haidvogel, D.B. Modeling South China Sea circulation: Response to seasonal forcing regimes. *J. Geophys. Res. Ocean.* **2006**, *111*. [CrossRef]

16. Wang, G.; Chen, D.; Su, J. Generation and life cycle of the dipole in the South China Sea summer circulation. *J. Geophys. Res. Ocean.* **2006**, *111*. [CrossRef]

17. Ning, X.; Chai, F.; Xue, H.; Cai, Y.; Liu, C.; Shi, J. Physical-biological oceanographic coupling influencing phytoplankton and primary production in the South China Sea. *J. Geophys. Res.* **2004**, *109*. [CrossRef]

18. Field, C.B.; Behrenfeld, M.J.; Randerson, J.T.; Falkowski, P. Primary production of the biosphere: Integrating terrestrial and oceanic components. *Science* **1998**, *281*, 237–240. [CrossRef] [PubMed]

19. Mahadevan, A. The Impact of submesoscale physics on primary productivity of plankton. *Ann. Rev. Mar. Sci.* **2016**, *8*, 161–184. [CrossRef]

20. Zhong, C.; Xiao, W.P.; Huang, B.Q. The response of phytoplankton to mesoscale eddies in western South China Sea. *Adv. Mar. Sci.* **2013**, *31*, 213–220.

21. Tang, D.; Kawamura, H.; Van Dien, T.; Lee, M. Offshore phytoplankton biomass increase and its oceanographic causes in the South China Sea. *Mar. Ecol. Prog. Ser.* **2004**, *268*, 31–41. [CrossRef]

22. Tang, D.L.; Kawamura, H.; Doan-Nhu, H.; Takahashi, W. Remote sensing oceanography of a harmful algal bloom off the coast of southeastern Vietnam. *J. Geophys. Res.* **2004**, *109*. [CrossRef]

23. Wang, J.J.; Tang, D.L. Phytoplankton patchiness during spring intermonsoon in western coast of South China Sea. *Deep Sea Res. II Top. Stud. Oceanogr.* **2014**, *101*, 120–128. [CrossRef]

24. Liang, Y.; Yuan, D.; Li, Q.; Lin, Q. Flow injection analysis of nanomolar level orthophosphate in seawater with solid phase enrichment and colorimetric detection. *Mar. Chem.* **2007**, *103*, 122–130. [CrossRef]

25. Wang, L.; Huang, B.; Chiang, K.-P.; Liu, X.; Chen, B.; Xie, Y.; Hu, J.Y.; Dai, M.H. Physical-biological coupling in the western South China Sea: The response of phytoplankton community to a mesoscale cyclonic eddy. *PLoS ONE* **2016**, *11*, e0153735. [CrossRef] [PubMed]

26. Ke, Z.; Tan, Y.; Huang, L.; Zhang, J.; Lian, S. Relationship between phytoplankton composition and environmental factors in the surface waters of southern South China Sea in early summer of 2009. *Acta Oceanol. Sin.* **2012**, *31*, 109–119. [CrossRef]

27. Liang, W.; Tang, D.; Luo, X. Phytoplankton size structure in the western South China Sea under the influence of a 'jet-eddy system'. *J. Mar. Syst.* **2018**, *187*, 82–95. [CrossRef]

28. Falkowski, P.G.; Barber, R.T.; Smetacek, V. Biogeochemical controls and feedbacks on ocean primary production. *Science* **1998**, *281*, 200–206. [CrossRef]

29. Hu, J.; Gan, J.; Sun, Z.; Zhu, J.; Dai, M. Observed three-dimensional structure of a cold eddy in the southwestern South China Sea. *J. Geophys. Res.* **2011**, *116*. [CrossRef]

30. Fang, W.D.; Fang, G.H.; Shi, P.; Huang, Q.Z.; Xie, Q. Seasonal structures of upper layer circulation in the southern South China Sea from in situ observations. *J. Geophys. Res.* **2002**, *107*, 1–12. [CrossRef]

31. Shaw, P.T.; Chao, S.Y. Surface circulation in the South China Sea. *Deep Sea Res. Part I Oceanogr. Res. Pap.* **1994**, *41*, 1663–1683. [CrossRef]

32. Yuan, Y.C.; Liu, Y.G.; Liao, G.H.; Lou, R.Y.; Su, J.L.; Wang, K.S. Calculation of circulation in the South China Sea during summer of 2000 by the modified inverse method. *Acta Oceanol. Sin.* **2005**, *24*, 14–30. [CrossRef]

33. Lin, H.; Hu, J.; Liu, Z.; Belkin, I.M.; Sun, Z.; Zhu, J. A peculiar lens-shaped structure observed in the South China Sea. *Sci. Rep.* **2017**, *7*, 478. [CrossRef] [PubMed]

34. Xu, Y.P. Nutrient Dynamics Associated with Mesoscale Eddies in the Western South China Sea. Master's Dissertation, Xiamen University, Xiamen, China, 2009.

35. Liu, Q.; Kaneko, A.; Su, J.L. Recent progress in studies of the South China Sea circulation. *J. Oceanogr.* **2008**, *64*, 753–762. [CrossRef]

36. Sun, J.; Liu, D.; Qian, S. A quantative research and analysis method for marine phytoplankton: An introduction to utermöhl method and its modification. *J. Oceanogr. Huanghai Bohai Seas* **2002**, *20*, 105–112.

37. Sun, J.; Liu, D.Y.; Feng, S.Z. Preliminary study on marine phytoplankton sampling and analysis strategy for ecosystem dynamic research in coastal waters. *Oceanol. Limnol. Sin.* **2003**, *34*, 224–232.

38. Sun, J.; Liu, D.Y. The preliminary notion on nomenclature of common phytoplankton in China seas waters. *Oceanol. Limnol. Sin.* **2002**, *33*, 271–286.

39. Dai, M.; Wang, L.; Guo, X.; Zhai, W.; Li, Q.; He, B.; Kao, S.-J. Nitrification and inorganic nitrogen distribution in a large perturbed river/estuarine system: The Pearl River estuary, China. *Biogeosciences* **2008**, *5*, 1227–1244. [CrossRef]

40. Pai, S.-C.; Yang, C.-C.; Riley, J.P. Effects of acidity and molybdate concentration on the kinetics of the formation of the phosphoantimonylmolybdenum blue complex. *Anal. Chim. Acta* **1990**, *229*, 115–120.

41. Ma, J.; Yuan, D.; Liang, Y.; Dai, M.H. A modified analytical method for the shipboard determination of nanomolar concentrations of orthophosphate in seawater. *J. Oceanogr.* **2008**, *64*, 443–449. [CrossRef]

42. Schlitzer, R. Ocean Data View. 2018. Available online: https://odv.awi.de (accessed on 2 March 2018).

43. Sun, J.; Liu, D.Y. The application of diversity indices in marine phytoplankton studies. *Acta Oceanol. Sin.* **2004**, *26*, 62–75.

44. RCRTeam. R: A Language and Environment for Statistical Computing, Version 3.6.1. Vienna, Austria. 2019. Available online: http://www.R-project.org/ (accessed on 5 July 2019).

45. Zhu, G.H.; Ning, X.R.; Cai, Y.M.; Liu, Z.L.; Liu, C.G. Studies on species composition and abundance distribution of phytoplankton in the South China Sea. *Haiyang Xuebao* **2003**, *25*, 8–23. (In Chinese)

46. Ke, Z.X.; Huang, L.M.; Tan, Y.H.; Yin, J.Q. Species composition and abundance of phytoplankton in the northern South China Sea in summer 2007. *J. Trop. Oceanogr.* **2011**, *30*, 131–143.

47. Li, X.; Sun, J.; Tian, W.; Wang, M. Phytoplankton community in the northern area of South China Sea in summer of 2009. *Mar. Sci.* **2012**, *36*, 33–39. (In Chinese)

48. Ma, W.; Sun, J. Characteristics of phytoplankton community in the northern South China Sea in summer and winter. *Acta Ecol. Sin.* **2014**, *34*, 621–632. (In Chinese)

49. Wang, X.; Wei, Y.; Wu, C.; Guo, C.; Sun, J. The profound influence of Kuroshio intrusion on microphytoplankton community in the northeastern South China Sea. *Acta Oceanol. Sin.* **2020**, *39*, 79–87. [CrossRef]

50. Zhong, Q.; Xue, B.; Noman, M.A.; Wei, Y.; Sun, J. Effect of river plume on phytoplankton community structure in Zhujiang River estuary. *J. Oceanol. Limnol.* **2020**, *39*, 550–565. [CrossRef]

51. Chen, G.; Xiu, P.; Chai, F. Physical and biological controls on the summer chlorophyll bloom to the east of Vietnam. *J. Oceanogr.* **2014**, *70*, 323–328. [CrossRef]

52. Martin, A.P.; Pondaven, P. On estimates for the vertical nitrate flux due to eddy pumping. *J. Geophys. Res. Ocean.* **2003**, *108*, 3359. [CrossRef]

53. McGillicuddy, D.J.; Anderson, L.A.; Bates, N.R.; Bibby, T.; Buesseler, K.O.; Carlson, C.; Carlson, C.; Davis, C.S.; Ewart, C.; Falkowski, P.G.; et al. Eddy/wind interactions stimulate extraordinary mid-ocean plankton blooms. *Science* **2007**, *316*, 1021–1026. [CrossRef] [PubMed]

54. Kumar, S.P.; Nuncio, M.; Ramaiah, N.; Sardesai, S.; Narvekar, J.; Fernandes, V.; Paul, J.T. Eddy-mediated biological productivity in the Bay of Bengal during fall and spring intermonsoons. *Deep Sea Res. Part I Oceanogr. Res. Pap.* **2007**, *54*, 1619–1640. [CrossRef]

55. Edwards, K.F.; Litchman, E.; Klausmeier, C.A. Functional traits explain phytoplankton community structure and seasonal dynamics in a marine ecosystem. *Ecol. Lett.* **2013**, *16*, 56–63. [CrossRef]

56. Boyd, P.W.; Hutchins, D.A. Understanding the responses of ocean biota to a complex matrix of cumulative anthropogenic change. *Mar. Ecol. Prog. Ser.* **2012**, *470*, 125–135. [CrossRef]

57. Edwards, M.; Richardson, A.J. Impact of climate change on marine pelagic phenology and trophic mismatch. *Nature* **2004**, *430*, 881–884. [CrossRef]

58. Zehr, J.P.; Capone, D.G. Changing perspectives in marine nitrogen fixation. *Science* **2020**, *368*. [CrossRef]

59. Xiao, W.; Wang, L.; Laws, E.; Xie, Y.; Chen, J.; Liu, X.; Chen, B.; Huang, B. Realized niches explain spatial gradients in seasonal abundance of phytoplankton groups in the South China Sea. *Prog. Oceanogr.* **2018**. [CrossRef]

60. Chen, C.-T.A.; Wang, S.-L.; Wang, B.-J.; Pai, S.-C. Nutrient budgets for the South China Sea basin. *Mar. Chem.* **2001**, *75*, 281–300. [CrossRef]

61. Zhang, Y.; Zhao, Z.; Sun, J.; Jiao, N. Diversity and distribution of diazotrophic communities in the South China Sea deep basin with mesoscale cyclonic eddy perturbations. *FEMS Microbiol. Ecol.* **2011**, *78*, 417–427. [CrossRef]

62. Moisander, P.H.; Beinart, R.A.; Voss, M.; Zehr, J.P. Diversity and abundance of diazotrophic microorganisms in the South China Sea during intermonsoon. *ISME J.* **2008**, *2*, 954–967. [CrossRef] [PubMed]

63. Bonnet, S.; Berthelot, H.; Turk-Kubo, K.; Cornet-Barthaux, V.; Fawcett, S.; Berman-Frank, I.; Barani, A.; Grégori, G.; Dekaezemacker, J.; Benavides, M.; et al. Diazotroph derived nitrogen supports diatom growth in the south West Pacific: A quantitative study using nanosims. *Limnol. Oceanogr.* **2016**, *61*, 1549–1562. [CrossRef]

64. Carpenter, E.J.; Montoya, J.P.; Burns, J.; Mulholland, M.; Subramaniam, A.; Capone, D.G. Extensive bloom of N_2-fixing diatom/cyanobacterial association in the tropical Atlantic Ocean. *Mar. Ecol. Prog. Ser.* **1999**, *185*, 273–283. [CrossRef]

65. Foster, R.A.; Zehr, J.P. Characterization of diatom-cyanobacteria symbioses on the basis of *nifH*, *hetR*, and 16S rRNA sequences. *Environ. Microbiol.* **2006**, *8*, 1913–1925. [CrossRef]

66. Foster, R.A.; Subramaniam, A.; Mahaffey, C.; Carpenter, E.J.; Capone, D.G.; Zehr, J.P. Influence of the Amazon River plume on distributions of free-living and symbiotic cyanobacteria in the western tropical North Atlantic Ocean. *Limnol. Oceanogr.* **2007**, *52*, 517–532. [CrossRef]

67. Foster, R.A.; Goebel, N.L.; Zehr, J.P. Isolation of *Calothrix Rhizosoleniae* (Cyanobacteria) strain SC01 from *Chaetoceros* (Bacillariophyta) spp. diatoms of the Subtropical North Pacific Ocean. *J. Phycol.* **2010**, *46*, 1028–1037. [CrossRef]

68. Foster, R.A.; Kuypers, M.M.M.; Vagner, T.; Paerl, R.W.; Musat, N.; Zehr, J.P. Nitrogen fixation and transfer in open ocean diatom-cyanobacterial symbioses. *ISME J.* **2011**, *5*, 1484–1493. [CrossRef]
69. Cheng, L.; Trenberth, K.E.; Fasullo, J.; Boyer, T.; Abraham, J.; Zhu, J. Improved estimates of ocean heat content from 1960 to 2015. *Sci. Adv.* **2017**, *3*, e1601545. [CrossRef]
70. Bopp, L.; Monfray, P.; Aumont, O.; Dufresne, J.-L.; Le Treut, H.; Madec, G.; Terray, L.; Orr, J.C. Potential impact of climate change on marine export production. *Glob. Biogeochem. Cycles* **2001**, *15*, 81–99. [CrossRef]
71. Reynolds, S.E.; Mather, R.L.; Wolff, G.A.; Williams, R.G.; Landolfi, A.; Sanders, R.; Woodward, E.M.S. How widespread and important is N_2 fixation in the North Atlantic Ocean? *Glob. Biogeochem. Cycles* **2007**, *21*. [CrossRef]
72. Huertas, I.E.; Rouco, M.; Lopez-Rodas, V.; Costas, E. Warming will affect phytoplankton differently: Evidence through a mechanistic approach. *Proc. R. Soc. B Biol. Sci.* **2011**, *278*, 3534–3543. [CrossRef]
73. Thomas, M.K.; Kremer, C.T.; Klausmeier, C.A.; Litchman, E. A global pattern of thermal adaptation in marine phytoplankton. *Science* **2012**, *338*, 1085–1088. [CrossRef] [PubMed]
74. Hobday, A.J.; Alexander, L.V.; Perkins, S.E.; Smale, D.A.; Straub, S.C.; Oliver, E.C.J.; Benthuysen, J.A.; Burrows, M.T.; Donat, M.G.; Feng, M.; et al. A hierarchical approach to defining marine heatwaves. *Prog. Oceanogr.* **2016**, *141*, 227–238. [CrossRef]
75. Oliver, E.C.J.; Donat, M.G.; Burrows, M.T.; Moore, P.J.; Smale, D.A.; Alexander, L.V.; Benthuysen, J.A.; Feng, M.; Sen Gupta, A.; Hobday, A.J.; et al. Longer and more frequent marine heatwaves over the past century. *Nat. Commun.* **2018**, *9*, 1324. [CrossRef] [PubMed]
76. Wernberg, T.; Bennett, S.; Babcock, R.C.; De Bettignies, T.; Cure, K.; Depczynski, M.; Dufois, F.; Fromont, J.; Fulton, C.J.; Hovey, R.K.; et al. Climate-driven regime shift of a temperate marine ecosystem. *Science* **2016**, *353*, 169–172. [CrossRef] [PubMed]
77. Bond, N.A.; Cronin, M.F.; Freeland, H.; Mantua, N. Causes and impacts of the 2014 warm anomaly in the NE Pacific. *Geophys. Res. Lett.* **2015**, *42*, 3414–3420. [CrossRef]
78. Irwin, A.J.; Finkel, Z.V.; Müller-Karger, F.E.; Ghinaglia, L.T. Phytoplankton adapt to changing ocean environments. *Proc. Natl. Acad. Sci. USA* **2015**, *112*, 5762–5766. [CrossRef]
79. Liang, Y.; Koester, J.A.; Liefer, J.D.; Irwin, A.J.; Finkel, Z.V. Molecular mechanisms of temperature acclimation and adaptation in marine diatoms. *ISME J.* **2019**, *13*, 2415–2425. [CrossRef] [PubMed]
80. Mao, Y.; Sun, J.; Guo, C.; Wei, Y.; Wang, X.; Yang, S.; Wu, C. Effects of typhoon Roke and Haitang on phytoplankton community structure in northeastern South China Sea. *Ecosyst. Health Sustain.* **2019**, *5*, 144–154. [CrossRef]
81. Liu, H.; Wu, C.; Xu, W.; Wang, X.; Thangaraj, S.; Zhang, G.; Zhang, X.; Zhao, Y.; Sun, J. Surface phytoplankton assemblages and controlling factors in the strait of Malacca and Sunda Shelf. *Front. Mar. Sci.* **2019**, *7*, 33. [CrossRef]
82. Moore, C.M.; Mills, M.M.; Arrigo, K.R.; Berman-Frank, I.; Bopp, L.; Boyd, P.W.; Galbraith, E.D.; Geider, R.J.; Guieu, C.; Jaccard, S.L.; et al. Processes and patterns of oceanic nutrient limitation. *Nat. Geosci.* **2013**, *6*, 701–710. [CrossRef]

MDPI

St. Alban-Anlage 66

4052 Basel

Switzerland

Tel. +41 61 683 77 34

Fax +41 61 302 89 18

www.mdpi.com

Water Editorial Office

E-mail: water@mdpi.com

www.mdpi.com/journal/water